MATERIAL SCIENCE LABORATORY

MATERIAL SCIENCE LABORATORY

Editor
Vincent Dias

DPH

DISCOVERY PUBLISHING HOUSE
INDIA

Published by:

DISCOVERY PUBLISHING HOUSE
4383/4B, Ansari Road, Darya Ganj
New Delhi-110 002 (India)
Phone : +91-11-23279245; 23253475; 43596065
Mobile : +91 9811179893 / +91 9871656464
E-mail : discoverybooksindia@gmail.com
orderdphbooks@gmail.com
namitwasan9@gmail.com
web : www.discoverypublishinggroup.com

Material Science Laboratory
Edited by: **Vincent Dias**

International Standard Book Number: **978-81-19365-73-9** (Hardback)

PREFACE

Materials science is an exciting and multidisciplinary field that delves into the structure, properties, synthesis, and performance of a vast array of materials. From metals and ceramics to polymers, semiconductors, composites, and biomaterials, materials science encompasses a wide range of substances that shape our modern world. The advancements and knowledge gained in this field have paved the way for remarkable technological breakthroughs across industries such as aerospace, electronics, medicine, and energy.

The aim of this book is to provide an extensive and practical resource for individuals interested in materials science and engineering, including students, researchers, and enthusiasts. Its purpose is to accompany your studies and research endeavors, offering a hands-on approach to comprehending the fundamental principles and applications of materials.

The book features a carefully curated selection of laboratory experiments that cover key concepts and techniques in material characterization, processing, and analysis. Each experiment is accompanied by clear instructions, safety guidelines, and background information to enrich your understanding of the underlying principles and methodologies. Whether you are a beginner embarking on your materials science journey or an experienced researcher seeking to deepen your knowledge and skills, you will find valuable insights and practical guidance within these pages.

In addition to the laboratory experiments, this book also includes research papers and reviews contributed by leading experts in the materials science field. These articles offer a broader perspective on cutting-edge research, emerging trends, and innovative applications. By exposing you to real-world research and the latest advancements, we aim to inspire you to explore new horizons and contribute to the ever-growing body of knowledge in materials science.

The readership of this laboratory book is diverse, catering to students, researchers, and professionals at various stages of their academic and professional journeys. For students, this book serves as a valuable resource for their coursework, bridging the gap between theoretical concepts and practical application. The detailed experimental procedures, coupled with insightful explanations, enable students to gain hands-on experience and develop critical thinking skills.

For researchers and professionals, this laboratory book provides a platform to expand their knowledge, explore novel techniques, and stay abreast of the latest research findings. The included research papers and reviews offer a comprehensive overview of various aspects of materials science, encompassing both fundamental concepts and advanced applications. It is our hope that this compilation serves as a catalyst for innovative research ideas and fosters collaborations within the materials science community.

As you engage with this book, you will discover the captivating world of materials science and its immense impact on our daily lives. Through practical experiments, in-depth explanations, and exposure to cutting-edge research, you will gain a deeper understanding of materials and their applications. Whether you are exploring materials science for the first time or seeking to expand your expertise, this book will be your guide on this thrilling journey of discovery and innovation. Embrace the opportunities it presents, and let your curiosity lead you to new frontiers in materials science.

The editor has collected reliable and scientifically proven information from various credible sources to provide readers with the latest and most accurate information available. The book is a compendium of contributions from various experts in the field, and the editor expresses gratitude to these contributors for sharing their knowledge and expertise.

Editor

CONTENTS

1 DESIGN OF EXPERIMENT: A RATIONAL AND TILL UNEXPLORED APPROACH TO INORGANIC MATERIALS' SYNTHESIS

Francesco Lamberti [1,2], Chiara Mazzariol [1], Federico Spolaore [1], Riccardo Ceccato [3], Luigi Salmaso [3], and Silvia Gross [1,2,4]

[1]*Department of Chemical Sciences, University of Padova, Via Marzolo 1, 35131 Padova, Italy*

[2]*Interdepartmental Centre Giorgio Levi Cases for Energy Economics and Technology, University of Padova, Via Marzolo 9, 35131 Padova, Italy*

[3]*Department of Management and Engineering, University of Padova, Stradella S. Nicola 3, 36100 Vicenza, Italy*

[4]*Karlsruher Institut für Technologie (KIT), Institut für Technische Chemie und Polymerchemie (ITCP), Engesserstr. 20, 76131 Karlsruhe, Germany*

ABSTRACT

This review was devoted to outlining the use and potential increasing application of the Design of Experiment (DoE) approach to the rational and planned synthesis of inorganic nanomaterials, with a particular focus on polycrystalline nanostructures (metal and alloys, oxides, chalcogenides, halogenides, etc.) produced by sustainable wet chemistry routes based on a multi-parameter experimental landscape. After having contextualised the stringent need for a rational approach to inorganic materials' synthesis, a concise theoretical background on DoE is provided, focusing on its statistical basis, shortly describing the different sub-methodologies, and outlining the pros and cons of each. In the second part of the review, a wider section is dedicated to the application of DoE to the rational synthesis of different kinds of chemical systems, with a specific focus on inorganic materials.

Keywords: *inorganic chemistry; Design of Experiment; inorganic synthesis*

1. INTRODUCTION

Inorganic materials' synthesis, also remarkably boosted by the requirement for robust, reproducible, and up-scalable approaches to materials' development, supporting the energetic transition and renewable energies' conversion and storage (e.g., catalysts, fuel cells, batteries, photovoltaics, etc.), is currently experiencing an actual renaissance aimed at matching a sustainable and green approach to synthesis with the demand for highly performing, stable, long-lasting, cost-effective and possibly multifunctional inorganic materials [1,2,3,4]. According to the paradigms of green chemistry [5] and to the emerging issues of circular chemistry [6,7,8,9,10,11,12,13], currently adopted synthetic requirements [14,15] encompass (1) a low temperature of processing (<200 °C, to pursue energy consumption reduction), (2) a short processing time (again for time and energy consumption reduction), (3) routes ensuring high yield and high throughput, (4) the use of environmentally friendly solvents (e.g., water, polyols) and of earth-abundant precursors (to address materials' criticality), in view of the application of these routes at an industrial scale, (5) the easy implementation of the synthetic procedure, (6) cost effectiveness, (7) reproducibility, and (8) up-scalability. This, in turn, implies the careful and informed optimisation of the whole experimental parameters' landscape, entailing several factors such as (1) pH, (2) temperature and time of processing, (3) the chemical nature of the solvent (i.e., dielectric constant, polarity, viscosity, density), (4) the structure and composition of the precursors, (5) molar ratios among the reagents, (6) the nature and amount of additives (i.e., stabilizing ligands, surfactants, structure-directing agents, porogens, etc.), (7) ionic strength, and (8) pressure, which are in many cases also intertwined with and interdependent of each other (e.g., dielectric constant and temperature/pressure). To avoid time- and resources-consuming approaches, a methodological paradigm shift in approaching this complexity is urgently needed. The current research in chemical synthesis is still chiefly relying on traditional inductive approaches such as "trial and error" methodologies for the development of new drugs, molecules, materials, or chemical protocols: this approach derives from the researcher's expertise, background, and knowledge. In many cases, the chemist approaches the research to optimise well-established and known recipes by continuously tuning and optimizing previous results and achievements. This safer approach, although simple and generally well accepted by literature, is, however, resource and time consuming and experimentally demanding, particularly when different parameters contribute in an interrelated fashion to affect the synthesis output, though their specific role is hardly identified. As a working example, the mentioned combined effect of temperature and pressure on dielectric constant, viscosity, and density of a suspension used for (even sub-critical) hydrothermal synthesis [16,17,18] would firstly require a deep understanding of the relation between temperature and pressure on the mentioned chemical–physical properties of the dispersing medium and secondly of the influence these experimental parameters have on the dissolution/reprecipitation phenomena occurring during the synthesis and eventually on the outcomes of the synthesis itself, in terms of crystallinity and built crystalline phase, composition, and yield of the product(s) [14,15,19,20]. The main issue derives from the fact that, by manually changing one parameter at time, the obtained result does not represent the unconditional outcome but only a relative maximum in efficiency, causing a reduced yield for the products. An additional relevant point pushing towards the reduction of the experimental trials is the need to reduce the use of critical raw materials [2,4,21,22], and specifically metals [23,24], very often involved in the development of technologically relevant inorganic nanomaterials, such as catalysts and devices for solar energy conversion, e-mobility, wind energy production, etc. [2,3]. To reduce the experimental effort required for the synthesis optimisation, instead of a time- and resources-demanding systematic screening of all involved experimental parameters, a more rational approach is thus required. In this regard, a well-established and promising methodology, based on statistics, is the Design of Experiments (DoE) [25,26,27,28,29] in which an impartial multi-variable analysis is carried out, thoroughly uncorrelated from user know-how, thus aseptic and focused only to boost the product yield in terms of sustainability (both environmental as well as economic) and efficiency, where the full experimental space can be truly explored. DoE is a well-known approach to optimise different processes and industrial manufacturing approaches, but its application to the specific field of inorganic materials chemistry is still relatively unexplored, at least at a scientific literature level, whereas its implementation at an industrial level is rapidly spreading.

In this review, we aimed at highlighting DoE main features by focusing on its potential application in the synthesis of inorganic materials, with a specific focus on wet chemistry and colloidal routes, which are the main methodological approaches of our research group at the University of Padova [14,15]. Related to this, we would like to explicitly point out that the choice of the examples reported concerning the DoE experimental methodology were in this case mainly related to wet inorganic chemistry, but the general considerations might apply also to solid-state and vapour-based inorganic synthesis methods. In particular, the DoE approach has been mostly implemented for optimizing the graphene synthesis [30,31] via chemical vapor deposition (CVD), although some examples concerning the growth of GaN rods [32] and silicon [33] through CVD could also be retrieved. DoE methodology has been also applied to support plasma-enhanced chemical vapor deposition (PECVD) synthesis (for example, of TiS_2 [34], SiN_x [35], and AlO_x [36]), as well as to physical vapor deposition (PVD) for the synthesis of TiAlN [37], ZrN, TiN [38], and yttria-stabilised zirconia [39]. Concerning solid-state synthesis, very few examples could be found. DoE has been successfully applied to accomplish a more rational preparation of alumina-based nanocomposites [40], ball-milling synthesis of electrolytes [41]. These few examples are neither comprehensive nor complete, and interested readers can refer to specialised literature on this. In addition, other approaches (for example, deep [42] and machine [43] learning, which will be mentioned later) have been also used to rationalise and optimise synthesis routes. The choice made in this review was made by the authors, also based on previous experiences, but does not represent a fully comprehensive view of the approaches to rationalise syntheses.

First, the theoretical framework underpinning DoE will be concisely described. In the second part of the review, some selected examples of the application of DoE to the synthesis of inorganic materials are provided and critically discussed. Eventually, perspective developments in this exciting field are highlighted.

2. THEORETICAL BACKGROUND OF DOE AND ITS APPLICATION TO INORGANIC CHEMISTRY

The DoE can be applied in chemistry as a methodology to statistically optimise processes and reaction conditions based on different experimental parameters, possibly but not necessarily interdependent (e.g., temperature and dielectric constant of a solvent). The first conceptualisation of DoE was introduced by Fisher in 1935 [29], who described a general problem of experiment design outlining which recurring factors' combination would best describe the outcome of an experiment and how the response is influenced by the single factors. Since the outcome of a chemical reaction is typically the result of multiple reaction conditions (factors), it is therefore possible, by screening the 'parameter space' of a process, to optimise the desired outcome. Specifically, DoE is a powerful tool that aims (1) to optimise and minimise the numbers of experiments through a statistical approach, (2) to identify the effects of the investigated parameters on the outcomes, and (3) to predict the response of untested conditions within the range of the explored experimental domains. The advantage of applying DoE to inorganic synthesis also lies in the possibility of assessing the synergistic effect of several parameters on the final result of the synthetic pathways, which would not be possible with a one-variable-at-a-time (OVAT) approach [25]. Therefore, a Design of Experiment approach will enable a rational planning of the synthetic efforts by careful screening the experimental parameters that most influence the final features, taking into account also interaction thereof, to identify the most promising combinations among them to obtain the desired product. Within this framework, in the field of inorganic synthesis, the possibility to apply the DoE approach to figure out the relationship among the experimental parameters, the outcomes of the reaction, and the final functional properties of the inorganic nanomaterials is extremely attractive as well as being an approach that can lead to a comprehensive exploration of the whole experimental factors. Even if in literature several examples of a DoE approach so far concern the optimisation of pharmaceutical processes [44,45,46], there are some studies related to the inorganic syntheses of different systems. A relevant disclaimer is that, being a statistical methodology, this approach is best applicable when large amounts of data can be retrieved. At the same time, it allows a more complete overview of the reaction conditions and it allows studying the synergistic and antagonistic interactions of the studied parameters, with an optimised (and limited) number of experiments, therefore reducing costs, time, and efforts. This approach can be used by researchers in two phases: (1) during an initial screening of a complex reaction with multiple parameters, thereby providing the understanding of the key factors impacting the outcome of a reaction, or (2) during the fine optimisation and tuning of a defined process, increasing yields or reducing costs.

2.1. Theoretical Background of DoE

In order to study the performance of a process/product and to try to systematically improve it, Design of Experiments (DoE) could be conducted. DoE deals with accurately planning, conducting, analyzing, and interpreting experiments so that valid and objective conclusions can be obtained about the process/product of interest.

In an experiment we deliberately manipulate some process/product variables, called factors, to investigate their effect on the desired output properties measured by response variables. It is worth noting that not all process/product variables are controllable by the experimenter. For this reason, controllable factors need to be distinguished from the uncontrollable ones (see Figure 1). For instance, in a synthesis based on the decomposition/reduction of tailor-made molecular precursor(s) to metal nanoparticles, controllable factors are, for instance, temperature and concentration of the precursor, whereas a not controllable factor is the oligomerisation degree and dynamics of the molecular precursor(s). An experiment is, therefore, a series of trials where the experimenter intentionally varies one or more controllable factors to observe changes in the output response and determine which factors affect the response variables.

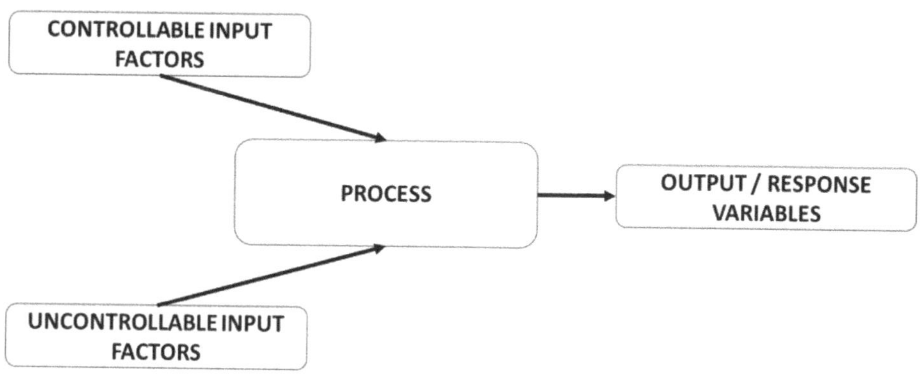

Figure 1.: *Experiment description.*

The experimenter needs to choose a few qualitative categories or numeric values, called levels, of the controllable factors to be tested in the experiment. Responses and factors are then measured or controlled on a randomised sample of experimental units. Each combination of factor levels tested on this sample represents an experimental condition, also called treatment or run. The entire set of runs is the design and its choice is a fundamental part in Design of Experiments.

The three basic principles [47] of experimental designs are:

- randomisation: both the allocation of the experimental units to treatments and the order in which the individual runs are to be performed are randomly determined;
- blocking: a technique for dealing with known and controllable nuisance factors (i.e., factors having an impact on the response, but of no interest to the experimenter), blocking out their potential effect on the response; and
- replication: each factor combination is, generally, assigned to more than one experimental unit. Replicates are indeed multiple independent executions of the same experimental conditions, which are processed individually in the experiment and should be run in random order.

It should be noted that a replicate is not a repeated or duplicate measurement of the response variable, which, on the other hand, is meant to reflect the inherent variability in the measurement system.

Let us now focus on the case where we are interested in evaluating the impact of more than one controllable factor on a specific response. A valuable approach is to plan a factorial experiment in which factors vary simultaneously, instead of one at a time. A factorial design indeed allows us to investigate whether each factor has an effect on the response (main effect) but also whether interactions between/among factors exist, i.e., if the effect of one factor on the response is not the same for all the levels of the other factor(s).

When the number of factors is particularly high, the experiment becomes too expensive or time and resources consuming to be performed because the number of required runs to investigate all possible treatments rapidly increases. In these cases, the experimenter can decide not to run all possible treatments, focusing on a suitable subset of the runs, by planning a fractional factorial design [48] instead of a full factorial design [49] (see Figure 2).

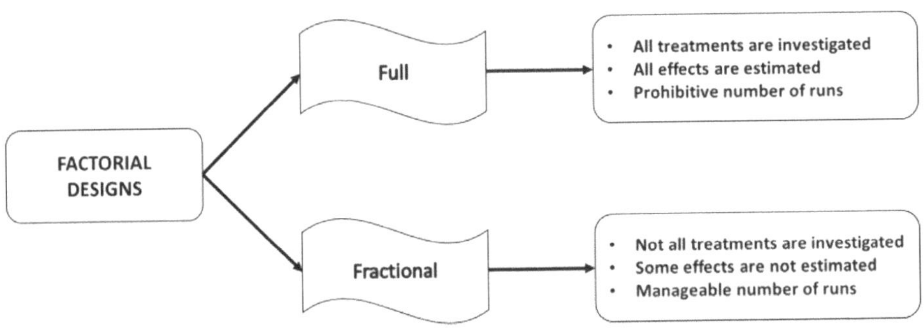

Figure 2.: *Advantages and disadvantages of full and fractional designs.*

It is worth noting that, in a full factorial design, all main effects and interactions can be estimated, but large numbers of factors and levels may result in a prohibitive number of runs. On the other hand, in a fractional factorial design the number of runs is reduced to a manageable size, minimizing time and cost, but some of the effects will be confounded (aliased).

Especially in the early stages of an experiment, a situation where many factors need to be investigated is quite common. Under such circumstances, experimenters commonly rely on screening experiments with factorial designs considering k factors, each at only two levels. As a working example, in a chemistry solution-based experiment, it can be chosen to vary pH and the temperature at two different levels, i.e., pH = 2 and pH = 5, and 100 °C and 150 °C, respectively. Screening experiments are widely used to identify the key factors affecting responses. After conducting them, the experimenter then proceeds with a general factorial design, increasing the number of levels to optimise factors' levels according to process/product targets in response variables.

Full factorials and fractional factorials are only two examples of designs adopted in the context of response surface methodology (RSM) [50,51]. RSM is an experimental methodology based on a collection of statistical designs and procedures and has two main objectives:

- understanding the effect of factors on the response variable(s), possibly identifying also nonlinear trends, and
- identifying the optimal configuration of the levels of the factors, typically by a maximisation or a minimisation of the response variables.

The sequential steps of the RSM procedure are [52]:

(1) Select factors (and their levels) and response variables. A screening experiment can be run to identify the most relevant factors.
(2) Select an appropriate experimental design (usually an optimal design) and collect data.
(3) Model the relationship between response variables and the factors.
(4) Evaluate the quality of the fitted model.
(5) Find optimal configurations of the factors' levels.

One of the most common designs used in the response surface methodology is the Central Composite Design (CCD). CCDs allow us to estimate first- and second-order terms and model the response with curvature by adding center and star points to a factorial design [52].

Another particularly relevant design in the chemical field is the mixture design [53]. It is used when the experiment involves a mixture of reagents and/or solvents and/or additives to form a solution/suspension/formulation. The goal is to determine if there is a combination of ingredients that improves the response variable according to pre-defined target values. In this case, a specific constraint needs to be considered differently from classical designs: The proportions (or molar fraction) all of the ingredients must add up to 100% (or 1, in the case of molar fraction, respectively). Then, in order to model the relationships between response variables and the factors, polynomial models are commonly adopted, but more complex techniques can be also used, such as machine learning methods [51].

Sometimes the number of response variables is larger than 1. Several different regression models are, therefore, required, and in the final optimisation step the desirability approach [52] is usually applied.

2.2. Selected Examples

The DoE methodology requires a specific and robust knowledge of the addressed topic. The reported examples focused on some of the most important subfields of modern inorganic chemistry that, although multidisciplinary, are at the interface between basic research and industrial research and development. For this reason, we will first illustrate the DoE application on a more simple topic such as solvent choice in a reaction and we will end the discussion by referring to outcomes obtained in the field of flow chemistry, competing with batch chemistry also in industrial research [54].

2.2.1. Solvent Optimisation

When referring to solution processing and wet chemistry routes to inorganic materials, solvents and dispersing media play an essential role in the outcome of the synthesis, as well as in the stabilisation of the resulting product (e.g., nanoparticles). Chemico-physical properties such as dielectric constant and polarity, presence of protons and hydrogen bonding properties, viscosity, and density all dramatically affect solubilisation, diffusion, and precipitation/crystallisation of both precursors and end products. Identifying the most suitable dispersing medium might be useful to optimise the synthetic pathway, accomplish polymorph selection, and maximise purity and yield of the targeted material.

The DoE approach can be conveniently used for the choice of a particular solvent (or mixture thereof) in a new synthetic route [55,56]. First of all, it is convenient to write down a list of the desired properties that the solvent needs to possess, not limited to chemical constraints but also including socio-economic issues: (1) the ecological impact on environment and health, (2) cost, (3) availability on the market, (4) chemical specifications such as boiling point, vapor pressure, and density, and (5) conditions for possible recycling/reuse, disposal, and/or (if needed) substitution with greener/cheaper alternatives. It is now possible to realise a principal component analysis (PCA) map in which tabulated and calculated variables for each solvent are inserted and analysed. The reported map in Figure 3, for instance, shows a PCA solvent map in which hydrogen bonding properties and polarity of 136 solvents are correlated, thus allowing the analysis to rationally build a tri-dimensional "solvent space" (a cuboid) in which at each vertex a solvent class is located: This class derives from the results of the previous correlation and links several species that in principle are not connected to each other or do not belong to the same chemical family. This intriguing analysis allows the researcher to systematically change the solvent with a rationale: As in this case, hazardous dichloromethane can easily be replaced with 1,4-dioxane or dimethyl carbonate and, likewise, carcinogenic and highly toxic benzene can be replaced with xylene. It is straightforward to understand that the choice would fall on the cheaper, the less harmful, and the more convenient one [57].

Another recent example of critical solvent choice using DoE was used by a group from the University of Bath in collaboration with the Paul Marray Catalysis Consulting Ltd. firm in the UK [58]. In this work, the authors applied DoE for a rational understanding of the best conditions for designing halide perovskite nanocrystals with tunable emissive properties by the realisation of an empiric model taking into account a reactant-to-solvent weight ratio, ligand concentration, ligand-to-metal molar ratio, and non-polar solvent polarizability, thus obtaining withstanding adherence to experimental data (Figure 3b).

Solvent selection for optimizing the solubilisation process under the desired optimal reaction temperature is, therefore, a significant challenge in processability [57,59]. One perspective evolution of DoE would be the use of machine learning for the automatisation of process development: In a recent paper [60], the authors, starting from a library of 459 solvents, divided them into 17 groups, each one characterised by a specific molecular descriptor, and obtained different recipes and better reaction conditions for the catalytic hydrogenation of a lactam.

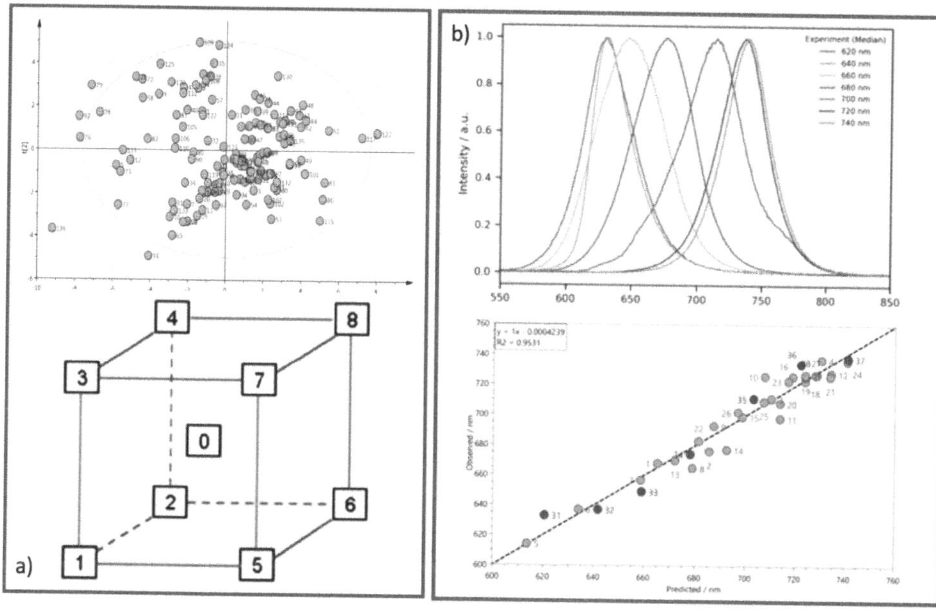

Figure 3.: *(a) The PCA solvent map (top panel) and the cuboid representing the solvent space (bottom panel) (reproduced from Ref. [57] with permission from the Royal Society of Chemistry). (b) The perovskite photoluminescence model validation experiment described in the text, where both S1-normalised PL emission (top panel) and observed vs. predicted wavelength PL, showing design experiments in green and validation experiments in blue (bottom panel), are presented (reproduced from Ref. [58] with permission from the Royal Society of Chemistry).*

The ultimate application of machine learning in chemistry design is the implementation of the Bayesian optimisation algorithm [61] in the exploration of reaction space: Within this approach, even literature and the operator's intuition that are already present in DoE boundary conditions are overcome, allowing us to balance the exploration of areas of uncertainty and the exploitation of available information. By doing so, Bayesian optimisation algorithms can be applied to different areas of research, enabling the selection of multiple experiments in parallel [62].

2.2.2. Synthesis of Inorganic Nanomaterials and Nanoparticles

The interest in nanomaterials [63] derives from the fact that, at this length scale, new properties are available and that these properties change drastically depending on nanomaterial size and/or shape. Nanoparticles are used in many chemical processes and often as catalysts with optimised dimensions to maximise efficiency and reduce volumes [64,65,66].

As far as a Design of Experiment is concerned, it is worth considering its application in the design of inorganic functional nanomaterials. Nowadays, the possibility to finely optimise and control the size, morphology, and size distribution of inorganic nanoparticles is becoming increasingly relevant to confer tuneable and desired final functional properties to the nanomaterials. In this regard, a Design of

Experiment approach applied to the inorganic synthesis of nanoparticles will enable assessing the relevant experimental parameters that most affect the outcomes of the reactions and, therefore, to determine which experimental conditions, e.g., temperature, concentration of precursor, pH, flow rate, etc., allow us to obtain the nanomaterials with the desired properties in terms of size, size distribution, etc. Going more into detail, Barglik-Chory et al. [67] reported a DoE approach for the adjustment and tuning of the band gap energies of CdS nanoparticles, exploiting a colloidal synthesis with cysteine and glutathione as biostabilisers. Indeed, it is well known that the size of the nanoparticles affects their band gap energies, and, therefore, it is of significant importance to understand the role played by the various experimental parameters on the final size of the nanoparticles. This can be evaluated, as an output response of the synthesis, either directly by determining the Feret diameter obtained by direct imaging (i.e., TEM or SEM images) or indirectly by analyzing UV-Vis absorption spectra.

The discussed study relates to the analysis of the influence of three different experimental parameters, i.e., the pH value and the relative amount of the stabiliser (cysteine or glutathione) and of the sulfur source (1,1,1,3,3,3-hexamethyl-disi-lathiane) on the band gap energies of CdS nanoparticles by measuring, as measurable output, the UV-Vis absorption spectra of samples synthesised with different reaction parameters. As a result, as an example of the cysteine-stabilised nanoparticles, it was possible to observe that the position of the band gap was significantly influenced by the three considered factors of pH value, the amount of the stabiliser, and the concentration of the sulfur source. In addition, it was found that the effect of pH on both cysteine and the sulfur source also affects the final band gap of the nanoparticles. These synergistic parameter interactions, which provide thus an exciting tool to tune synthesis outcomes, can be identified only by the simultaneous variation of parameter values as provided by DoE.

A further example of the application of DoE in the inorganic synthesis can be found in the work of Sadat-Shojai et al. [68]. The study focused on the optimisation of the hydrothermal synthesis of hydroxyapatite (HAp) nanoparticles, which is a widely exploited biomaterial due to its biocompatibility, bioactivity, and favorable osteointegration [69]. In particular, the synthesis of particles with optimised size, morphology, crystallinity, and stoichiometry is becoming a focus of interest in the biomedical fields. Within this framework, and exploiting a DoE approach, this paper reported the controlled growth of HAp nanoparticles with tuneable size, morphology, crystallinity, and stoichiometry by adjusting the experimental parameters. Reactant concentration, pH, temperature, time of the hydrothermal treatment, and presence of urea in the reaction medium were considered as the process variables that may influence the characteristics of HAp nanoparticles. As a conclusion, the paper presented a general methodology to predict the suitable conditions for the hydrothermal synthesis of shape-controlled HAp nanoparticles by performing a systematic evaluation of the most influencing variables in the synthesis of HAp by the hydrothermal method and of their effects on the final HAp properties.

In a further paper, the synthesis of iron oxide nanoparticles to be used as anode materials in Li-ion batteries via coprecipitation was studied by exploiting a factorial design of experiment methodology to investigate the influence of pH, medium temperature, Fe^{3+}/Fe^{2+} ratio, and reaction time on the crystallite size, taken as an output factor [70]. X-ray diffraction evidenced as the crystallite size decreased with increased pH and the Fe^{3+}/Fe^{2+} molar ratio, therefore enabling the rapid optimisation of a robust and reproducible protocol for the preparation of monodispersed maghemite nanoparticles.

Analogously, a DoE approach was implemented to prepare pure and chitosan-coated Fe_3O_4 nanoparticles [71] for the extraction and green removal of chromium (VI) from water samples. The optimised experimental parameters' space encompassed pH, contact time, adsorbent dosage, and agitation speed, assessed and evaluated by DoE method and investigated by response surface methodology (RSM).

The same kind of magnetic nanoparticles deposited on a potato peel (PP) was used for remediation from Pb2+. The L16 (4(boolean AND)4) method of Taguchi DOE was used for the optimisation of the adsorption condition, showing that at pH 6 with 10 min of contact time and a dose of 15 g/L can give more than 90% removal efficiency of Pb2+. Contour maps, Taguchi response analysis, and analysis of variance (ANOVA) suggested that pH has a dominant contribution in the removal of Pb2+ [72].

DoE can also be used to enhance synthesis reproducibility. A method for the synthesis of silver nanoparticles (AgNP) using gallic acid as reductant was optimised using DoE strategies based on response surface methodologies. Fractional factorial design was used in the screening stage. The obtained AgNP presented improved repetitivity and reproducibility of photophysical properties between batches compared to the synthesis method reported in literature [73].

Nanosized titanium oxide (TiO_2) powders prepared by conventional and microwave hydrothermal methods by forced hydrolysis of $TiOCl_2$ were followed by DoE and different characterisation methods. The effects of mixing time, HPC, and TiO_2 concentration and their mutual interactions on shear stress were evaluated with a Design of Experiment (DoE) approach [74].

In a further study, a one-step custom Design of Experiment was employed for the simultaneous screening and optimisation of raw yield (Y-1) and percent yield (Y-2) with heating temperature T (190–220 °C) and reaction time t (1.0–2.5 h) as the predictors in a conventional solvothermal magnetite nanoparticles' (NPs) synthesis [75].

A further application of DoE to rapid and high-throughput screening of inorganic nanomaterials relates to the development of new heterogeneous catalysts in industrial research. The field of catalysts' design and upscale production is characterised by an extremely large parameter space, implying that an increase in throughput in the preparation and testing of candidate substances is mandatory in order to identify a new and effective catalyst in a reasonable timeframe. In a paper by Duff et al. [76], the authors developed a workflow in which 10,000 substances per day were synthesised and their activity tested in a heterogeneously catalysed gas phase reaction (alkene epoxidation). The synthesis was based on a careful dosing of precursor solutions onto a single substrate, using one ink-jet printer technology and subsequent thermal treatment. For the activity testing, the product stream of each candidate was carried out through a detection layer, where the target product was converted into a fluorescent substance, then detected and assessed by locally resolved fluorescence spectroscopy. Since even an effective throughput of 10,000 substances per day does not allow mapping the whole parameter space in a practicable amount of time, a combination of evolutionary optimisation and data mining was explored.

2.2.3. Precious Metal NP Optimisation for Exhaust Gas After-Treatment

An everyday example of the application of nanomaterials can be taken from the heterogeneous reactions happening in the exhaust after-treatment systems of fuel vehicles, where precious metal nanoparticles are used to abate harmful gases resulting from the fuel combustion inside the engine. Being scarce, expensive, and recognised as critical raw materials by the European Commission [21], these metals are already used in the form of nanoparticles (NPs) to maximise the surface available for the reactions, while at the same time reducing their amount. Nevertheless, smaller does not necessarily mean better, and for certain applications stability towards sintering/coalescence and durability contribute to the overall performance throughout time. Within this context, a DoE was used by our group, within a collaboration with Umicore AG&Co. KG, to optimise the reaction conditions of a hydrothermal colloidal synthesis of Pt NPs as a possible approach to tune the resulting particle size [77].

Indeed, relevant physical properties of water vary under hydrothermal treatments, such as viscosity, ionic product, and dielectric constant, that change inside the autoclaves used in the synthesis upon temperature (and consequent autogenous pressure) increase [16,18,19]. From prior art and researcher experience, the resulting reagent solubilisation and product crystallisation were known to be functions of the temperature, pH, and molar ratio of the reagents (precious metal complexes and polymeric stabiliser). Nevertheless, with the reagents, set of factors, and reaction set up having never been applied before, it was important to investigate the interplay among the different experimental parameters involved and to single out the most relevant ones in ruling the synthesis output in terms of targeted size.

The reduction of the number of experiments to be performed was particularly important in this case because of the high criticality of the precious metals involved (Pd, Pt) [2,21,22].

In case of a larger number of factors, the experimental cost required tends to be high as all possible combinations of factors must be taken into account. Usually, fractional factorial designs are preferred in this case to reduce the size of the design matrix. Various applications of model-based experimental designs can be found in the review by Franceschini et al. [78]. When the number of parameter combination leads to an affordable number of experiments, the so-called full-factorial method can be used, where the additive effects on a response for each of the input factors, as well as the interaction between them, are determined. Before setting the experiment and the DoE matrix, a few experiments were performed to understand the feasibility of the applied reaction conditions in order to choose upper and lower values in the DoE matrix that would not alter the stability of the colloidal suspensions. The range of temperatures was known to span between 120 °C (temperature needed to provide enough energy to the system to reduce the Pt metal salt) but lower than about 200 °C, which is the deformation temperature of the Teflon liner (polytetrafluoroethylene (PTFE)) inside the autoclave. The pH values below 8 showed the formation of precipitates (aquo-hydroxo complexes of the chosen Pt precursor) whereas the upper value of 10 was given by the dilution of the basic Pt solution. Finally, the chosen ranges for the atomic ratio between the Pt and the polymeric stabiliser (polyvinylpyrrolidone (PVP)) were selected in order to obtain stable suspensions.

Once the applicable parameter space for the hydrothermal syntheses was defined, to screen the impact of the three selected parameters (temperature, pH, and molar ratio between precious metal complexes and polymeric stabiliser), each with two possible values (low or high) in a full-factorial experiment design mode, eight (=23) experiments were conducted. By measuring the Feret diameter of at least 300 particles for each experiment (which can be considered as internal replicates of each set of analyzed experimental condition), the average NPs' diameter was used as the outcome. A model was constructed, based on an Analysis of Variance (ANOVA) approach, that was able to fit for the selected DoE design the reaction outcomes [79].

As outlined in Figure 4 (Pareto Chart), only one of the factors resulted as statistically relevant in determining a size variation of the resulting NPs, whereas the effect of the other parameters was negligible. Despite the reduced number of experiments, the chosen DoE approach allowed the identification of the main parameter (the pH, terms B in the Pareto Chart in Figure 4) responsible for the particle growth under the tested experimental conditions.

Instead, temperature and molar ratio between precious metal complexes and polymeric stabiliser were evidenced to be less relevant; therefore, their contribution reported in the Pareto Chart of Figure 4 resulted as below the threshold of statistically relevant parameters. Likewise, also the mutual interactions pH–temperature, temperature–molar ratio, and pH–molar ratio did not remarkably affect the reaction output.

As general comments for the performed DoE, the goodness of the generated model to fit the outcomes was confirmed by the linearity of the regression as reported in the residual plots in Figure 4. The model assumptions, such as linearity of relations and normal distribution of residuals, were adequate even if the low number of experiments could not provide strong evidence for the homoscedasticity (i.e., when all the aleatory variables have a similar variance) and the independency of errors. Overall, the standard variance for the values, the random distribution against the sequence of experiments, and the normal distribution of the residuals resulted in a good regression analysis. At the end of the study, the Pt NPs obtained with the optimised reaction parameters (Figure 4, TEM B) resulted as bigger than those obtained at the beginning of the study (Figure 4, TEM A).

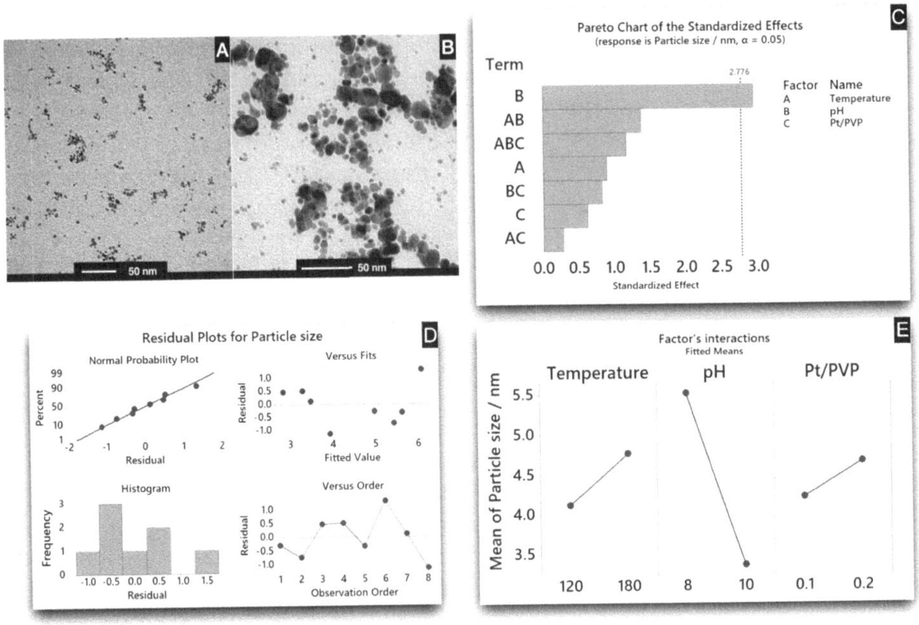

Figure 4.: *DoE for the investigation of the main reaction parameter responsible for the nanoparticle size increase under hydrothermal synthesis conditions. Minitab was used for the data analysis and model generation. (A) TEM micrograph of Pt NPs obtained with the optimised reaction parameters. (B) TEM micrograph of Pt NPs obtained at the beginning of the study. (C) The Pareto Chart is reported, ranking the significance of the different effect as analysed by Minitab. (D) The Residual plot is also reported, providing an overview of the generated model to fit the experimental results and presenting the homoscedasticity of the residuals vs different variables. (E) The Factor's interaction plot is reported, providing an overview of the impact of the different factors on the overall mean particle size. (Portions of information contained in this publication/book are printed with permission of Minitab, LLC. All such material remains the exclusive property and copyright of Minitab, LLC. All rights reserved. Minitab, LLC (2021),* www.minitab.com*) (accessed on 29 December 2021).*

2.2.4. Application to Flow Chemistry

Microfluidic and, generally speaking, flow chemistry are branches of research that are projected for industrial evolution because of the reduction in the experimental volumes and, accordingly, of the used reactants [80,81]. The opportune designing of starting experimental conditions is notably well accepted by entrepreneurs who can improve the quality and the quantity of goods to be sold. Several examples have already been reviewed and presented in literature [82].

Its application to the field of inorganic chemistry synthesis is largely less impactful than that to the field of organic chemistry and drug design, but relevant contributions are emerging in the state of the art [83,84,85,86,87,88,89]. Indeed, microfluidic approaches offer a promising route for the synthesis of high-quality and highly monodisperse inorganic nanoparticles. In microfluidic reactors, fluids flow in a parallel fashion, providing mixing of the reactants by diffusion of the molecules across the interface between the fluids [90]. The limited dimension of the reactor and the large ratio of surface to volume ensures thermal homogeneity across the reactor as well as a tight control on the mass and heat transfer. Therefore, it can be assumed that the reaction mixture is homogeneous with respect to the concentration and temperature [81]. In addition, microfluidic approaches, suitably optimised, allow the temporal separation of the two steps of nucleation and growth of nanoparticles, since the nucleation initiates with the efficient rapid mixing of the reactants in the microreactor and subsequently the growth of particles occurs, allowing the generation of nanoparticles with highly precise sizes and shapes [80,81].

In relation to microfluidics, and within a colloidal regime framework, DoE approaches are also useful for smartly controlling the nanoscale interactions of colloidal building blocks in order to obtain high-throughput functional materials. In particular, supraparticles (i.e., complex spherical structures obtained by self-assembling of evaporated emulsion droplets in flow reactors) yield is strongly affected by different physico-chemical properties (size, dispersity, porosity, and so on) that need rationalisation. In a recent paper [91], the authors showed the possibility of up-scaling the continuous flow production of these structures in a defined range size (from 0.1 to 10 micron) by carefully tuning the involved experimental parameters.

A further contribution [92], concerning a design of experiment approach for flow synthesis, was applied to the synthesis of citrate functionalised calcium phosphate CaP NPs aided by sonication using a continuous flow wet chemical precipitation, and the effect of some of the most relevant process factors (i.e., reactant flow rate, sonication amplitude, and maturation time) on the physico-chemical properties of the NPs were evaluated. From the statistical data analysis, the authors evidenced that CaP NP dimensions are influenced by the reactor flow rate, while the crystalline domain dimensions and product purity are influenced by the maturation process.

Even metal-organic frameworks (MOFs) are promising candidates for DoE application: Indeed, the daily productivity (defined as kg of MOF $m^{-3}d^{-1}$) by microfluidics can be boosted by several orders of magnitude by proper optimisation of synthesis parameters such as residence time, linker concentration, metal/linker volumetric ratio, and solvent choice [93].

Another interesting example is the effort to optimise the parameters in a reactor flow during the synthesis of biocatalysts exploiting 3D printed scaffolds [94]: In such a way, the system automatically chooses the best result of the DoE analysis and runs the experiments by remote controlling the equipment, saving time and avoiding human intervention.

Other works concerning the application of DoE for flow chemistry reactions, not necessarily related to inorganic chemistry, are present in the literature: one work reports on a liquid-phase dehydrogenation, which is strongly influenced by temperature and liquid flee flow [95]. A further study in which the amount of carcinogenic reagents for performing a reaction of hydrogenation is strongly reduced [96] was reported. A simultaneous self-optimisation strategy combining IR and mass spectrometry [97] or exotic nanomole-scale reaction screening optimisation for pharmaceutical purposes [98] were reported. Finally, continuous flow synthesis of an anticancer molecule in which an interesting synergic double application of DoE and high-throughput experimentation methodologies was presented [99].

3. CONCLUSIONS AND PERSPECTIVES

The increasing contribution of inorganic materials chemistry for strategically relevant technologies such as CO2 reduction and valorisation, heterogeneous catalysis, energy conversion and storage, and electric mobility urgently requires the implementation of rapid, up-scalable, and cost-effective synthetic approaches to highly performing functional inorganic materials. This in turn relies on the systematic screening of several (and in many cases interrelated) experimental parameters and, with the experimental parameters' landscape dramatically influencing size, size distribution, shape, and morphology, a rational and rapid exploration of these parameters becomes a ruling factor in determining a possible industrial implementation of a given synthesis route. In this framework, DoE can rapidly become a relevant methodological tool supporting inorganic materials' design and optimisation, also complying with the growing requirement of reducing the water and carbon footprint of the preparation approaches. Nevertheless, it should be pointed out that, particularly in complex synthetic systems and in the presence of tightly interrelated and interdependent parameters, even robust chemical intuition and knowledge are not always sufficient to support a reliable rationalisation. Since not every reaction is suited to be simplified in such a manner, whether a high-throughput screening approach is effective or not has to be considered on a case-by-case basis. In some cases, high throughput becomes only possible at the cost of abstraction and simplification with a concomitant reduction in knowledge gain per individual experiment. Finally, from the multidisciplinary point of view, it is worth noting that robust chemical knowledge and chemical intuition might not be sufficient to rationally address modern inorganic materials' synthesis, and support from statisticians is required to implement reliable and fruitful DoE.

Author Contributions

All authors to conceptualization, writing and reviewing tasks. All authors have read and agreed to the published version of the manuscript.

Acknowledgments

The University of Padova is gratefully acknowledged for administrative and logistic support to the research activity. S.G. additionally thanks DFG for a Mercator Fellow position at KIT funded by the DFG within the SFB 1441—Project-ID 426888090 and DAAD (Germany) for a visiting professorship at JLU. F.L. and S.G. acknowledge the Interdepartmental Centre Giorgio Levi Cases for Energy Economics and Technology of the University of Padova for the funding of the project AMON-RA.

REFERENCES

1. Gatti, T.; Lamberti, F.; Mazzaro, R.; Kriegel, I.; Schlettwein, D.; Enrichi, F.; Lago, N.; Di Maria, E.; Meneghesso, G.; Vomiero, A.; et al. Opportunities from Doping of Non-Critical Metal Oxides in Last Generation Light-Conversion Devices. Adv. Energy Mater. 2021, 11, 2101041.
2. European Commission. Critical Raw Materials for Strategic Technologies and Sectors in the EU—A Foresight Study. 2020.
3. Kloo, L. Inorganic chemistry for renewable energy conversion and storage. Dalton Trans. 2014, 43, 14924–14925.
4. Salviulo, G.; Lavagnolo, M.C.; Dabalà, M.; Bernardo, E.; Polimeno, A.; Sambi, M.; Bonollo, F.; Gross, S. Enabling Circular Economy: The Overlooked Role of Inorganic Materials Chemistry. Chem.—Eur. J. 2021, 27, 6676–6695.
5. Anastas, P.; Warner, J. Green Chemistry: Theory and Practice; Oxford University Press Inc.: Oxford, UK, 2000.
6. Keijer, T.; Bakker, V.; Slootweg, J.C. Circular chemistry to enable a circular economy. Nat. Chem. 2019, 11, 190–195.
7. Kümmerer, K.; Clark, J.H.; Zuin, V.G. Rethinking chemistry for a circular economy. Science 2020, 367, 369–370.
8. Clark, J.H. Green biorefinery technologies based on waste biomass. Green Chem. 2019, 21, 1168–1170.
9. Clark, J.H.; Farmer, T.J.; Herrero-Davila, L.; Sherwood, J. Circular economy design considerations for research and process development in the chemical sciences. Green Chem. 2016, 18, 3914–3934.
10. Linder, M. Ripe for disruption: Reimagining the role of green chemistry in a circular economy. Green Chem. Lett. Rev. 2017, 10, 428–435.
11. Loste, N.; Roldán, E.; Giner, B. Is Green Chemistry a Feasible Tool for the Implementation of a Circular Economy? Environ. Sci. Pollut. Res. 2020, 27, 6215–6227.
12. Thomas, J.M.; Raja, R. Designing Catalysts for Clean Technology, Green Chemistry, and Sustainable Development. Annu. Rev. Mater. Sci. 2005, 35, 315–350.
13. Everts, S. GREEN CHEMISTRY Environmentally friendly synthesis of niacin generates less inorganic waste. Chem. Eng. News 2008, 86, 15.
14. Bretos, I.; Diodati, S.; Jiménez, R.; Tajoli, F.; Ricote, J.; Bragaggia, G.; Franca, M.; Calzada, M.L.; Gross, S. Low-Temperature Solution Crystallization of Nanostructured Oxides and Thin Films. Chem.—Eur. J. 2020, 26, 9157–9179.
15. Diodati, S.; Dolcet, P.; Casarin, M.; Gross, S. Pursuing the Crystallization of Mono- and Polymetallic Nanosized Crystalline Inorganic Compounds by Low-Temperature Wet-Chemistry and Colloidal Routes. Chem. Rev. 2015, 115, 11449–11502.
16. Einarsrud, M.-A.; Grande, T. 1D oxide nanostructures from chemical solutions. Chem. Soc. Rev. 2013, 43, 2187–2199.
17. Rabenau, A. The Role of Hydrothermal Synthesis in Preparative Chemistry. Angew. Chem. Int. Ed. 1985, 24, 1026–1040.

18. Byrappa, K.Y.M. Handbook of Hydrothermal Technology; Elsevier: New York, NY, USA, 2013.
19. Walton, R.I. Subcritical solvothermal synthesis of condensed inorganic materials. Chem. Soc. Rev. 2002, 31, 230–238.
20. Modeshia, D.R.; Walton, R.I. Solvothermal synthesis of perovskites and pyrochlores: Crystallisation of functional oxides under mild conditions. Chem. Soc. Rev. 2010, 39, 4303–4325.
21. European Commission. Report on Critical Raw Materials and the Circular Economy. 2018.
22. European Commission. List of Critical Raw Materials for the EU (COM/2017/0490 Final). 2017.
23. Hagelüken, C. Bedeutung des EU Kreislaufwirtschaftspakets für das Metallrecycling. Chem. Ing. Tech. 2017, 89, 17–28.
24. Espinoza, L.T.; Schrijvers, D.; Chen, W.-Q.; Dewulf, J.; Eggert, R.; Goddin, J.; Habib, K.; Hagelüken, C.; Hurd, A.J.; Kleijn, R.; et al. Greater circularity leads to lower criticality, and other links between criticality and the circular economy. Resour. Conserv. Recycl. 2020, 159, 104718.
25. Benedetti, B.; Caponigro, V.; Ardini, F. Experimental Design Step by Step: A Practical Guide for Beginners. Crit. Rev. Anal. Chem. 2020, 1–14.
26. Leardi, R. Experimental design in chemistry: A tutorial. Anal. Chim. Acta 2009, 652, 161–172.
27. Leardi, R. D-Optimal Designs. In Encyclopedia of Analytical Chemistry; John Wiley & Sons, Ltd.: Chichester, UK, 2018; pp. 1–11.
28. Fitzpatrick, D.; Ley, S. Engineering chemistry for the future of chemical synthesis. Tetrahedron 2018, 74, 3087–3100.
29. Fisher, R.A. The Design of Experiments. Nature 1936, 137, 252–254.
30. Papon, R.; Pierlot, C.; Sharma, S.; Shinde, S.M.; Kalita, G.; Tanemura, M. Optimization of CVD parameters for graphene synthesis through design of experiments. Phys. Status Solidi B 2016, 254, 1600629.
31. San, F.G.B.; Dursun, S.; Yazici, M.S. Optimization of the PEMFC operating parameters for cathode in the presence of PtCo/CVD graphene using factorial design. Int. J. Energy Res. 2019, 43, 4506–4519.
32. Arízaga, G.G.C.; Herrera, G.S.; Fischer, A.; López, O.E.C. Influence of reaction conditions on the growth of GaN rods in an ammono-CVD reactor. J. Cryst. Growth 2011, 319, 19–24.
33. Ramadan, Z.; Im, I.-T.; Park, C.W. Process Optimization and Modeling of the Silicon Growth in Trichlorosilane-Hydrogen Gas Mixture in a Planetary CVD Reactor. IEEE Trans. Semicond. Manuf. 2021, 34, 1–8.
34. Hößler, D.; Ernst, M. Optimization of a TiSi2 formation based on PECVD Ti using DoE methodology. Solid-State Electron. 2019, 158, 51–58.

35. Bucio, T.D.; Khokhar, A.Z.; Lacava, C.; Stankovic, S.; Mashanovich, G.Z.; Petropoulos, P.; Gardes, F.Y. Material and optical properties of low-temperature NH3 -free PECVD SiNx layers for photonic applications. J. Phys. D Appl. Phys. 2017, 50, 025106.

36. Kim, K.; Borojevic, N.; Winderbaum, S.; Duttagupta, S.; Zhang, X.; Park, J.; Hameiri, Z. Investigation of industrial PECVD AlOx films with very low surface recombination. Sol. Energy 2019, 186, 94–105.

37. Yu, D.; Wang, C.; Cheng, X.; Zhang, F. Optimization of hybrid PVD process of TiAlN coatings by Taguchi method. Appl. Surf. Sci. 2008, 255, 1865–1869.

38. Chou, W.-J.; Sun, C.-H.; Yu, G.-P.; Huang, J.-H. Optimization of the deposition process of ZrN and TiN thin films on Si(100) using design of experiment method. Mater. Chem. Phys. 2003, 82, 228–236.

39. Yang, J.; Cheng, Z.; Zhong, X.; Shao, F.; Zhao, H.; Zhuang, Y.; Sheng, J.; Ni, J.; Tao, S. Deposition Behavior of PS-PVD Yttria Partially Stabilized Zirconia Coatings. J. Therm. Spray Technol. 2021, 1–12.

40. Araújo, A.J.; Alves, H.P.; Andrade, R.M.; Campos, L.F.; Macedo, D.A.; Pinho, A.L.; Nascimento, R.M.; Paskocimas, C.A. Designing experiments for the optimization of solid-state synthesis and characterization of alumina-based composites. Ceram. Int. 2019, 45, 8525–8532.

41. Chable, J.; Martin, A.; Bourdin, A.; Body, M.; Legein, C.; Jouanneaux, A.; Crosnier-Lopez, M.-P.; Galven, C.; Dieudonné, B.; Leblanc, M.; et al. Fluoride solid electrolytes: From microcrystalline to nanostructured tysonite-type La0.95Ba0.05F2.95. J. Alloy. Compd. 2017, 692, 980–988.

42. Ryan, K.; Lengyel, J.; Shatruk, M. Crystal Structure Prediction via Deep Learning. J. Am. Chem. Soc. 2018, 140, 10158–10168.

43. Gzyl, A.S.; Oliynyk, A.O.; Adutwum, L.A.; Mar, A. Solving the Coloring Problem in Half-Heusler Structures: Machine-Learning Predictions and Experimental Validation. Inorg. Chem. 2019, 58, 9280–9289.

44. Draheim, C.; De Crécy, F.; Hansen, S.; Collnot, E.-M.; Lehr, C.-M. A Design of Experiment Study of Nanoprecipitation and Nano Spray Drying as Processes to Prepare PLGA Nano- and Microparticles with Defined Sizes and Size Distributions. Pharm. Res. 2015, 32, 2609–2624.

45. Parikh, K.J.; Sawant, K.K. Comparative Study for Optimization of Pharmaceutical Self-Emulsifying Pre-concentrate by Design of Experiment and Artificial Neural Network. AAPS PharmSciTech 2018, 19, 3311–3321.

46. Turk, C.T.S.; Oz, U.C.; Serim, T.M.; Hascicek, C. Formulation and Optimization of Nonionic Surfactants Emulsified Nimesulide-Loaded PLGA-Based Nanoparticles by Design of Experiments. AAPS PharmSciTech 2013, 15, 161–176.

47. Box, J.F.; Fisher, R.A. Design of Experiments. Am. Stat. 1980, 34, 1–7.

48. Antony, J. Fractional Factorial Designs. In Design of Experiments for Engineers and Scientists, 2nd ed.; Elsevier: Amsterdam, The Netherlands, 2014; pp. 87–112.

49. Antony, J. Full Factorial Designs. In Design of Experiments for Engineers and Scientists, 2nd ed.; Elsevier: Amsterdam, The Netherlands, 2014; pp. 63–85. ISBN 9780080994178.

50. Box, G.E.P.; Wilson, K.B. On the Experimental Attainment of Optimum Conditions. J. R. Stat. Soc. Ser. B 1951, 13, 1–45.
51. Arboretti, R.; Ceccato, R.; Pegoraro, L.; Salmaso, L. Design of Experiments and machine learning for product innovation: A systematic literature review. Qual. Reliab. Eng. Int. 2021, 38, 1131–1156.
52. Bezerra, M.A.; Santelli, R.E.; Oliveira, E.P.; Villar, L.S.; Escaleira, L.A. Response surface methodology (RSM) as a tool for optimization in analytical chemistry. Talanta 2008, 76, 965–977.
53. Eriksson, L.; Johansson, E.; Wikström, C. Mixture design-Design generation, PLS analysis, and model usage. Chemom. Intell. Lab. Syst. 1998, 43, 1–24.
54. Holtze, C.; Boehling, R. Batch or flow chemistry?—A current industrial opinion on process selection. Curr. Opin. Chem. Eng. 2022, 36, 100798.
55. Winterton, N. The green solvent: A critical perspective. Clean Technol. Environ. Policy 2021, 23, 2499–2522.
56. Zhang, C.; Amar, Y.; Cao, L.; Lapkin, A.A. Solvent Selection for Mitsunobu Reaction Driven by an Active Learning Surrogate Model. Org. Process Res. Dev. 2020, 24, 2864–2873.
57. Murray, P.M.; Bellany, F.; Benhamou, L.; Bučar, D.-K.; Tabor, A.B.; Sheppard, T.D. The application of design of experiments (DoE) reaction optimisation and solvent selection in the development of new synthetic chemistry. Org. Biomol. Chem. 2015, 14, 2373–2384.
58. Baker, R.W.; Forfar, L.; Liang, X.; Cameron, P.J. Using design of experiment to obtain a systematic understanding of the effect of synthesis parameters on the properties of perovskite nanocrystals. React. Chem. Eng. 2021, 6, 709–719.
59. Byrne, F.P.; Jin, S.; Paggiola, G.; Petchey, T.H.M.; Clark, J.H.; Farmer, T.J.; Hunt, A.J.; McElroy, C.R.; Sherwood, J. Tools and techniques for solvent selection: Green solvent selection guides. Sustain. Chem. Process. 2016, 4, 7.
60. Amar, Y.; Schweidtmann, A.M.; Deutsch, P.; Cao, L.; Lapkin, A. Machine learning and molecular descriptors enable rational solvent selection in asymmetric catalysis. Chem. Sci. 2019, 10, 6697–6706.
61. Shields, B.J.; Stevens, J.; Li, J.; Parasram, M.; Damani, F.; Alvarado, J.I.M.; Janey, J.M.; Adams, R.P.; Doyle, A.G. Bayesian reaction optimization as a tool for chemical synthesis. Nature 2021, 590, 89–96.
62. Roussel, R.; Gonzalez-Aguilera, J.P.; Kim, Y.-K.; Wisniewski, E.; Liu, W.; Piot, P.; Power, J.; Hanuka, A.; Edelen, A. Turn-key constrained parameter space exploration for particle accelerators using Bayesian active learning. Nat. Commun. 2021, 12, 5612.
63. Rauscher, H.; Roebben, G.; Mech, A.; Gibson, N.; Kestens, V.; Linsinger, T.P.J.; Riego Sintes, J. An Overview of Concepts and Terms Used in the European Commission's Definition of Nanomaterial, EUR 29647 EN; Publications Office of the European Union: Luxembourg, 2019.
64. Liu, L.; Corma, A. Metal Catalysts for Heterogeneous Catalysis: From Single Atoms to Nanoclusters and Nanoparticles. Chem. Rev. 2018, 118, 4981–5079.

65. Schmid, G. (Ed.) Nanoparticles; Wiley-VCH Verlag GmbH & Co. KGaA: Weinheim, Germany, 2010.
66. Stark, W.J.; Stoessel, P.R.; Wohlleben, W.; Hafner, A. Industrial applications of nanoparticles. Chem. Soc. Rev. 2015, 44, 5793–5805.
67. Barglik-Chory, C.; Remenyi, C.; Strohm, H.; Müller, G. Adjustment of the Band Gap Energies of Biostabilized CdS Nanoparticles by Application of Statistical Design of Experiments. J. Phys. Chem. B 2004, 108, 7637–7640.
68. Sadat-Shojai, M.; Atai, M.; Nodehi, A. Design of experiments (DOE) for the optimization of hydrothermal synthesis of hydroxyapatite nanoparticles. J. Braz. Chem. Soc. 2011, 22, 571–582.
69. Pu'Ad, N.M.; Haq, R.A.; Noh, H.M.; Abdullah, H.; Idris, M.; Lee, T. Synthesis method of hydroxyapatite: A review. Mater. Today Proc. 2020, 29, 233–239.
70. Golmohammad, M.; Mirhabibi, A.; Golestanifard, F.; Kelder, E.M. Optimizing Synthesis of Maghemite Nanoparticles as an Anode for Li-Ion Batteries by Exploiting Design of Experiment. J. Electron. Mater. 2015, 45, 426–434.
71. Pourmortazavi, S.M.; Sahebi, H.; Zandavar, H.; Mirsadeghi, S. Fabrication of Fe3O4 nanoparticles coated by extracted shrimp peels chitosan as sustainable adsorbents for removal of chromium contaminates from wastewater: The design of experiment. Compos. Part B Eng. 2019, 175, 107130.
72. Jalees, M.I. Synthesis and application of magnetized nanoparticles to remove lead from drinking water: Taguchi design of experiment. J. Water Sanit. Hyg. Dev. 2020, 10, 56–65.
73. Núñez, R.N.; Veglia, A.V.; Pacioni, N.L. Improving reproducibility between batches of silver nanoparticles using an experimental design approach. Microchem. J. 2018, 141, 110–117.
74. Baldassari, S.; Corradi, A.B.; Bondioli, F.; Ferrari, A.; Romagnoli, M.; Villa, C. DOE analyses on aqueous suspensions of TiO2 nanoparticles. J. Eur. Ceram. Soc. 2008, 28, 2665–2671.
75. Sani, S.; Adnan, R.; Iqbal, M.A.M. One-step statistical design of experiment for the screening and optimization of magnetite nanoparticles yields from solvothermal synthesis. Microporous Mesoporous Mater. 2020, 312, 110775.
76. Duff, D.G.; Ohrenberg, A.; Voelkening, S.; Boll, M. A Screening Workflow for Synthesis and Testing of 10,000 Heterogeneous Catalysts per Day– Lessons Learned. Macromol. Rapid Commun. 2003, 25, 169–177.
77. Spolaore, F. Size- and Shape-Controlled Syntheses of Metal and Alloy Nano-Particles by Sustainable and Green Colloidal and Wet Chemistry Routes for Automotive Applications. Ph.D. Thesis, University of Padova, Padua, Italy, 2020.
78. Franceschini, G.; Macchietto, S. Model-based design of experiments for parameter precision: State of the art. Chem. Eng. Sci. 2008, 63, 4846–4872.
79. Khan, R.M. Problem Solving and Data Analysis Using Minitab; John Wiley & Sons, Ltd.: Chichester, UK, 2013.
80. DeMello, J.; DeMello, A. Microscale Reactors: Nanoscale Products. Lab Chip. 2004, 4, 11–15.

81. Elvira, K.S.; Solvas, X.C.; Wootton, R.C.R.; de Mello, A.J. The Past, Present and Potential for Microfluidic Reactor Technology in Chemical Synthesis. Nat. Chem. 2013, 5, 905–915.

82. Taylor, C.J.; Baker, A.; Chapman, M.R.; Reynolds, W.R.; Jolley, K.E.; Clemens, G.; Smith, G.E.; Blacker, A.J.; Chamberlain, T.W.; Christie, S.D.R.; et al. Flow chemistry for process optimisation using design of experiments. J. Flow Chem. 2021, 11, 75–86.

83. Tajoli, F.; Dengo, N.; Mognato, M.; Dolcet, P.; Lucchini, G.; Faresin, A.; Grunwaldt, J.-D.; Huang, X.; Badocco, D.; Maggini, M.; et al. Microfluidic Crystallization of Surfactant-Free Doped Zinc Sulfide Nanoparticles for Optical Bioimaging Applications. ACS Appl. Mater. Interfaces 2020, 12, 44074–44087.

84. Dengo, N.; Faresin, A.; Carofiglio, T.; Maggini, M.; Wu, L.; Hofmann, J.P.; Hensen, E.J.M.; Dolcet, P.; Gross, S. Ligand-free ZnS nanoparticles: As easy and green as it gets. Chem. Commun. 2020, 56, 8707–8710.

85. Abou-Hassan, A.; Sandre, O.; Cabuil, V. Microfluidics in Inorganic Chemistry. Angew. Chem. Int. Ed. 2010, 49, 6268–6286.

86. Chan, E.M.; Mathies, A.R.A.; Alivisatos, A.P. Size-Controlled Growth of CdSe Nanocrystals in Microfluidic Reactors. Nano Lett. 2003, 3, 199–201.

87. Nightingale, A.M.; de Mello, J.C. Microscale synthesis of quantum dots. J. Mater. Chem. 2010, 20, 8454–8463.

88. Nightingale, A.M.; Bannock, J.H.; Krishnadasan, S.H.; O'Mahony, F.T.F.; Haque, S.A.; Sloan, J.; Drury, C.; McIntyre, R.; Demello, J.C. Large-scale synthesis of nanocrystals in a multichannel droplet reactor. J. Mater. Chem. A 2013, 1, 4067–4076.

89. Edel, J.B.; Fortt, R.; Demello, J.C.; Demello, A.J. Microfluidic routes to the controlled production of nanoparticlesElectronic supplementary information ESI available: Image of the central portion of the micromixer chip.

90. Whitesides, G.M. The origins and the future of microfluidics. Nature 2006, 442, 368–373.

91. Plunkett, A.; Eldridge, C.; Schneider, G.A.; Domènech, B. Controlling the Large-Scale Fabrication of Supraparticles. J. Phys. Chem. B 2020, 124, 11263–11272.

92. Degli Esposti, L.; Dotti, A.; Adamiano, A.; Fabbi, C.; Quarta, E.; Colombo, P.; Catalucci, D.; De Luca, C.; Iafisco, M. Calcium Phosphate Nanoparticle Precipitation by a Continuous Flow Process: A Design of Experiment Approach. Crystals 2020, 10, 953.

93. Bagi, S.D.; Yuan, S.; Rojas-Buzo, S.; Shao-Horn, Y.; Román-Leshkov, Y. A continuous flow chemistry approach for the ultrafast and low-cost synthesis of MOF-808. Green Chem. 2021.

94. Valotta, A.; Maier, M.C.; Soritz, S.; Pauritsch, M.; Koenig, M.; Brouczek, D.; Schwentenwein, M.; Gruber-Woelfler, H. 3D printed ceramics as solid supports for enzyme immobilization: An automated DoE approach for applications in continuous flow. J. Flow Chem. 2021, 11, 675–689.

95. Ashraf, M.A.; Tan, J.; Davidson, M.G.; Bull, S.; Hutchby, M.; Mattia, D.; Plucinski, P. Continuous-flow liquid-phase dehydrogenation of 1,4-cyclohexanedione in a structured multichannel reactor. React. Chem. Eng. 2018, 4, 27–40.

96. Örkényi, R.; Beke, G.; Riethmüller, E.; Szakács, Z.; Kóti, J.; Faigl, F.; Éles, J.; Greiner, I. Environmentally Friendly Synthesis of Indoline Derivatives using Flow-Chemistry Techniques. Eur. J. Org. Chem. 2017, 2017, 6525–6532.

97. Fath, V.; Lau, P.; Greve, C.; Weller, P.; Kockmann, N.; Röder, T. Simultaneous self-optimisation of yield and purity through successive combination of inline FT-IR spectroscopy and online mass spectrometry in flow reactions. J. Flow Chem. 2021, 11, 285–302.

98. Perera, D.; Tucker, J.W.; Brahmbhatt, S.; Helal, C.J.; Chong, A.; Farrell, W.; Richardson, P.; Sach, N.W. A platform for automated nanomole-scale reaction screening and micromole-scale synthesis in flow. Science 2018, 359, 429–434.

99. Biyani, S.A.; Qi, Q.; Wu, J.; Moriuchi, Y.; Larocque, E.A.; Sintim, H.O.; Thompson, D.H. Use of High-Throughput Tools for Telescoped Continuous Flow Synthesis of an Alkynylnaphthyridine Anticancer Agent, HSN608. Org. Process Res. Dev. 2020, 24, 2240–2251.

2 GREEN SYNTHESIS OF DE NOVO BIOINSPIRED POROUS IRON-TANNATE MICROSTRUCTURES WITH AMPHOTERIC SURFACE PROPERTIES

Hemali Rathnayake, Sheeba Dawood, Gayani Pathiraja, Kelvin Adrah and Olubunmi Ayodele

Nanoscience/Joint School of Nanoscience and Nanoengineering, University of North Carolina at Greensboro, Greensboro, NC 27401, USA

ABSTRACT

Bioinspired porous microstructures of iron-tannate (Fe(III)-TA) coordination polymer framework were synthesized by catenating natural tannic acid with iron(II), using a scalable aqueous synthesis method in ambient conditions. The chemical composition, morphology, physiochemical properties, and colloidal stability of microstructures were elucidated. The surface area (S_{BET}) and the desorption pore volume were measured to be 70.47 m^2/g and 0. 44 cm^3/g, respectively, and the porous structure was confirmed with an average pore dimension of ~27 nm. Microstructures were thermally stable up to 180 °C, with an initial weight loss of 13.7% at 180 °C. They exhibited high chemical stability with pH-responsive amphoteric properties in aqueous media at pH levels ranging from 2 to 12. Supporting their amphoteric sorption, microstructures exhibited rapid removal of Pb^{+2} from water, with 99% removal efficiency, yielding a maximum sorption capacity of 166.66 mg/g. Amphoteric microstructures of bioinspired metal–phenolate coordination polymers remain largely unexplored. Additionally, natural polyphenols have seldomly been used as polytopic linkers to construct both porous and pH-responsive amphoteric coordination polymer frameworks with a robust structure in both acidic and basic media. Thus, this de novo porous microstructure of Fe(III)-TA and its physiochemical surface properties have opened new avenues to design thermally and chemically stable, eco-friendly, low-cost amphoteric sorbents with multifunctionality for adsorption, ion exchange, separation, storage, and sensing of both anions and cations present in heterogeneous media.

Keywords: *green synthesis; amphoteric surface properties; biopolymers; natural polyphenols; tannic acid*

1. INTRODUCTION

Bioinspired metal–organic coordination polymers with outstanding physiochemical properties are expected to be of intense interest in the development of advanced multifunctional hybrid materials—for example, environmentally benign hybrid materials with amphoteric natures that display cationic behavior below the isoelectric point and anionic behavior at higher pH and offer unique properties, such as high surface activities and selectivity for analytes, a wide isoelectric range, bioactivity, and high interface interactions [1]. Because of these physiochemical properties, amphoteric materials have been attractive as mixed-mode ion-exchange sorbents in the fields of cosmetics, chromatography, oil recovery, electrochemistry, nanoscience, polymer chemistry, and wastewater treatment [1,2]. Specifically, amphoteric sorbents are intended to enhance the sensitivity and selectivity of solid-phase, extraction-based methods, thereby enabling increasingly complex matrixes to be handled. These sorbents are either silica-based or organic polymer-based, with the ion selectivity arising from the presence of ionizable functional groups able to interact with anionic and cationic species [3,4]. To date, however, no bio-based amphoteric porous microstructures with robust coordination polymer frameworks have been developed from low-cost, readily available biomass precursors or biopolymers.

Considering the potential stability of metal–phenolate coordination bonds—together with the growing global demand to broaden the use of renewable resources—synthesis of natural polyphenol-based sorbents with amphoteric characteristics has appeal for the solid-phase extraction of acidic and basic analytes [5,6]. They have the ability to act as multifunctional rigid spacers between cations, contributing to the porosity of the resulting coordination polymer framework, with additional functionalities, such as phenolic, carbonyls, esters, and cyclic ether groups. Therefore, they show great promise as a rich supply of precursors for green chemistry and strong ion exchangers that could charge across the entire pH range of operation, providing opportunities for the development of bioinspired sorbents for environmental remediation.

Coordination polymers (CPs) and their subclass, metal–organic frameworks (MOFs) are highly porous self-assembled nanostructures with high surface areas and defined pore dimensions. They have been used in a wide range of applications, such as gas storage and separation, drug delivery and storage, chemical separation, sensing, catalysis, and bio-imaging [7]. The MOFs precisely position metal-based nodes by connecting them with functional organic ligands via coordination bonds. Compared with traditional porous solids, such as zeolites, activated carbon and mesoporous silica, MOFs allow for designed framework structures and tailored pore environments at the molecular level. Their size, shape, and self-assembly can be carefully controlled via an effective covalent synthesis method to yield three-dimensional (3-D) hierarchical architectures. They offer a new method for fine-tuning structural nodes with known geometries and coordination environments. Their porous structures and the geometric arrangement of inorganic and organic components enable rational design for high surface area platforms [8]. Exploring the potential design and synthesis of naturally abundant and environmentally friendly novel classes of coordination polymer frameworks similar to MOFs' robust metal–carboxylate framework, herein, for the first time, we developed a de novo bio-based coordination polymer framework with amphoteric properties by coordinating Fe+3 metal nodes with a highly branched polytopic organic ligand (tannic acid) to yield a porous and robust structure of iron-tannate coordination polymer framework (Fe(II)-TA). Its amphoteric properties were revealed for the first time, and its potential applicability for heavy metal adsorption from water was demonstrated.

Tannic acid (TA) is a natural polyphenol and one of the cheapest natural abundant functional materials [9,10]. Its five pyrogallol and five catechol groups provide multiple bonding sites with diverse interactions, such as ionic, coordination, hydrogen bonding, and hydrophobic interactions [11,12,13]. Its oxygen-rich functional groups, which include catechol and pyrogallol groups, can coordinate with variety of transition metal ions, forming coordination complexes. They possess metal–phenolic coordination and have been explored in the past [14,15,16,17,18]. A variety of metal–tannate coordination complexes and polymers have been applied as either thin films or particles with tailored properties [14] or in the formation of novel metallogels [17]. The use of TA as either porogen or an additive component in materials science has attracted great attention. This is due not only to its cheap, environmentally friendly, and nontoxic nature, but also its ability to be used as a non-surfactant template. For example, past research has demonstrated that TA can be used as a porogen to tune the porosity of other inorganic particles and make mesoporous materials with tunable mesopore sizes ranging from 6 to 13 nm [19,20,21,22,23]. The coordination complex formation between tannic acid and iron(III) was demonstrated recently, albeit in a very few literature reports [24,25,26,27,28]. A supramolecular assembly of iron(III)-tannic acid metal–organic complex was developed as an antimicrobial spray nanocoating. It has demonstrated its utility as an antimicrobial coating in shoe insoles and fruits [24]. Additionally, a molecular nanoparticle-based Fe(III)-tannic acid complex was previously synthesized using a low-cost, reproducible method and its physicochemical properties (including its capability of inducing autophagy effect in two selected liver cell lines) were studied [25,26,27,28]. Furthermore, a self-assembly of metal ions with tannic acid on various substrates was demonstrated using coordination programming to prepare multifunctional thin films and capsules [29,30,31,32,33].

Despite the aforementioned research advances in bio-based metal–tannic acid coordination materials, very few studies have reported using iron-tannate coordination complexes either as an adsorbent or an adsorptive catalyst for removing the fluorides, organic dyes, and nitrogen compounds present in water resources [34,35,36]. In all these prior studies, metal–tannate was a catenated organometallic complex with no robust coordination framework where the metal nodes precisely coordinated onto the catechol units to make a robust metal–phenolate coordination geometry that controlled the porosity of hierarchical microstructures of Fe(III)-TA coordination polymer frameworks. Thus, our microstructure of Fe(III)-TA coordination polymer frameworks was de novo and possessed porous hierarchical coordination geometry with a robust structural framework similar to MOFs architectures, but constructed from a polytopic organic ligand and metal nodes [7,8]. Additionally, the porous microstructures of iron-tannate coordination polymer frameworks with tailored amphoteric surface properties were not reported, nor were their potential applicability as sorbents to remove heavy metals from water demonstrated. Thus, the work described herein contributed to the first synthesis of a de novo bioinspired sorbent with amphoteric properties.

2. MATERIALS AND METHODS

2.1. Materials

Tannic acid ($C_{76}H_{52}O_{46}$, Molar Mass = 1701.19 g mol^{-1}), iron(II) acetate hexahydrate, and anhydrous ethanol were obtained from Sigma-Aldrich (St Louis, MO, USA). Unless otherwise stated, all chemicals were used as received.

2.2. Procedure for the Preparation of Fe(III)-TA Microstructures

The self-assembled microstructures were prepared at room temperature by dissolving 1:10 molar ratio of tannic acid (0.10 g, 0.05 mmol) to iron(II) acetate (0.10 g, 0.50 mmol) in DI water (each 10 mL in volume). Initial pH of both solutions were recorded. The initial pH of the tannic acid and Fe^{+2} solutions were 3.28 and 5.51, respectively. The iron(II) acetate solution was added into the tannic acid solution in a 50 mL beaker dropwise (0.5 mL/min) with rapid stirring at room temperature. Initially, light purple (Lilac color),) precipitate was formed; gradually, it turned dark purple on complete addition of Fe^{+2} solution. The nanostructure formation was monitored using UV-visible spectroscopy by taking an aliquot from the reaction mixture just after the addition of the ion solution after stirring for one hour. The darker purple suspension that resulted was imaged under TEM. The suspension was transferred into a 50 mL falcon tube and sonicated for one additional hour to break any aggregation. After sonication, 1 mL of sample was taken for morphology analysis using TEM, SEM, and UV-vis. The dark purple precipitate was collected by centrifugation at 10,000 RPM for 10 min. The sample was dried under vacuum at −25 Torr at room temperature to yield dark purple powder (80% yield). The structural composition was analyzed by FT-IR and micro-elemental analysis and oxidation states of each element were confirmed by elemental analyzer and XPS, respectively. FT-IR stretching (cm^{-1}): 3421 (broad peak, –OH), 1689 (sharp peak, metal ion-coordinated carbonyls in pyrogallol units), 1608–1446 (weak multiple peaks, C=C aromatic), 1330 (sharp peak, C-O-C=O), 1106, and 1036 (sharp peaks, C-O-C); elemental composition by XPS: C (45.33%), Fe (11.62%), and O (39.05%); micro-elemental analysis (experimental): C (45.79%), and H (3.19); calculated for empirical formula of $Fe_4C_{72}H_{40}O_{46}$—C (46.38%), Fe (11.98%), H (2.16%), and O (39.47%).

2.3. Characterization

The chemical composition and functional groups were analyzed using Fourier transform infrared spectroscopy (FTIR-Varian 670-IR spectrometer, Agilent Technologies, Santa Clara, CA, USA). The morphology of the microstructures was analyzed using transmission electron microscopy (TEM Carl Zeiss Libra 120 and HR-TEM JEOL2100PLUS with STEM/EDS capability at 120 kV and 200 kV, respectively) and scanning electron microcopy (Carl Zeiss Auriga FIB-FESEM, Carl

Zeiss, Jena, Germany). Bulk elemental analysis was also obtained using energy dispersive X-ray spectroscopy (EDS) from JEOL IT800 SEM/EDS. The thermal stability of the microstructures was analyzed using the thermogravimetric analyzer (Q500). Samples were heated up to 700 °C at increments of 10 °C/min in nitrogen gas flow. The elemental compositions for C and H were obtained from the micro-elemental analyzer. Chemical oxidation states of each element were obtained from X-ray photon spectroscopy (XPS-Escalab Xi+-Thermo Scientific, Thermo Fisher Scientific, Waltham, MA, USA). Elemental compositions for Fe, C, O were also obtained from XPS elemental survey analysis. Full N_2-isotherm with micropore analysis (Physi-13) for single point BET, BET surface area, micropore volume area using t-plot, and micropore volume distribution using DFT were conducted using Micromeritics 3Flex Gas Analyzer by Micromeritics Instrument Corp (Norcross, GA, USA). The sample was degassed under vacuum at 180 °C for 12 h, prior to collection of the N_2-isotherms. The Zeta potential of the microstructures was measured using the Zeta sizer dip cell kit (Malvern Nano ZS, Horiba, London, UK). Zeta potential measurements: microstructures (0.050 g) were dispersed in ethanol (99.5%) and DI water, respectively, and ultrasonicated for 15 min to get a colloidal suspension. The natural pH of sorbents was measured to be 6.22 in ethanol and 3.8 in water. The zeta potential measurements were performed for six different pH solutions (pH = 2, 4, 6, 7, 10 and 12) in ethanol and water at 20 °C using the Zeta sizer dip cell kit (Malvern Nano ZS, UK). The pH of each solution was adjusted by dropwise addition of appropriate amounts of 0.1 M NaOH or 0.1 M HCl. Three measurements of zeta potentials were taken for one sample and the median was recorded. The level of lead(II) ions in aqueous solutions, before and after being soaked with Fe(III)-TA sorbents, was analyzed using a simultaneous inductively coupled plasma optical emission spectrometer (ICP-OES) with trace element analysis capability for wavelengths from 177 to 785 nm (Varian 710-ES ICP Spectrophotometer, Agilent Technologies, Santa Clara, CA, USA) in 3% (v/v) nitric acid solution.

2.4. Lead (Pb^{+2})Adsorption Experiments

The Pb+2 ions adsorption was studied using a series of experiments at varied initial heavy metal ion concentrations ranging from 20 to 500 mg/L. In each test, the microstructures (20.0 mg) were dispersed and soaked in 10 mL of heavy metal ion solutions for 15 min to reach equilibrium. The initial and equilibrium pH were 5.88 and 3.72, respectively. The suspensions were filtered through 0.45 μm membrane filters and analyzed for heavy metal ion concentration using an ICP-OES in 2.5% (v/v) nitric acid solution. The amount of heavy metal ions adsorbed onto the sorbent at equilibrium (qe) was calculated by mass balance equation of heavy metal ions before and after the sorption, using Equation (1):

$$q_e = \frac{(C_i - C_t)V}{W} \tag{1}$$

where q_e is the equilibrium sorption capacity, and C_i and C_t are initial and final heavy metal concentration, respectively, at equilibrium.

3. RESULTS

3.1. Synthesis

The chemistry for the formation of Fe(III)-TA coordination polymers with hexacoordinated Fe(III)-phenolate coordination framework is depicted in **Scheme 1**. As described in the experimental procedure for the preparation of iron-tannate microstructures, our synthesis followed green chemistry principles. We utilized a renewable feedstock, a natural polyphenol as the precursor, water as the solvent, preventing toxic and hazardous chemicals, and room temperature chemical synthesis, providing an energy efficient synthesis approach, to produce bio-based Fe(III)-TA sorbent. In our synthesis, dropwise slow addition of aqueous solution of Fe^{+2} at a rate of 0.5 mL/min into the tannic acid aqueous solution while the reaction was stirring at room temperature yielded a lilac-colored precipitate which gradually turned to dark purple upon complete addition of iron acetate solution (**Figure 1**a,b). The gradual color change from the initial yellow color to purple and eventually dark purple during and after the addition of Fe^{+2} demonstrated that the complex formation was very rapid. Product formation was successful within an hour (**Figure 1**a,b). The final pH of the reaction mixture was 4.99.

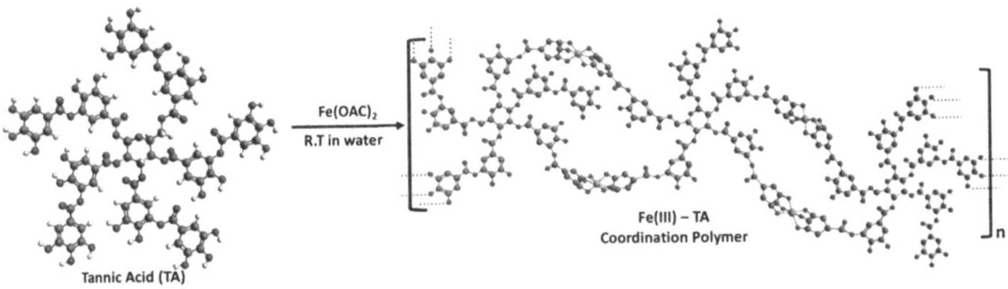

***Scheme 1.**: Synthesis scheme for the formation of Fe(III)-TA by catenating tannic acid with Fe^{+2}.*

The product formation was monitored by collecting UV-visible spectra at the initial stage, during the reaction, and after one hour of reaction time, and comparing them with the UV-visible spectra taken for washed and purified nanoparticles after redispersal in ethanol (Figure 1c,d). The UV-visible spectrum of TA typically exhibited two absorption vibrionic peaks at 214 nm and 276 nm, which corresponded to outer gallic acid units (catechol) and inner gallic acid units bonded to glucose moiety (pyrogallols), respectively. With the addition of Fe^{+2} solution, two characteristic vibrational peaks corresponding to TA showed a gradual red shift, to 241 nm and 326 nm, with an additional broader peak at 577 nm, demonstrating the formation of an oxidized Fe(III)-TA coordination framework. The broader peak at 577 nm corresponded to the charge transfer (CT) band, arising due to complex formations between Fe(III) and TA and confirming the coordination bonding between Fe(III) and hydroxy groups of gallic acids [25].

3.2. Characterization

As depicted in Figure 2a, the FT-IR spectra confirmed the formation of Fe-O bonds from the presence of $\nu_{(Fe\text{-}O)}$ stretch at 660 cm^{-1} and the shift of ester carbonyl of the catechol units, $\nu_{(C=O)}$ from 1720 cm^{-1} to 1693 cm^{-1}. The chemical composition and monomer empirical formulae of Fe(III)-TA were deduced by combining surface analysis data obtained from the XPS elemental survey and micro-elemental analysis, supporting the presence of Fe, C, H, and O with the empirical formula $Fe_4C_{72}H_{40}O_{46}$. Since iron composition was obtained from the XPS elemental survey, a surface analysis method, EDS analysis (Figure S1) on the bulk sample was also performed to support the iron content present in Fe-TA microstructures. The iron content obtained from the EDS analysis closely agreed with both XPS elemental analysis and theoretical elemental composition. Table 1 summarizes the XPS elemental survey and micro-elemental analysis, along with the bonding type and oxidation states of each element.

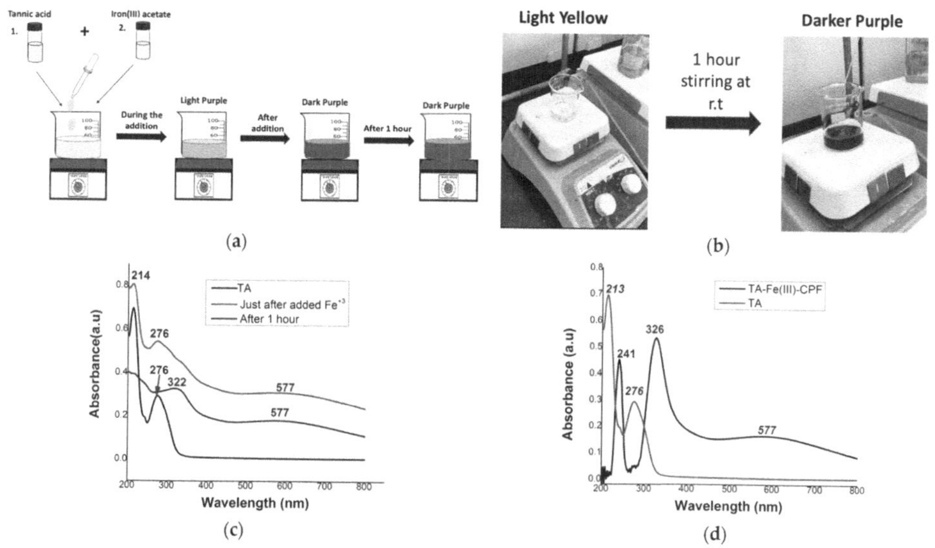

Figure 1.: *(a) Color changes during the reaction progress. (b) Images taken before addition of Fe+3 solution and after 1 h of reaction time. (c) Respective UV-visible spectral traces collected during the reaction progress. (d) A comparison of UV-visible spectra of tannic acid (TA) and purified and redispersed Fe(III)-TA microstructures in ethanol.*

Figure 2b–d represents the binding energy spectra for each element, confirming their oxidation states. The binding energy spectrum of Fe 2p, depicted in Figure 2b, showed that the oxidation state of Fe was +3 state from the corresponding two peaks, at 710.9 eV for Fe $2p_{3/2}$ and 725.9 eV for Fe $2p_{1/2}$, confirming that the metal–phenolate bonds were indeed Fe(III)-O in $[Fe(-CO)_6]$ metal oxide clusters. The accompanied two (less intense) satellite peaks at 714.8 and 735.5 of Fe $2p_{3/2}$ and Fe $2p_{1/2}$ were consistent with the presence of tri-valent Fe, rather than divalent Fe [25,26]. The presence of two well-resolved characteristic binding energy peaks for C 1s at 284.1 eV and 287.6 eV corresponded to C-C and O-C=O bonding, confirming the presence of tannic acid (Figure 2c). In the O 1s spectrum depicted in Figure 2d, Fe(III)-O bonding was confirmed from the peak at 530.2 eV, with a larger FWHM (4.44 eV), which further supported the oxygen chemical bonding state of O^{-2}, corresponding to the phenolate form.

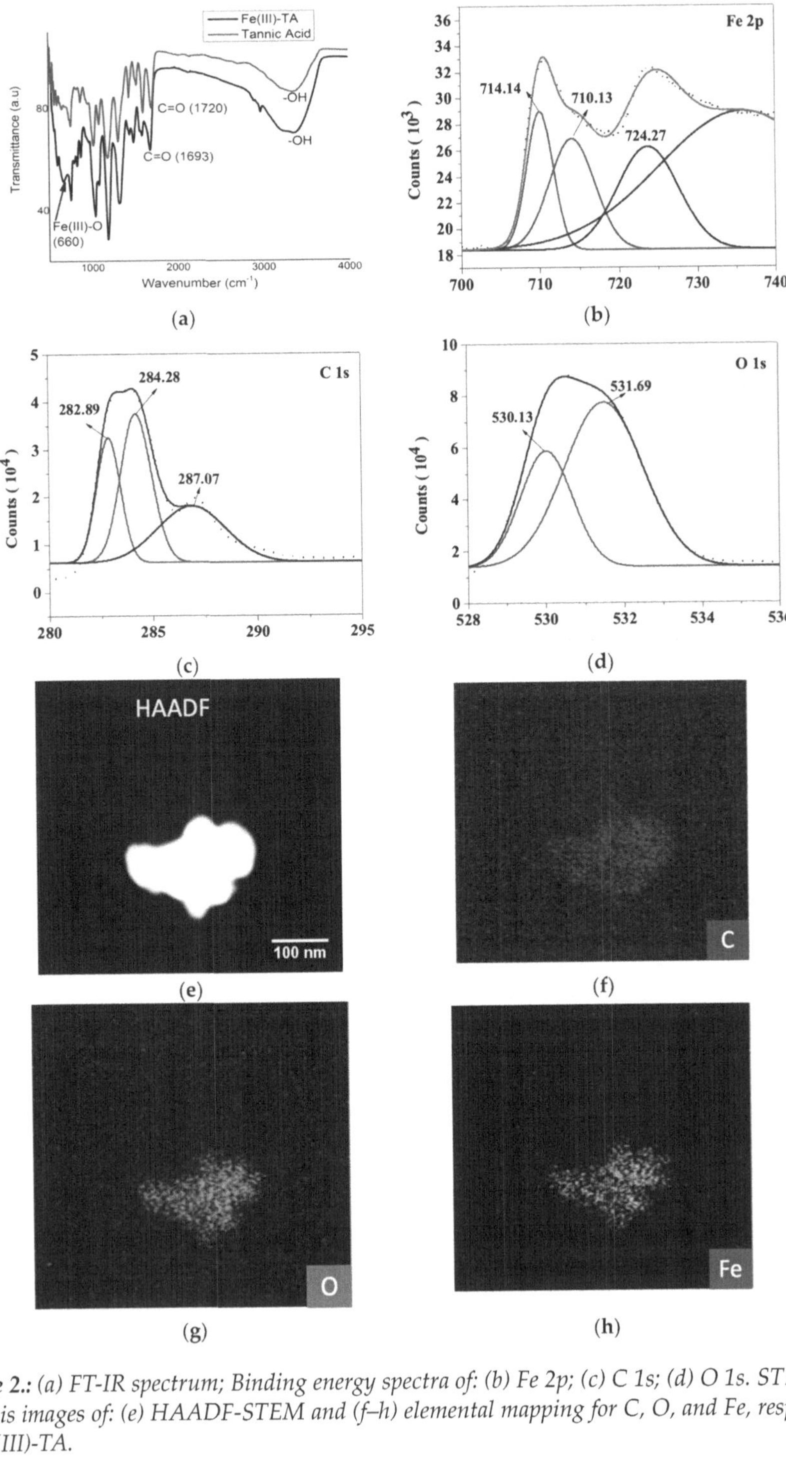

Figure 2.: *(a) FT-IR spectrum; Binding energy spectra of: (b) Fe 2p; (c) C 1s; (d) O 1s. STEM/EDS analysis images of: (e) HAADF-STEM and (f–h) elemental mapping for C, O, and Fe, respectively, for Fe(III)-TA.*

Table 1.: *XPS survey analysis and elemental compositions and binding energies of Fe(III)-TA.*

Elemental Transition Peak	Binding Energy (eV)	Chemical and Oxidation State	Elemental Analysis Experimental	Theoretical
C 1s	284.1, 287.9	C-C (sp^3), O-C=O (sp^2)	45.79	46.28
Fe 2p $2p_{3/2}$ & $2p_{1/2}$	701.9, 714.8, 729.9, 735.5, 723.6	Fe^{+3}, Fe^{+2}	11.78 [a] / 12.20 [b]	11.96
O 1s	530.2, 531.0, 532.4	Fe-O-C, C-O (sp^3), C=O (sp^2)	39.24 [a]	39.39
H	-	-	3.19	2.16

[a] Obtained from XPS elemental survey analysis; [b] Obtained from EDS bulk elemental analysis.

The binding energies of 531.0 eV and 532.4 eV, which corresponded to sp^3 C-O and sp^2 C=O, further supported the presence of glucose core units of the tannic acid. The bulk elemental distribution mapping obtained from the EDS analysis further supported the distribution of iron and the chemical composition of the microstructure (Figure 2e–h) and Figures S1 and S2 in the Supplementary Materials).

3.3. Morphology and Surface Properties of Fe(III)-TA Microstructures

Morphologies of Fe(III)-TA microstructures were studied using scanning electron microscopy (SEM) and transmission electron microscopy (TEM), and are depicted in Figure 3. The SEM image (Figure 3a) of particles exhibited raspberry-like particles with aggregative clusters in a wide size distribution, ranging from an average size of ~50 nm in diameter for small particles to 100–300 nm in diameter for large particles and clusters. The TEM images of microstructures showed large cavities (10–25 nm in diameter, Figure 3b,c) along with nanopores <2 nm (in Figure 3d). The porosity, pore volume distribution and the N_2-adsorption Brunauer–Emmett–Teller Surface area (S_{BET}) were analyzed, utilizing Barret–Joyner–Halenda (BJH) and Brunauer–Emmett–Teller analyses. The BET adsorption-desorption linear isotherm and BJH desorption pore volume plots for BJH desorption cumulative pore volume and BJH desorption dV/dw are depicted in Figure S3 in the Supplementary Materials. Type II N2 isotherm (Figure S3a in the Supplementary Materials), typical for microporous materials, was observed, with a S_{BET} of 70.47 m^2/g and the micropore area of 7.04 m^2/g. The BJH desorption average pore diameter was 27.2 nm, with a BJH desorption cumulative pore volume of 0.44 cm^3/g (Figure S3b,c).

3.4. Thermal and Physiochemical Properties

The thermal stability of Fe(III)-TA, obtained from thermogravimetric analysis of the powder samples, exhibited three distinct weight losses at 180 °C, 350 °C, and 390 °C (Figure S4 in the Supplementary Materials). The initial weight loss of 13.7% at 180 °C could be ascribed to the decomposition of residual hydroxyl groups of peripheral pyrogallol units in the framework. The second and third weight losses at 350 °C and 390 °C demonstrated further decomposition of the pyrogallol and glucose units of the tannic acid, contributing to the partial decomposition of organic content of the framework at 350 °C and the collapsing of the framework and complete decomposition of the organic content at 390 °C, respectively. The experimental weight loss of organic content was ~85%, comparable to the theoretical weight loss of organic content. There was no further weight loss after 390 °C, confirming the formation of carbonaceous products, which included iron oxide and char, contributing to the experimental weight loss of 14.98%. This weight loss was comparable to the theoretical weight loss for iron oxide and char yielded from organic content.

Figure 3.: Morphology of Fe(III)-TA microstructure: (a) SEM image; (b–d) HR-TEM images representing pore distribution.

The pH-dependent surface properties and chemical stability of the microstructures in a colloidal solution of water and ethanol were evaluated by soaking the microstructures in a series of aqueous solutions of HCl and NaOH. The structure of microstructures remained intact over a large pH range: $2 < pH < 12$. The interfacial surface properties of the colloidal solution of Fe(III)-TA in water exhibited negative zeta potentials at all pH levels, reflecting negatively charged particles' surfaces. These contributed to the deprotonation of hydroxyl groups in peripheral pyrogallol units of tannic acid (Table 2 and Figure 4). The magnitude of the zeta potential gradually increased as the pH increased from 2 to 7, and colloids became very stable at pH = 7, exhibiting the highest zeta potential of −45.0 mV. This was evidence of the surface charge stabilization via electrostatic repulsions between particles, thereby avoiding flocculation and providing very good colloidal stability in water at pH 7 and above. In an aqueous solution of alcohol (70% ethanol), the colloidal solution of Fe(III)-TA exhibited amphoteric surface properties, with a positively charged particle surface at pH < 4 and a negatively charged surface at pH > 4.

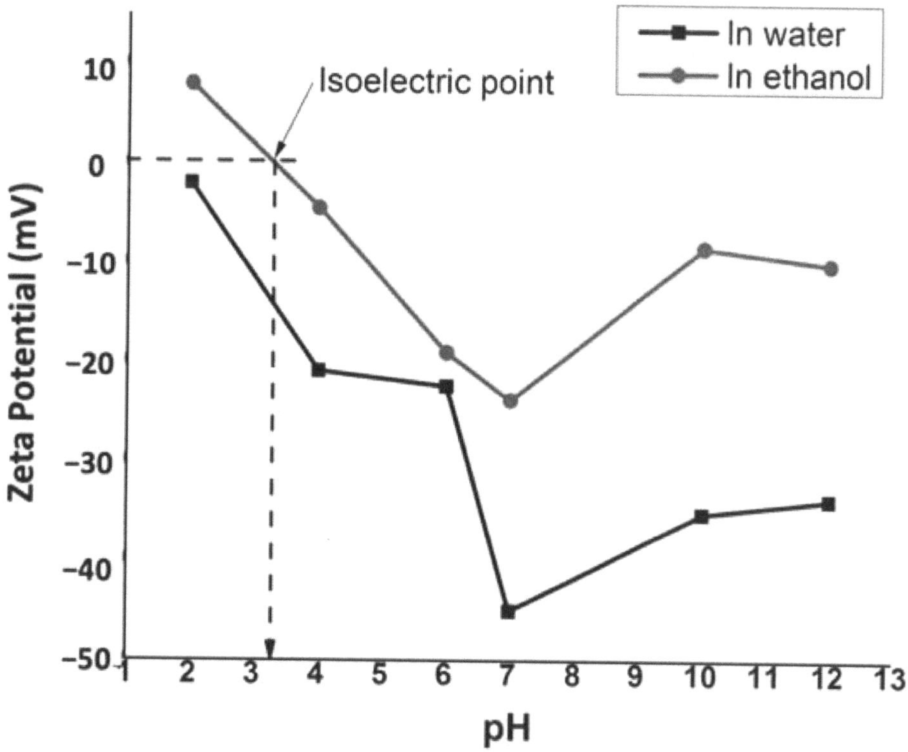

Figure 4.: *Zeta potential vs. pH of Fe(III)-TA colloids in water and ethanol.*

Table 2.: *The physiochemical properties of a colloidal solution of microstructures.*

pH of the Solution	Zeta Potential (eV)	
	in Water	in Ethanol
2	−2.0	8.0

pH of the Solution	Zeta Potential (eV)	
	in Water	in Ethanol
4	−20.9	−4.5
6	−22.4	−19.0
7	−45.0	−23.8
10	−35.2	−8.4
12	−33.8	−10.1

3.5. Heavy Metal Adsorption Studies

To demonstrate the heavy metal sorption properties of our bio-based sorbent, Pb^{+2} ions adsorption efficiency was investigated using water samples with Pb^{+2} ion concentrations ranging from 20 ppm to 500 ppm. The equilibrium sorption capacity, with respect to the Pb^{+2} ion concentration, was calculated using Equation (1), and the respective equilibrium sorption curve and adsorption isotherm are depicted in Figure 5. The equilibrium sorption plot shown in Figure 5a exhibited excellent Pb^{+2} adsorption with 99% efficiency removing lead from water. The adsorption isotherm shown in Figure 5b, which was established using the Langmuir model, exhibited a linear adsorption relationship, as stated by its linear form, Equation (2):

$$\frac{1}{q_e} = \frac{1}{q_m} + \frac{1}{q_m k_l c_e} \tag{2}$$

where q_e is the equilibrium sorption capacity, q_m is the maximum sorption capacity of the sorbent (saturation point), C_e is the concentration of heavy metal ion in solution at equilibrium [mg/L], and k_l is the Langmuir adsorption constant [L/mg].

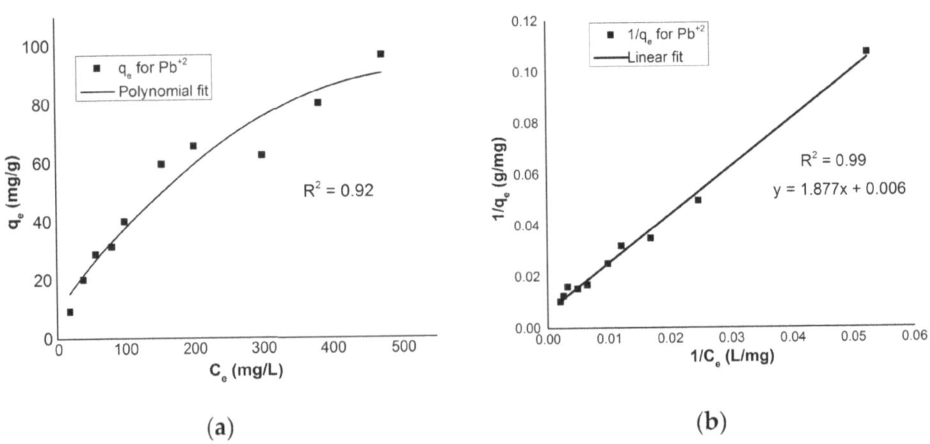

Figure 5.: (a) Equilibrium sorption curve for Pb^{+2} and (b) adsorption isotherm curve, established for Pb^{+2} using Langmuir model.

The experimental sorption data fit very well to the Langmuir equation, with the highest average regression coefficients (R^2 = 0.99), denoting homogeneous adsorption and supporting the clear concept of monomolecular adsorption on homogenous charged surfaces [37]. The Langmuir adsorption constant (k_l), which was related to the apparent energy of sorption, was calculated to be 3.19 × 10^{-3}, and the maximum adsorption capacity (q_m) for Pb^{+2} was 166.66 mg/g. This initial study demonstrated that the porous microstructures of Fe(III)-TA showed energetically favorable adsorption of heavy metals. However, further investigation will be necessary to understand the heavy metal ion–sorbent interactions. Thus, future studies will focus on understanding the host–guest chemistry of heavy metal ion interactions with Fe(III)-TA by investigating the pH-dependent kinetic adsorption isotherms to reveal its adsorption mechanism.

4. DISCUSSION

In summary, as the first bioinspired coordination polymer with a rigid framework synthesized from a polytopic natural polyphenol, the porous microstructure of the Fe(III)-TA coordination polymer demonstrated the use of biomass building blocks to construct environmentally benign, bioinspired porous microstructures using green synthesis, on a large scale, under ambient conditions in water. In a recent work, we reported the crystal structure (CCDC 2108217–2108218) of a dimer of iron(III)-tannate, confirming that Fe^{+3} coordinated to phenolate groups of each catechol unit by catenating up to four catechol units per one tannic acid molecule, supporting our empirical formula for the monomer unit of the coordination polymer framework [38]. In the dimer form, the coordination number for Fe^{+3} was six in its slightly disordered FeO_6 octahedral clusters, resulting in an overall charge-neutral coordination complex. In its dimer form, all three phenolate groups in four catechol units were presumed to be deprotonated, constructing metal–phenolate FeO_6 clusters, supporting the coordination geometry and stoichiometry of the metal–phenolate coordination nodes in the coordination polymer of iron(III)-tannate, constructing a robust framework. Microporosity, amphoteric surface properties, and colloidal stability in water at a wider range of pH, as well as Pb^{+2} adsorption, highlighted the potential of microstructures of Fe(III)-TA and their potential utility as an amphoteric sorbent material for separation, extraction, and upcycling of toxic heavy metals, valuable minerals, and organic contaminants from water resources. For instance, the physiochemical surface properties of microstructures could provide greater insight into interfacial interactions with a variety of impurities present in water, enabling their use as sorbents for contaminant removal and separation from aqueous environments. Thus, our future work will focus on studying the absorption/desorption kinetics of microstructures for the purpose of removing a variety of heavy metal contaminants and other valuable minerals from a variety of water resources.

5. CONCLUSIONS

For the first time, we demonstrated a versatile green synthesis method to make porous iron-tannate microstructures (Fe(III)-TA) and their physiochemical properties. The Fe(III)-TA microstructures were synthesized by catenating the peripheral catechol groups in tannic acid with iron(II)acetate, using an environmentally friendly aqueous synthesis method under ambient conditions. This environmentally benign synthesis method facilitated a cost-effective, easy scale-up process. The structural and chemical compositions were analyzed, confirming the formation of an iron-tannate coordination polymer network with the empirical formula of $[Fe_4C_{76}H_{40}O_{46}]_n$. The surface area, porosity distribution, adsorption/desorption pore volume, and pH-responsive amphoteric behaviors of the iron-tannate microstructures were assessed, confirming their potential as a bioinspired porous material with amphoteric characteristics. Given the oxygen-rich functionality and highly porous robust framework, Fe(III)-TA microstructures have demonstrated the ability to extract Pb^{+2} present in water, with 99% extraction efficiency. The maximum adsorption capacity of the sorbent for Pb^{+2} was found to be 166.66 mg/g. The adsorption was energetically favorable and followed the linear form of the Langmuir isotherm model. Moreover, future studies on understanding and controlling the properties of highly porous, natural, polyphenol-based coordination polymer frameworks, including Fe(III)-TA, have great potential for the development of safe and innovative bio-based sorbents and will advance the host–guest chemistry of natural, amphoteric, polyphenol-based coordination polymer frameworks.

Author Contributions

The manuscript was written by H.R., with contributions from S.D. (65%), G.P. (15%), K.A. (15%), and O.A. (5%). All authors have read and agreed to the published version of the manuscript.

Acknowledgments

This work was performed at the Joint School of Nanoscience and Nanoengineering, a member of the South-eastern Nano-technology Infrastructure Corridor (SENIC) and National Nanotechnology Coordinated Infrastructure (NNCI), supported by the NSF (Grant ECCS-1542174). Authors also acknowledge the DOD HBCU/MSI instrumentation award (Contract #: W911NF1910522) used to acquire HR-TEM (JEOL 2100PLUS) with STEM/EDS capability and USDA NIFA equipment awarded to acquire Field-Emission SEM (JEOL IT800SEM) with EDS capability. Additionally, authors acknowledge Pravalika Butreddy for performing the EDS analysis.

REFERENCES

1. Sarkar, R.; Pal, A.; Rakshit, A.; Saha, B. Properties and applications of amphoteric surfactant: A concise review. J. Surfactants Deterg. 2021, 24, 709–730.
2. Eredyuk, V.; Alami, E.; Nydén, M.; Holmberg, K.; Peresypkin, A.V.; Menger, F.M. Micellization and Adsorption Properties of Novel Zwitterionic Surfactants. Langmuir 2001, 17, 5160–5165.

3. Li, Y.; Huang, C.; Yang, J.; Peng, J.; Jin, J.; Ma, H.; Chen, J. Multifunctionalized mesoporous silica as an efficient reversed-phase/anion exchange mixed-mode sorbent for solid-phase extraction of four acidic nonsteroidal anti-inflammatory drugs in environmental water samples. J. Chromatogr. A 2017, 1527, 10–17.
4. Fontanals, N.; Borrull, F.; Marcé, R.M. Mixed-mode ion-exchange polymeric sorbents in environmental analysis. J. Chromatogr. A 2020, 1609, 460531.
5. Andrade-Eiroa, A.; Canle, M.; Leroy-Cancellieri, V.; Cerdà, V. Solid-phase extraction of organic compounds: A critical review. Part II. TrAC Trends Anal. Chem. 2016, 80, 655–667.
6. Madikizela, L.M.; Ncube, S.; Chimuka, L. Recent developments in selective materials for solid phase extraction. Chromatographia 2019, 82, 1171–1189.
7. Houk, R.J.T.; Jacobs, B.W.; Gabaly, F.E.; Chang, N.N.; Talin, A.A.; Graham, D.D.; House, S.D.; Robertson, I.M.; Allendorf, M.D. Silver Cluster Formation, Dynamics, and Chemistry in Metal–Organic Frameworks. Nano Lett. 2009, 9, 3413–3418.
8. Burtch, N.C.; Jasuja, H.; Walton, K.S. Water stability and adsorption in metal–organic frameworks. Chem. Rev. 2014, 114, 10575–10612.
9. Auriemma, M.; Piscitelli, A.; Pasquino, R.; Cerruti, P.; Malinconico, M.; Grizzuti, N. Blending poly (3-hydroxybutyrate) with tannic acid: Influence of a polyphenolic natural additive on the rheological and thermal behavior. Eur. Polym. J. 2015, 63, 123–131.
10. Makris, D.P.; Boskou, G.; Andrikopoulos, N.K. Polyphenolic content and in vitro antioxidant characteristics of wine industry and other agri-food solid waste extracts. J. Food Compos. Anal. 2007, 20, 125–132.
11. Shutava, T.; Prouty, M.; Kommireddy, D.; Lvov, Y. pH Responsive Decomposable Layer-by-Layer Nanofilms and Capsules on the Basis of Tannic Acid. Macromolecules 2005, 38, 2850–2858.
12. Erel-Unal, I.; Sukhishvili, S.A. Hydrogen-Bonded Multilayers of a Neutral Polymer and a Polyphenol. Macromolecules 2008, 41, 3962–3970.
13. Fan, H.; Wang, L.; Feng, X.; Bu, Y.; Wu, D.; Jin, Z. Supramolecular Hydrogel Formation Based on Tannic Acid. Macromolecules 2017, 50, 666–676.
14. Ejima, H.; Richardson, J.J.; Liang, K.; Best, J.P.; van Koeverden, M.P.; Such, G.K.; Cui, J.; Caruso, F. One-step assembly of coordination complexes for versatile film and particle engineering. Science 2013, 341, 154–157.
15. Li, W.; Bing, W.; Huang, S.; Ren, J.; Qu, X. Mussel Byssus-Like Reversible Metal-Chelated Supramolecular Complex Used for Dynamic Cellular Surface Engineering and Imaging. Adv. Funct. Mater. 2015, 25, 3775–3784.
16. Rahim, M.A.; Kempe, K.; Müllner, M.; Ejima, H.; Ju, Y.; van Koeverden, P.M.; Suma, T.; Braunger, J.A.; Leeming, M.G.; Abrahams, B.F.; et al. Surface-Confined Amorphous Films from Metal-Coordinated Simple Phenolic Ligands. Chem. Mater. 2015, 27, 5825–5832.
17. Rahim, M.A.; Björnmalm, M.; Suma, T.; Faria, M.; Ju, Y.; Kempe, K.; Müllner, M.; Ejima, H.; Stickland, A.D.; Caruso, F. Metal–Phenolic Supramolecular Gelation. Angew. Chem. Int. Ed. 2016, 55, 13803–13807.

18. Rahim, M.A.; Björnmalm, M.; Bertleff-Zieschang, N.; Besford, Q.; Mettu, S.; Suma, T.; Faria, M.; Caruso, F. Rust-Mediated Continuous Assembly of Metal–Phenolic Networks. Adv. Mater. 2017, 29, 1606717.

19. Erwin, W.R.; Zarick, H.F.; Talbert, E.M.; Bardhan, R. Light trapping in mesoporous solar cells with plasmonic nanostructures. Energy Environ. Sci. 2016, 9, 1577–1601.

20. Chen, Y.; Shi, J. Chemistry of Mesoporous Organosilica in Nanotechnology: Molecularly Organic–Inorganic Hybridization into Frameworks. Adv. Mater. 2016, 28, 3235–3272.

21. Gao, Z.; Zharov, I. Large Pore Mesoporous Silica Nanoparticles by Templating with a Nonsurfactant Molecule, Tannic Acid. Chem. Mater. 2014, 26, 2030–2037.

22. Jiang, Y.; Sun, W.; Zhou, L.; Ma, L.; He, Y.; Gao, J. Improved Performance of Lipase Immobilized on Tannic Acid-Templated Mesoporous Silica Nanoparticles. Appl. Biochem. Biotechnol. 2016, 179, 1155–1169.

23. Gao, J.; Lei, H.; Han, Z.; Shi, Q.; Chen, Y.; Jiang, Y. Dopamine functionalized tannic-acid-templated mesoporous silica nanoparticles as a new sorbent for the efficient removal of Cu2+ from aqueous solution. Sci. Rep. 2017, 7, 45215.

24. Park, J.H.; Choi, S.; Moon, H.C.; Seo, H.; Kim, J.Y.; Hong, S.-P.; Lee, B.S.; Kang, E.; Lee, J.; Ryu, D.H.; et al. Antimicrobial spray nanocoating of supramolecular Fe(III)-tannic acid metal-organic coordination complex: Applications to shoe insoles and fruits. Sci. Rep. 2017, 7, 6980.

25. Saowalak, K.; Titipun, T.; Somchai, T.; Chalermchai, P. Iron(III)-Tannic Molecular Nanoparticles Enhance Autophagy effect and T1 MRI Contrast in Liver Cell Lines. Sci. Rep. 2018, 8, 6647.

26. Krungchanuchat, S.; Thongtem, T.; Thongtem, S.; Pilapong, C. Characterization and cellular studies of molecular nanoparticle of iron (III)-tannic complexes; toward a low cost magnetic resonance imaging agent. Biointerphases 2017, 12, 021005.

27. Kim, H.J.; Kim, D.-G.; Yoon, H.; Choi, Y.-S.; Yoon, J.; Lee, J.-C. Polyphenol/FeIII Complex Coated Membranes Having Multifunctional Properties Prepared by a One-Step Fast Assembly. Adv. Mater. Interfaces 2015, 2, 1500298.

28. Yang, L.; Han, L.; Ren, J.; Wei, H.; Jia, L. Coating process and stability of metal-polyphenol film. Colloids Surf. A Physicochem. Eng. Asp. 2015, 484, 197–205.

29. Rahim, M.A.; Ejima, H.; Cho, K.L.; Kempe, K.; Müllner, M.; Best, J.P.; Caruso, F. Coordination-Driven Multistep Assembly of Metal–Polyphenol Films and Capsules. Chem. Mater. 2014, 26, 1645–1653.

30. Nguyen, N.T.T.; Furukawa, H.; Gándara, F.; Trickett, C.A.; Jeong, H.M.; Cordova, K.E.; Yaghi, O.M. Three-Dimensional Metal-Catecholate Frameworks and Their Ultrahigh Proton Conductivity. J. Am. Chem. Soc. 2015, 137, 15394–15397.

31. Yang, S.J.; Antonietti, M.; Fechler, N. Self-Assembly of Metal Phenolic Mesocrystals and Morphosynthetic Transformation toward Hierarchically Porous Carbons. J. Am. Chem. Soc. 2015, 137, 8269–8273.

32. Zhang, P.; Li, H.; Veith, G.M.; Dai, S. Soluble Porous Coordination Polymers by Mechanochemistry: From Metal-Containing Films/Membranes to Active Catalysts for Aerobic Oxidation. Adv. Mater. 2015, 27, 234–239.

33. Wei, J.; Liang, Y.; Hu, Y.; Kong, B.; Zhang, J.; Gu, Q.; Tong, Y.; Wang, X.; Jiang, S.P.; Wang, H. Hydrothermal Synthesis of Metal–Polyphenol Coordination Crystals and Their Derived Metal/N-doped Carbon Composites for Oxygen Electrocatalysis. Angew. Chem. 2016, 128, 12658–12662.
34. Aktar, J.; Ray, M. Iron-polyphenol nanocluster removes fluoride and methylene blue dye from water and promotes plant growth. SSRN 2022, 32.
35. Elmorsi, T.M. Synthesis of Nano-Titanium Tannate as an Adsorbent for Crystal Violet Dye, Kinetic and Equilibrium Isotherm Studies. J. Environ. Prot. 2015, 6, 1454–1471.
36. Zhang, R.; Li, L.; Liu, J. Synthesis and characterization of ferric tannate as a novel porous adsorptive-catalyst for notrogen removal from wastewater. RSC Adv. 2015, 5, 40785–40791.
37. Chen, J.P. Decontamination of Heavy Metals: Processes, Mechanisms, and Applications; CRC Press: Boca Raton, FL, USA, 2012.
38. Rathnayake, H. CCDC 2108217: Experimental Crystal Structure Determination; The Cambridge Crystallographic Data Centre (CCDC): Cambridge, UK, 2021.

3

EFFECT OF FINE STRUCTURE AND PORE VOLUME IN THE MARIMO-LIKE CARBON CATHODE MATERIAL ON THE OXYGEN REDUCTION REACTION FOR THE POLYMER ELECTROLYTE FUEL CELL

M. Shiraishi[1], M. Inamoto[2], K. Nakagawa[3], T. Ando[4] and M. Nishitani-Gamo[5]

[1] *Graduate School of Science and Engineering, Course of Applied Chemistry, Toyo University, Saitama, Japan.*

[2] *Research Institute of Industrial Technology, Toyo University, Saitama, Japan.*

[3] *Saitama Industrial Technology Center (SAITEC), Saitama, Japan.*

[4] *Department of Chemical Engineering, Kansai University, Osaka, Japan.*

[5] *National Institute for Materials Science (NIMS), Ibaraki, Japan. 6 Department of Applied Chemistry, Toyo University, Saitama, Japan.*

ABSTRACT

We measured the rotating ring-disk electrode (RRDE) voltammetry curves of a Pt catalyst supported on Marimo-like carbon (MC) to clarify the effect of carbon nanofilaments (CNFs) morphology and fine structures consisting of MC cathode material on the oxygen reduction reaction (ORR) activity for Polymer electrolyte fuel cells. The specific surface area and pore volume of MCs are influenced by their morphology and fine structures. MCs grown with Ni–Cu bimetal catalyst (represented as Ni80Cu20MC), having an octopus-like morphology, yielded four times larger total pore volume than that measured from the MC grown with Ni catalyst (Ni100MC). In the case of Ni80Cu20MC, coin-stacked graphenes perpendicular to the filament axis formed CNFs. Ni100MC consists of cup-stacked CNFs. The difference in the graphene stacked fine structure resulted in a different number of graphene edges covering the CNF surface; however, both supported Pt particle sizes and their distributions were similar. Linear sweep voltammetry with a RRDE revealed that the Pt/Ni80Cu20MC ORR onset potential was higher than that observed from the Pt/Ni100MC. This value is suggested to be correlates with the ORR activity. The origin of the ORR difference was discussed on the basis of the structural difference in the MCs.

Keywords: *Marimo-like carbon, Carbon nanofilament, Polymer electrolyte fuel cell, Cathode, Pt, RRDE, ORR,Catalyst support*

1 INTRODUCTION

Polymer electrolyte fuel cells (PEFC) generate clean electric power from water formation energy with an efficiency ratio of more than 80%. The redox reaction occurs at the Pt electrode catalyst, consisting of a membrane electrode assembly (MEA) [1]. Carbon black is widely used as a Pt catalyst support material. A higher crystallinity of the support material is desirable for a long-range lifetime and a higher power generation performance [2], and prompt water removal from the system is also important [3].

We reported a durable, higher electrocatalytic activity on Marimo-like carbon (MC) supported Pt for a PEFC electrode catalyst [4]. The MC is a novel spherical carbon, consisting of carbon nanofilaments (CNFs) formed by the catalytic decomposition of hydrocarbons over transition metal catalysts supported on oxidized diamond powder [5]. CNFs are fixed on the diamond support surface, and a space volume formed by the CNFs is maintained to allow for water drainage. The filament fine structure is a cup-stacked or coin-stacked structure formed by many graphenes, namely, the crystallinity of CNF is much higher than that of carbon black. Several graphene edges exist on the CNF surface, and they play an indispensable role in highly dispersed Pt support sites. We assumed and studied whether the unique structural characteristics of MC, consisting of several graphene edges on the CNF surface and an appropriate space volume formed by the CNFs for the material transport such as fuel gas and water drain, could improve the MEA performance compared to that obtained using a commercially available Pt catalyst supported on the carbon black [6].

The morphology and fine structure of a fibrous carbon nanomaterial are strongly affected by the physicochemical states of the catalyst, such as the type of transition metal, its chemical states, and the addition of the second elements [7]. Catalytic growth of carbon filament using bimetal catalysts, such as Ni–Cu [8,9,10], Fe–Cu [11], and Co–Cu [12] was used and the effect of Cu addition on the carbon filament growth was discussed. Bernardo and coworkers reported that the Ni–Cu bimetal catalyst yielded "Octopus carbon," such as multiple CNF growths from a single catalyst particle [9]. We found that adding 20 wt% of Cu to the Ni catalyst resulted in an MC with several CNFs generated from a single catalyst particle, resembling an octopus, and established growth conditions to control the number of CNFs grown from a single catalyst particle [13]. This growth process can realize a well-controlled space volume formed by the CNFs consisting of MC, and we expect that it can assist to develop the PEFC performance, as the space plays an indispensable role in supporting gas diffusion and removing water produced by the redox reaction.

In this study, we measured the rotating ring-disk electrode (RRDE) voltammetry curves of Pt catalyst supported on the MCs, one is grown with Ni catalyst, and the other with Ni–Cu catalyst to determine the effect of CNF morphology and fine structures, consisting of the MC as a Pt catalyst support on the oxygen reduction reaction (ORR) activity.

2 EXPERIMENTAL

2.1 MC consists of a different type of CNFs for Pt support material

The details of the MC synthesis procedure were described in a previous report [13]. Our systematic study revealed that the oxidized diamond-supported Ni catalyst produced a cup-stacked structure, whereas Ni–Cu catalyst, with an 80 wt% Ni and 20 wt% Cu ratio produced four or five multiple CNF growths from a single catalyst particle, namely, an octopus-like CNF, with a platelet structure (also known as a coin-stacked structure). The reaction gas was methane, and the reaction time was 3 h. The reaction temperature was 550 °C for the Ni catalyst and 600 °C for the Ni80Cu20 catalyst. In this study, MC synthesized using oxidized diamond-supported Ni100 and Ni80–Cu20 catalysts are referred to as Ni100MC and Ni80Cu20MC, respectively.

The structure of the MC depended on the composition of the catalyst used for the synthesis. Scanning electron microscopy (SEM; Regulus 8230, Hitachi High-Tech Corporation, Tokyo, Japan) was used to examine the morphology of the obtained MC. The SEM images of the synthesized MCs are shown in Fig. 1. The brighter points represent the metal catalyst particles. In the case of Ni100MC, single carbon nanofilament growth from a single catalyst particle is shown in Fig. 1a. For Ni80Cu20MC shown in Fig. 1b, multiple carbon nanofilaments grow from a single catalyst particle-like as an octopus. The role of added Cu for the multiple CNF growths is suggested to yield an inhomogeneous composition of Ni–Cu particles, and we speculate that the particle having a partially covered surface with Cu, resulted in the multiple CNFs growths [13].

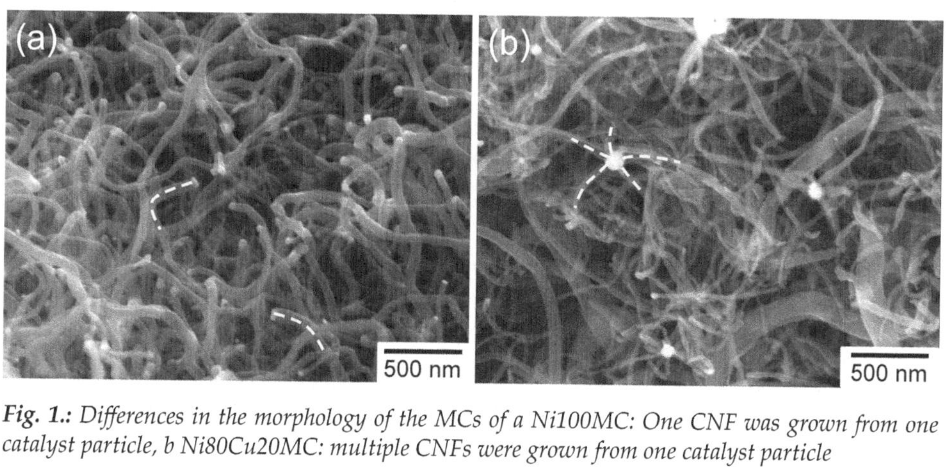

***Fig. 1.**: Differences in the morphology of the MCs of a Ni100MC: One CNF was grown from one catalyst particle, b Ni80Cu20MC: multiple CNFs were grown from one catalyst particle*

Numerous spaces were observed among the CNFs shown in Fig. 1. The space volume, a pore volume formed by the CNFs, is critical to water drainage and gas diffusion, as stated in a previous section. The specific surface area and porous characterization of MCs using N2 adsorption at 77 K were performed using Belsorp Mini (Microtrac BEL Corp.) [14]. The Brunauer–Emmett–Teller (BET) equations were used to determine the apparent surface area (SBET), and the Barrett–

Joyner–Halenda method was used to calculate the pore volume. The morphology, fine structure, surface area, and three types of pore volumes of the Ni100MC and Ni80u20MC are summarized in Table 1. The SBET of Ni80Cu20MC is almost 1.8 times higher than that of Ni100MC. The total pore volume (Vtotal) of Ni80Cu20MC is four times larger than that of Ni100MC. The macropore, having 50 nm or more in size, and volume (Vmacro) of Ni80Cu20MC is 7.6 times larger than that of Ni100MC. The mesopore, ranging from 2 to 50 nm in size, and the volume (Vmeso) of Ni80Cu20MC is also 2.7 times larger than that of Ni100MC. These pore volumes affected the larger Vtotal of Ni80Cu20MC. These pores of Ni80Cu20MC correlate with the octopus-like morphology. This larger space volume of Ni80Cu20MC can promote the transport of product water and fuel gas because the Ni100MC is used for the Pt support material to fabricate MEA in the PEFC [15, 19].

Table 1.: *The characteristics of the Marimo-like carbons for the RRDE measurement*

	Used catalyst **Carbon yield/ mol/mol Ni**	**Mor-phology**	**Orienta-tion of graphite**	S_{BET} **(m²/g)**	V_{micro} **(mL/g)**	V_{meso} **(mL/g)**	V_{macro} **(mL/g)**	V_{total} **(mL/g)**
Ni100MC	Ni100(5wt%)/O-dia 531.1	Whisker carbon	Cup-stacked	81	0.02	0.09	0.05	0.16
Ni80Cu20MC	Ni80Cu20(5wt%)/O-dia 623.4	Octopus carbon	Platelet (Coin-stacked)	148	0.02	0.24	0.38	0.64

S_{BET}: BET surface area, V_{micro}: Micropore volume (<2 nm), V_{meso}: Mesopore volume (2–50 nm), V_{macro}: Macropore volume (>50 nm), V_{total}: Total pore volume

We revealed a difference in a graphene-stacked fine structure of CNFs consisting of the MCs using transmission electron microscopy (TEM; JEM-2100, JEOL Ltd.). The nanostructures of the MCs synthesized from these different catalysts, differ from each other. The TEM images of the synthesized MCs are shown in Fig. 2. The white lines represent a part of the stacked graphene. CNFs grown from Ni100 catalysts had a cup-stacked structure (Fig. 2a); the oriented graphene sheets indicated around 35 degrees to the fiber axis, indicated as $sin\theta_{Ni100}$. However, the CNFs grown from Ni80Cu20 catalysts had a coin-stacked structure; the graphene sheets were oriented at an angle of 90 degrees with respect to the fiber axis ($sin\theta_{Ni-80Cu20}$). Therefore, CNFs grown from Ni80Cu20 catalysts have more graphene edges per unit surface area than Ni100. The distance between graphene edges is $d/sin\theta$, where θ is the angle between the fiber axis and graphene and d is the face spacing of graphite. Therefore, the number of graphene edges per unit length, n is $sin\theta/d$. The ratio of the values of n for Ni100 and for Ni80Cu20 is $n_{Ni80Cu20}/n_{Ni100} = 1/sin\theta_{Ni100} = 1/sin35° = 1.7$. The estimated number of edges included in Ni80Cu20MC per unit surface area was almost 1.7 times larger than those in Ni100MC. The graphene edges are critical to Pt particle support sites and realize a highly dispersed, narrower Pt diameter distribution, as well.

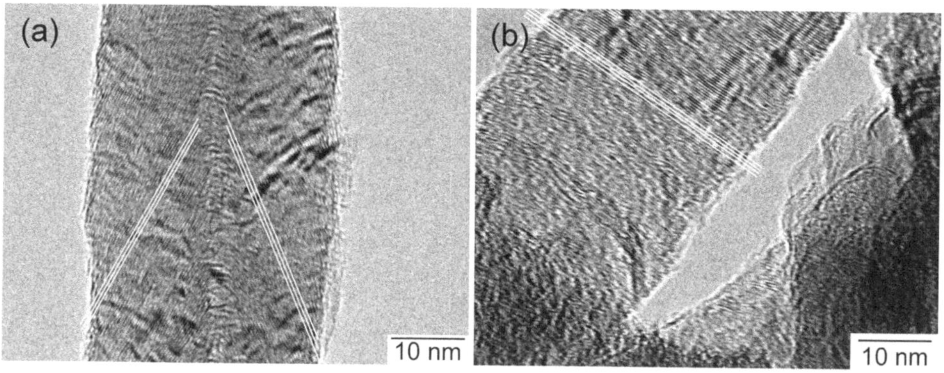

Fig. 2.: TEM images of the MC nanostructure. a Ni100MC consists of CNFs with a cup-stacked structure. b Ni80Cu20MC consists of CNFs with a platelet, coin-stacked structure. This structure difference was mainly due to a Cu addition to the Ni catalyst for the MC growth [13]

2.2 Preparation of the Pt-supported MC

The modified nanocolloidal method was used to load Pt particles onto the MC. The preparation flow of the Pt-supported MC is shown in Fig. 3. The actual number of Pt loadings was determined using thermogravimetry (Exstar TG/DTA6200, Seiko Instruments Inc.). Rapid combustion of the sample causes a powder to float in the platinum pan, resulting in an artifact effect in the TG curve. To prevent such failure, the temperature increase rate was decreased in the duration of a large weight loss rate. For Ni100MC, the temperature was raised from room temperature to 480 °C at 5 °C/min, from 480 to 640 °C at 1 °C/min, and finally from 640 to 900 °C again at 5 °C/min. For the Ni80Cu20MC, the temperature was increased from room temperature to 380 °C at 5 °C/min, then from 380 to 560 °C at 1 °C/min, and finally from 560 to 900 °C again at 5 °C/min. The remaining weight after the measurement was considered the amount of Pt, and the weight ratio of Pt loadings to MCs was calculated.

2.3 Electrochemical measurements with RRDE

For the half-cell ORR measurements, the catalyst ink was prepared by ultrasonically irradiating 10 mg of the MC-supported Pt catalyst, 0.45 mL of a propanol solution, and 32 μL of 5 wt% Nafion solution for 15 min. The ratio of Nafion to catalyst was determined using a previous report [15]. We also used a commercially available catalyst (TEC10E50E; Tanaka Kikinzoku Kogyo K.K. Pt loadings 46.7 wt%), which is a Pt catalyst supported on carbon black for an MEA electrode to prepare a catalyst ink. It was prepared with 0.45 mL of propanol solution and 54-μL of 5 wt% Nafion solution using 10 mg of the TEC10E50E catalyst. A Glassy carbon disk electrode was polished, ultrasonically cleaned in water, and dried. A 5.04-μL catalyst ink was pipetted onto the disk electrode and dried at 55 °C for 1 h to form a catalyst thin film followed by drying at room temperature for 15 min. Table 2 summarizes the amount of Pt on the disk electrode used for the RRDE measurements.

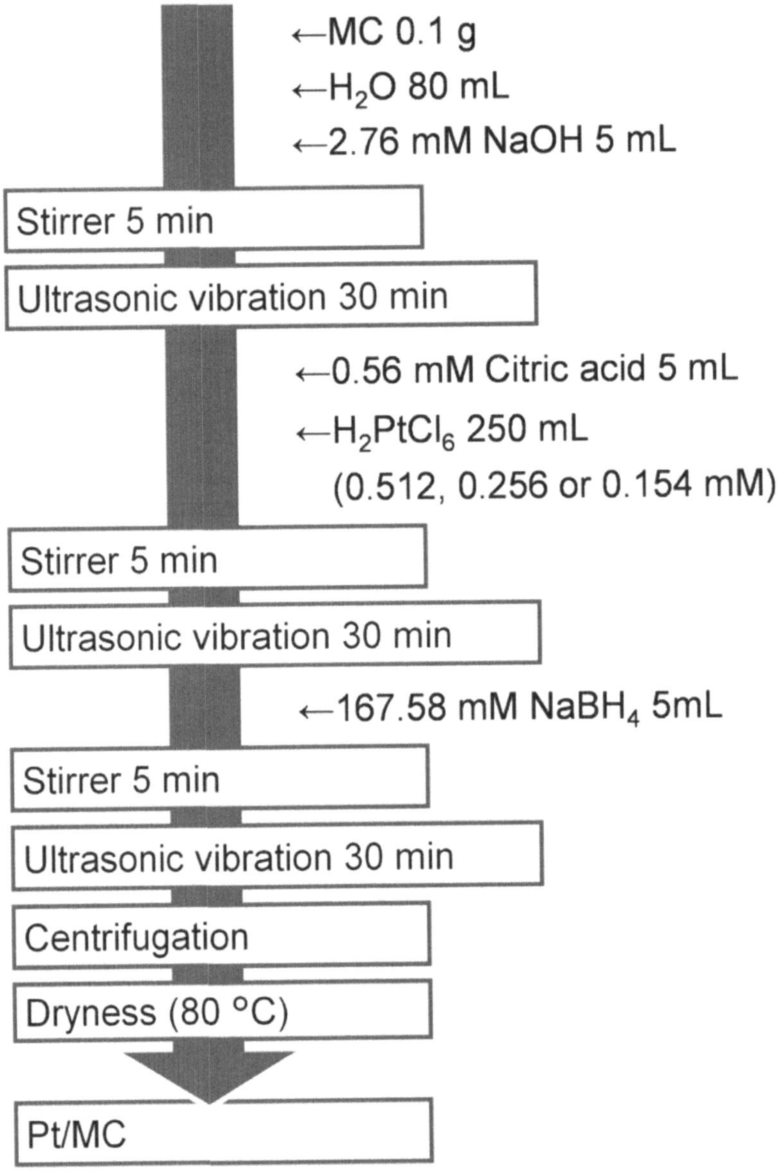

Fig. 3.: Preparation flow of the MC-supported Pt catalyst

Table 2.: *The amount of Pt on the disk electrode used for the RRDE mesurements*

Sample	The amount of Pt on the disk electrode (μg)	The concentration of Pt on the disk electrode (μg/cm²)
Pt(7.2 wt%)/Ni100MC	7.53	38.41
Pt(14.8 wt%)/Ni80Cu20MC	15.32	78.17

Sample	The amount of Pt on the disk electrode (μg)	The concentration of Pt on the disk electrode (μg/cm²)
Pt(11.9 wt%)/Ni80Cu20MC	12.57	64.12
Pt(6.9 wt%)/Ni80Cu20MC	7.43	37.92
TEC10E50E(Pt 46.7 wt%)	46.23	235.88

For the RRDE measurements, a Hg/Hg_2SO_4 electrode (MSE) in saturated K_2SO_4 solution was used as the reference electrode, the counter electrode was platinum wire, and the electrolyte solution was O_2 saturated 0.1-M $HClO_4$. The ORR current was measured by sweeping the potential from 0.4 to −0.59 V (vs MSE) at 10 mV/s at the rotation rates of 100, 400, 900, 1600, and 2500 rpm. All potentials in this study, however, refer to that of the reversible hydrogen electrode.

3 RESULTS AND DISCUSSIONS

3.1 Pt particles supported states over the MCs

Figure 4 shows TEM images of the MC-supported Pt catalyst. The effect of CNF fine structure on the supported states of the Pt particles when the amount of Pt is almost the same, and (a) Ni100MC is a cup-stacked structure, (b) Ni80Cu20MC is a coin-stacked one, is shown in Fig. 4a and b, respectively. Although the CNF fine structure between Ni100MC and Ni80Cu20MC differs, the Pt particle size and the loaded density are comparable. The average Pt diameter is estimated in the range of 2–4 nm. The loaded Pt ratio increased twice higher in the case of 14.8 wt% Pt-loaded Ni80Cu20MC (Fig. 4c) than in the case of 6.9 wt% (Fig. 4b), and the Pt loaded increased density and particle average size are comparable (Fig. 4a and b). We measured the size and number of Pt particles in the TEM images and described histograms to better understand the Pt particle diameter distribution. The histograms of the loaded Pt particle diameter distribution obtained from TEM images of (a) Pt(7.2 wt%)Ni100MC, (b) Pt(6.9 wt%)Ni80Cu20MC, and (c) Pt(14.8 wt%)Ni80Cu20MC, are shown in Fig. 5. These histograms show a similar profile and a peak in the range of 2–4 nm. These Pt-supported states were similar, but the MC fine structure and the amount of supported Pt differ. Both Ni100MC and Ni80Cu20MC include several graphene edges on the CNF surface. These edges provide Pt support sites for achieving uniform, dispersive Pt support states on the CNF surface.

3.2 ORR measurements using the RRDE method

Figure 6 shows the ring electrode currents of rotating ring-disk electrode voltammetry for the ORR of the samples shown in Table 2. We also measured the TEC10E50E (Tanaka Kikinzoku Kogyo K.K. Pt loadings 46.7 wt%) at the same time as a reference. The ring electrode currents of Pt/Ni80Cu20MC was very small in the same level of TEC10E50E, and while the Pt/Ni100MC indicated two times higher currents, the value is also small, as shown in Fig. 6. The measured ring currents suggest that the side reaction can be negligible in this study. We calculate number of electrons transferred of Pt/MCs and TEC10E50E catalysts with using

Koutecky-Levich equation, as shown in Table 3. The value is in the range from 3.3 to 3.8 obtained from Pt/MCs and around at 3.6 calculated from TEC10E50E. These results suggest that a four-electron reaction mainly occurred with negligible two-electron reaction.

Figure 7 shows the RRDE voltammetry curves for the ORR of Pt(7.2 wt%) Ni100MC (red line), and Pt(6.9 wt%)Ni80Cu20MC (blue line). We determined the ORR starting potential as the measured current density reached 5% of the limited current density, calculated as 4.0 mA/cm^2. In the case of Pt(7.2 wt%)Ni100MC, the ORR start potential was 0.87 V, and in the case of Pt(6.9 wt%)Ni80Cu20MC, it was 0.90 V. The higher ORR starting potential indicates a higher Pt activity. The Pt catalyst loaded on the Ni80Cu20MC is more active than in the case of Ni100MC on the ORR. The Pt particle size distributions of Pt(7.2 wt%)Ni100MC and Pt(6.9 wt%) Ni80Cu20MC were similar (Fig. 5).

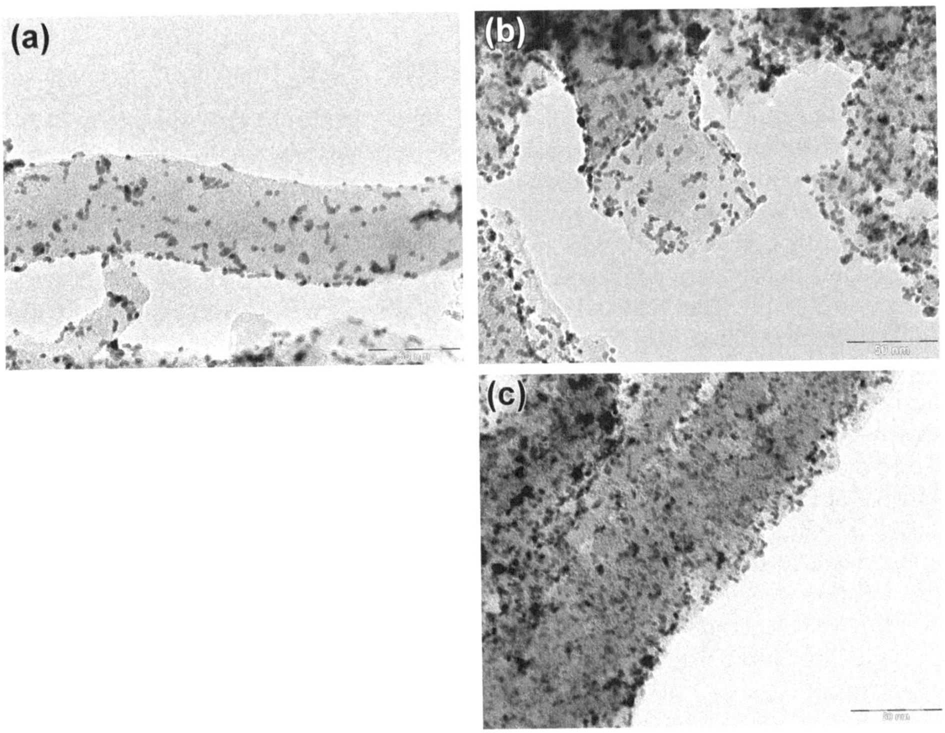

Fig. 4.: TEM images of the Pt/MCs. a Pt(7.2 wt%)Ni100MC, b Pt(6.9 wt%)Ni80Cu20MC, and c Pt(14.8 wt%)Ni80Cu20MC

The difference in the ORR starting potential between Pt(7.2 wt%)Ni100MC and Pt(6.9 wt%)Ni80Cu20MC is not explained by the Pt particle size distribution, but rather by a fine structure difference in Ni100MC and Ni80Cu20MC, particularly, the number of graphene edges. Tsuji and coworkers reported that the PtRu catalyst particle size did not affect the oxidation reactivity of methanol when different structures of CNFs were used [16].

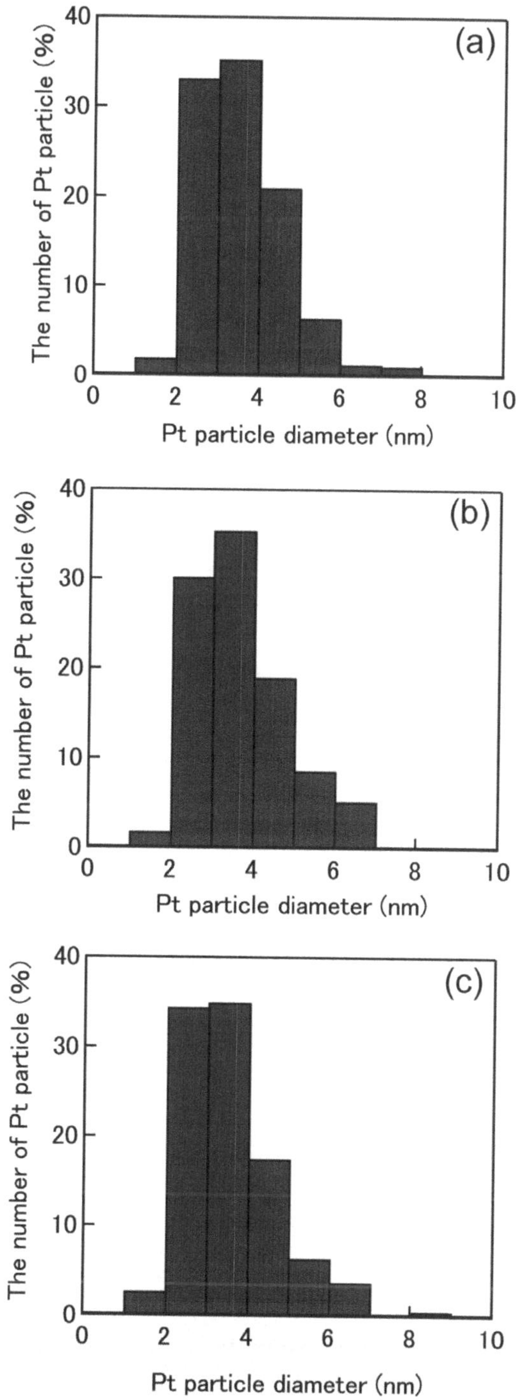

Fig. 5.: *A diameter distribution of the Pt particles supported on the MCs. a Pt(7.2 wt%) Ni100MC, b Pt(6.9 wt%)Ni80Cu20MC, and c Pt(14.8 wt%)Ni80Cu20MC*

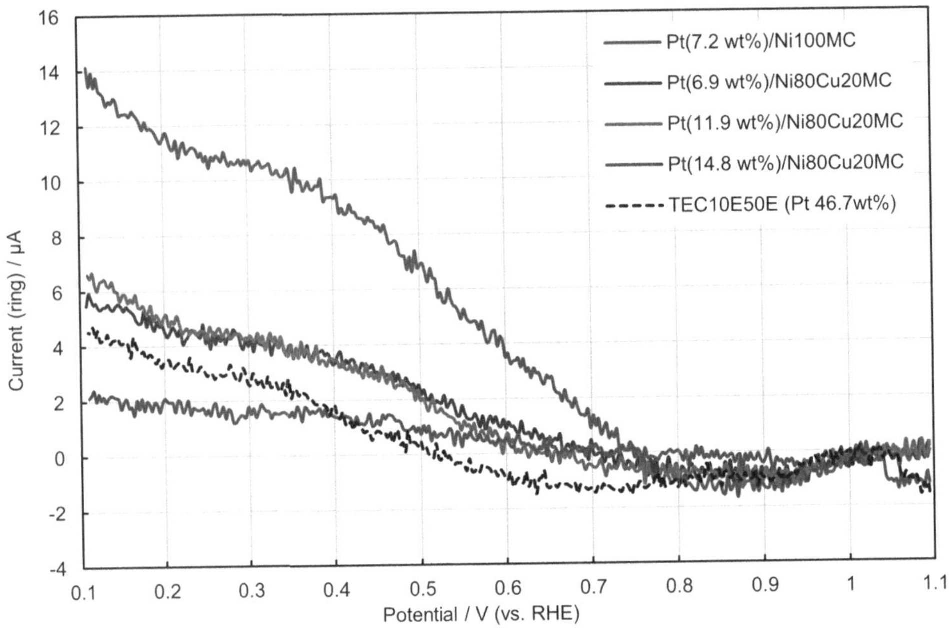

Fig. 6.: *The ring electrode currents of rotating ring-disk electrode voltammetry for the oxygen reduction reaction in an O_2-saturated 0.1-M $HClO_4$ solution over the samples shown in Table 2. The scan rate is 10 mV/s at 900 rpm. The results obtained from Pt-loaded MCs as Pt(7.2 wt%)Ni100MC (red line), Pt(6.9 wt%)Ni80Cu20MC (blue line), Pt(11.9 wt%)Ni80Cu20MC (green line), and Pt(14.8 wt%)Ni80Cu20MC (purple line) are indicated. The TEC10E50E (Tanaka Kikinzoku Kogyo K.K. Pt loadings 46.7 wt%) result is also shown as a reference with a black dotted line*

Table 3.: *Number for electrons transferred of Pt/MCs and commercially avairable Pt/C catalyst*

Sample	**Number of electrons transferred**
Pt(7.2 wt%)/Ni100MC	3.67
Pt(6.9 wt%)/Ni80Cu20MC	3.79
Pt(11.9 wt%)/Ni80Cu20MC	3.63
Pt(14.8 wt%)/Ni80Cu20MC	3.34
TEC10E50E (Pt loadings 46.7 wt%)	3.59

Bessel and coworkers also studied the effect of a support material structure on the Pt catalyst activity in methanol oxidation [17]. They reported that a fine support material structure, such as a stacked graphene orientation of CNFs, caused a variation in Pt crystallinity, implying that the Pt catalyst activity should be varied. Zheng and coworkers revealed that the electronic state of CNF-supported Pt catalysts is affected by the difference in the CNF fine structure using X-ray photoelectron spectroscopy [18]. They speculated that the CNF fine structure influenced the interaction between Pt catalysts and the support material, with a stronger interaction, resulting in higher Pt catalytic activity for the ORR. We also observed a higher ORR onset potential which correlates to the catalytic activity using Ni80Cu20MC as a support material. The number of graphene edges of Ni80Cu20MC is estimated to be 1.7 times larger than that of Ni100MC, and there may be a stronger interaction between Pt particles and Ni80Cu20MC, and the interaction possibly affected the higher ORR activity.

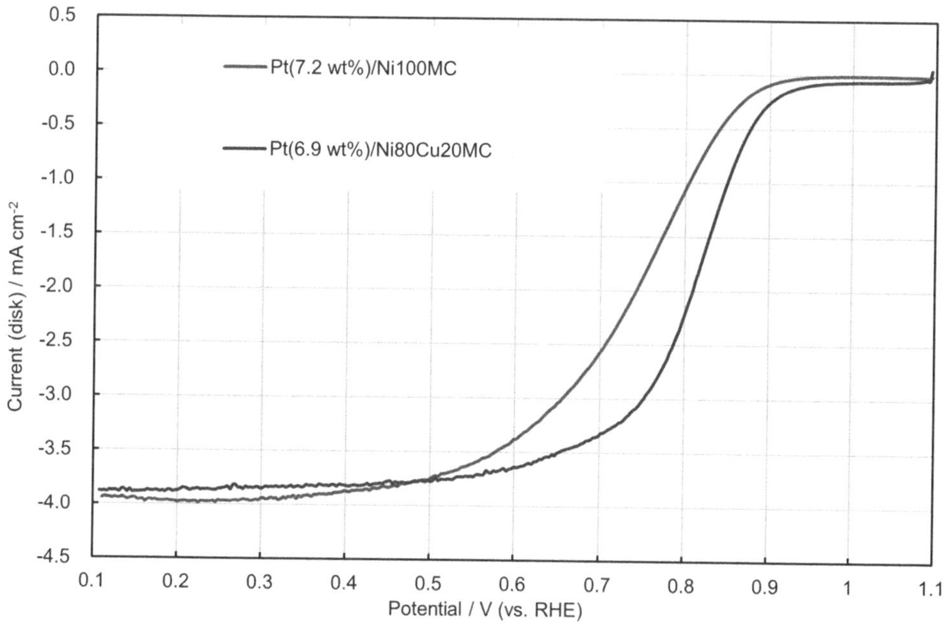

Fig. 7.: *Rotating ring-disk electrode voltammograms for the oxygen reduction reaction in an O_2-saturated 0.1-M $HClO_4$ solution over Pt-loaded Ni100MC (Pt(7.2 wt%)Ni100MC; red line) and Pt(6.9 wt%)Ni80Cu20MC (blue line). The scan rate is 10 mV/s at 900 rpm*

The possible higher Pt activity for the ORR using Ni80Cu20MC support also correlates with the specific surface area and porous characteristics of the MC shown in Table 1. The specific surface area of Ni80Cu20MC is 1.8 times larger than that of Ni100MC; simultaneously, the total pore volume of Ni80Cu20MC is almost four times higher than that of Ni100MC. We have reported that the MEA fabricated with the Ni100MC supported Pt catalyst had a good cell performance in the PEFC compared to the case of using amorphous carbon supported Pt catalyst, especially in the high current density range; the pores consist of CNFs can assist material transfer in the electrode catalyst [19]. By analogy from the MEA performance fabricated with the Pt/Ni100MC, the pores with tens of nanometers included in Ni80Cu20MC can aid in material transfer, and the Pt catalyst loaded over Ni80Cu20MC can produce a higher ORR starting potential. We expect that such a pore structure of Ni80Cu20MC can effectively enhance not only gas diffusion but also prompt water removal from the PEFC system, as used for the Pt support material in the MEA cathode. We have been fabricating MEA using Ni80Cu20MC as a Pt support material to measure its cell performance expecting for a durable and higher electric power generation performance.

The amount of loaded Pt over Ni80Cu20MC affected the ORR starting potential. The linear sweep voltammetric results for the different amounts of Pt loaded over Ni80Cu20MC are shown in Fig. 8. The results obtained from the TEC10E50E (Tanaka Kikinzoku Kogyo K.K. Pt loadings 46.7 wt%) are also shown as a reference in Fig. 8. In the case of loaded Pt at 6.9 wt%, the ORR starting potential was 0.90 V. As the Pt amount increases, the potential increased as 11.9 wt% Pt yielded 0.92 V, and 14.8 wt% Pt yielded 0.95 V. The increase in loaded Pt over Ni80Cu20MC is attributed to an increase in the ORR starting potential or the ORR activity. The TEC10E50E produced an ORR starting potential of 0.91 V; the potential of Pt(14.8 wt%)/Ni80Cu20MC was 0.95 V and it was comparable to that obtained using the TEC10E50E. Although the amount of loaded Pt over Ni80Cu20MC was almost a 1/3 less than that of Pt included in the TEC10E50E, the ORR activity of Pt(14.8 wt%)/Ni80Cu20MC is superior to that of TEC10E50E. The Pt/Ni80Cu20MC catalyst suggests realizing a PEFC with less Pt and higher electrochemical performance.

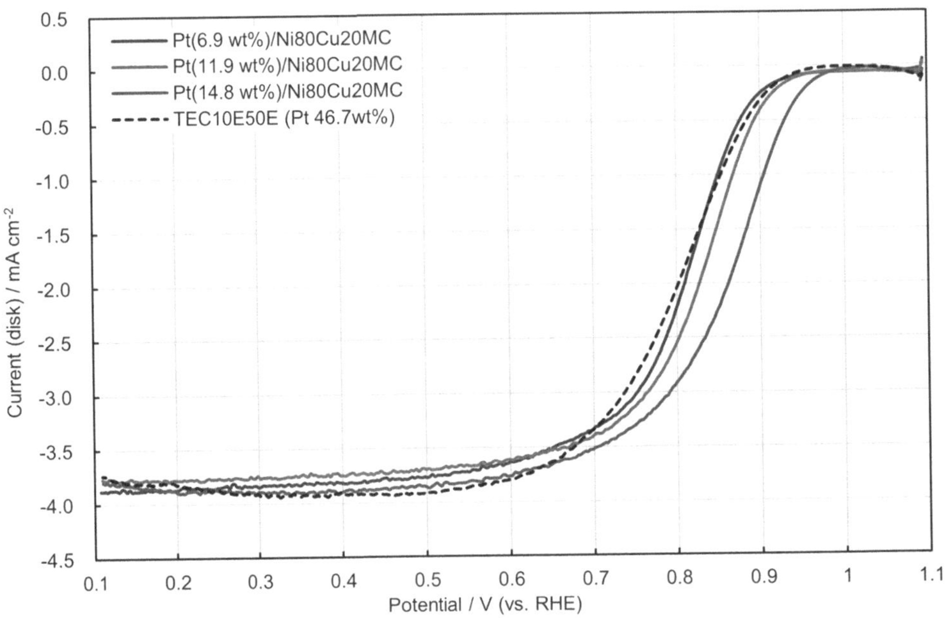

Fig. 8.: *Rotating ring-disk electrode voltammograms for the oxygen reduction reaction over various Pt ratios of Pt/Ni80Cu20MC in an O_2-saturated 0.1-M $HClO_4$ solution. The scan rate is 10 mV/s at 900 rpm*

4 SUMMARY

In this study, we performed linear sweep voltammetry using the RRDE for the Pt/Ni100MC and Pt/Ni80Cu20MC to determine the effect of CNF morphology and the fine structure consisting of the MCs as a Pt catalyst support material on the ORR onset potential values which reflect the ORR activity. FESEM analysis of the Ni80Cu20MC reveals an octopus-like morphology. TEM analysis revealed that the Ni80Cu20MC CNF fine structure was coin-stacked, whereas that of Ni100MC was a cup-stacked structure. Both fine structures provide several graphene edges on the CNF surface and a highly dispersed and narrow-diameter distribution of Pt particles supported on the MCs. A strong interaction between Pt particles and Ni80Cu20MC may exist because of more quantity of graphene edges, and the interaction can yield a higher ORR starting potential than in the case of Pt/Ni100MC. The specific surface area and porous characterization of MCs by N2 adsorption at 77 K revealed that the pores with tens of nanometers in the Ni80Cu20MC are more than four times greater than that in Ni100MC. These pores of Ni80Cu20MC correlate with the octopus-like morphology. They can aid in material transfer, and the Pt catalyst supported on Ni80Cu20MC can produce a higher ORR starting potential which possibly represents a higher ORR activity. Such pore structures of Ni80Cu20MC can enhance not only gas diffusion but also prompt water removal from the PEFC system as used in the Pt support material in the cathode. The Pt/Ni80Cu20MC catalyst is suggested to be better candidate for MEA in PEFC. We have been fabricating MEA using the Ni80Cu20MC as a Pt support material to realize a superior PEFC with a durable and higher electric power generation performance. At the same time, to clarify the electrochemical characteristics of MCs, such as a Pt mass activity and ECSA, is necessary and important for fabricating and characterizing MEA.

Acknowledgements

The study was supported by a Grant from Japan Keirin Autorace foundation (JKA) and the INOUE ENRYO Memorial Grant, TOYO University. The authors would like to thank Crimson Interactive Pvt. Ltd. for a careful review in English proofreading.

REFERENCES

1. Cameron DS (1978) Fuel cell energy generators. Platinum catalyst used in alternative energy source. Platin Met Rev 22:38–46
2. Kharwar PS, Akula S, Sahu AK, Ramanujam K (2020) Highly durable Pt-based catalyst supported on carbon deriverd from Tamarind seeds for oxygen reduction reaction in PEM fuel cell. J Electrochem Soc 167:104515
3. Wang YJ, Fang B, Li H, Bi XT, Wang H (2016) Progress in modified carbon support materials for Pt and Pt-alloy cathode catalysts in polymer electrolyte membrane fuel cells. Prog Mater Sci 82:445–498
4. Baba K, Nishitani-Gamo M, Ando T, Eguchi M (2016) Durable Marimo-like carbon support for platinum nanoparticle catalyst in polymer electrolyte fuel cell. Electrochim Acta 213:447–451
5. Nakagawa K, Oda H, Yamashita A, Okamoto M, Sato Y, Gamo H, Nishitani-Gamo M, Ogawa K, Ando T (2009) A novel spherical carbon. J Mater Sci 44:221–226

6. Eguchi M, Okubo A, Yamamoto S, Kikuchi M, Uno K, Kobayashi Y, Nishitani-Gamo M, Ando T (2010) Preparation of catalyst for a polymer electrolyte fuel cell using a novel spherical carbon support. J Power Sources 195:5862–5867
7. Rodriguez NM (1993) A review of catalytically grown carbon nanofibers. J Mater Res 8:3233–3250
8. Nishiyama Y, Tamai Y (1974) Carbon formation on copper-nickel alloys from benzene. J Catal 33:98–107
9. Bernardo CA, Alstrup I, Rostrup-Nielsen JR (1985) Carbon deposition and methane steam reforming on silica-supported Ni–Cu catalysts. J Catal 96:517–534
10. Kim MS, Rodriguez NM, Baker RTK (1991) The interaction of hydrocarbons with copper-nickel and nickel in the formation of carbon filaments. J Catal 131:60–73
11. Krishnankutty N, Rodriguez NM, Baker RTK (1996) Effect of copper on the decomposition of ethylene over an iron catalyst. J Catal 158:217–227
12. Chambers A, Rodriguez NM, Baker RTK (1995) Modification of the catalytic behavior of cobalt by the addition of copper. J Phys Chem 99:10581–10589
13. Shiraishi M, Nakagawa K, Ando T, Nishitani-Gamo M (2022) The effect of copper on the multiple carbon nanofilaments growths by the methane decomposition over the oxidized diamond-supported nickel–copper bimetallic catalyst. SN Appl Sci 4:126
14. Akiyama S, Nakagawa K (2021) Effect of mesopores in the Marimo nanocarbon anode material on the power generation performance of direct glucose fuel cell. Carbon Trends 4:100058
15. Baba K, Iwasawa K, Eguchi M, Kobayashi Y, Kobori M, Nishitani-Gamo M, Ando T (2013) Interfacial nanostructure of the polymer electrolyte fuel cell catalyst layer constructed with different ionomer contents. Jpn J Appl Phys 52:06GD06
16. Tsuji M, Kubokawa M, Yano R, Miyamae N, Tsuji T, Jun M-S, Hong S, Lim S, Yoon SH, Mochida I (2007) Fast preparation of PtRu catalysts supported on carbon nanofibers by the microwave-polyol method and their application to fuel cells. Langmuir 23:387–390
17. Bessel CA, Laubernds K, Rodriguez NM, Baker RTK (2001) Graphite nanofibers as an electrode for fuel cell applications. J Phys Chem B 105:1115–1118
18. Zheng JS, Wang XZ, Fu R, Yang DJ, Li P, Lv H, Ma JX (2012) Microstructure effect of carbon nanofibers on Pt/CNFs electrocatalyst for oxygen reduction. Int J Hydrogen Energy 37:4639–4647
19. Eguchi M, Baba K, Iwasawa K, Nishitani-Gamo M, Ando T (2013) The Marimo carbon as a polymer electrolyte fuel cell catalyst support. Trans Mat Soc Jpn 38(3):349–352

4 SYNTHESIS OF KETJENBLACK DECORATED PILLARED NI(FE) METAL-ORGANIC FRAMEWORKS AS PRECURSOR ELECTROCATALYSTS FOR ENHANCING THE OXYGEN EVOLUTION REACTION

Thi Hai Yen Beglau, Lars Rademacher, Robert Oestreich and Christoph Janiak

Institut für Anorganische Chemie und Strukturchemie, Heinrich-Heine-Universität Düsseldorf, 40204 Düsseldorf, Germany

ABSTRACT

Metal-organic frameworks (MOFs) have been investigated with regard to the oxygen evolution reaction (OER) due to their structure diversity, high specific surface area, adjustable pore size, and abundant active sites. However, the poor conductivity of most MOFs restricts this application. Herein, through a facile one-step solvothermal method, the Ni-based pillared metal-organic framework [$Ni_2(BDC)_2DABCO$] (BDC = 1,4-benzenedicarboxylate, DABCO = 1,4-diazabicyclo[2.2.2]octane), its bimetallic nickel-iron form [$Ni(Fe)(BDC)_2DABCO$], and their modified Ketjenblack (mKB) composites were synthesized and tested toward OER in an alkaline medium (KOH 1 mol L^{-1}). A synergistic effect of the bimetallic nickel-iron MOF and the conductive mKB additive enhanced the catalytic activity of the MOF/mKB composites. All MOF/mKB composite samples (7, 14, 22, and 34 wt.% mKB) indicated much higher OER performances than the MOFs and mKB alone. The Ni-MOF/mKB14 composite (14 wt.% of mKB) demonstrated an overpotential of 294 mV at a current density of 10 mA cm^{-2} and a Tafel slope of 32 mV dec^{-1}, which is comparable with commercial RuO_2, commonly used as a benchmark material for OER. The catalytic performance of Ni(Fe)MOF/mKB14 (0.57 wt.% Fe) was further improved to an overpotential of 279 mV at a current density of 10 mA cm^{-2}. The low Tafel slope of 25 mV dec^{-1} as well as a low reaction resistance due to the electrochemical impedance spectroscopy (EIS) measurement confirmed the excellent OER performance of the Ni(Fe)MOF/mKB14 composite. For practical applications, the Ni(Fe)MOF/mKB14 electrocatalyst was impregnated into commercial nickel foam (NF), where overpotentials of 247 and 291 mV at current densities of 10 and 50 mA cm^{-2}, respectively, were realized. The activity was maintained for 30 h at the applied current density of 50 mA cm^{-2}. More importantly, this work adds to the fundamental understanding of the in situ transformation of Ni(Fe)DMOF *into OER-active α/β-$Ni(OH)_2$, β/γ-NiOOH, and FeOOH with residual porosity inherited from the MOF structure, as seen by powder X-ray diffractometry and N_2 sorption analysis. Benefitting from the porosity structure of the MOF precursor, the nickel-iron catalysts outperformed the solely Ni-based catalysts due to their synergistic effects and exhibited superior catalytic activity and long-term stability in OER. In addition, by introducing mKB as a conductive carbon additive in the MOF structure, a homogeneous conductive network was constructed to improve the electronic conductivity of the MOF/mKB composites. The electrocatalytic system consisting of earth-abundant Ni and Fe metals only is attractive for the development of efficient, practical, and economical energy conversion materials for efficient OER activity.*

Keywords: *metal-organic frameworks (MOFs); Ketjenblack; electrocatalysis; oxygen evolution reaction (OER); nickel; iron*

1. INTRODUCTION

In recent decades, issues caused by the burning of fossil fuels have become a global concern. Therefore, it is important to replace fossil fuels with clean, sustainable, and renewable resources [1,2].

In the past years, water splitting, with the use of renewable electricity for hydrogen generation, seems to be one of the most promising technologies of electrical energy conversion for storage and other uses of H_2 [3,4]. The electrolytic water-splitting process consists of two half-cell reactions: the hydrogen evolution at the cathode (HER) and the oxygen evolution at the anode (OER). The oxygen evolution reaction consists of three main processes: (i) the adsorption of H_2O/OH^- on the electrocatalyst surface; (ii) the formation of reaction intermediates; (iii) the release of O_2 molecules. OER passes through a four-electron transfer process coupled with the breaking of the O–H bond and the formation of the O–O bond, which is kinetically slow and requires a high overpotential to overcome the energy barrier. Noble metals such as Ru, Pt, and Ir are highly-active electrocatalysts for the water splitting process [5,6]. However, their commercialization on a large scale is hampered by their scarcity and high cost. Therefore, finding advanced low-cost electrocatalysts that also have long-term stability to replace noble metals for the OER is the primary requirement to enable large-scale electrocatalytic water splitting [7,8,9].

Numerous earth-abundant transition metal (e.g., Fe, Co, Ni, and Cu)-based materials including oxides, hydroxides, nitrides, phosphides, sulfides, and selenides are considered promising toward OER [10,11]. Nickel-based materials have received considerable attention [12]. The optimization of the morphology and porosity of nickel-based oxides and hydroxides [13,14], perovskites [15], nitrides [16], phosphides [17], sulfides [18,19], selenides [20,21], and mixed-metal oxides [22,23] have been published in various extensive works. Subbaraman et al. [24] ranked the electrocatalytic efficiency of 3d M^{2+} (M = Fe, Co, Ni and Mn) ions in the following order: Ni > Co > Fe > Mn. This order is inverse to the strength of the HO–M^{2+} bond [24]. $Ni(OH)_2$-based OER catalysts have been identified as good OER catalysts. Stern et al. reported enhanced activity in OER for β-$Ni(OH)_2$ nanoparticles with different morphologies, with the low overpotential of 299 mV for $Ni(OH)_2$ catalysts [25]. Because of their facile synthesis and excellent OER activities, amorphous metal (oxide) hydroxide materials have consistently outperformed crystalline metal oxides as the preferred catalyst in an alkaline medium [25]. However, it is still a challenge to overcome the low stability of $Ni(OH)_2$. Many strategies have been devoted to enhance the stability of $Ni(OH)_2$ such as forming the $Ni(OH)_2$/NiOOH pair with a small amount of iron [26,27,28,29,30,31] or using a ternary Ni–Co–Fe mixed-metal hydroxide [32,33,34]. The combination of iron into composite materials containing amorphous nickel- or cobalt-based catalysts can be beneficial for electrocatalytic water oxidation [32,33]. The presence of iron promotes the nickel oxidation from +2 to +3, which is an active state and leads to fast reaction kinetics and enhanced conductivity [32,33]. As reported by Yu et al. [30], the electrocatalyst with the optimal Ni/Fe composition of 32/1 performed with the highest activity due to the small reaction resistance and superior intrinsic activity. However, these materials usually have low surface areas and are multi-step synthesis procedures also involving high-temperature treatments. It is still necessary to develop a facile approach to synthesize highly porous and stable nickel hydroxide materials with simple and energy efficient synthesis procedures.

Metal-organic frameworks (MOFs) are porous and crystalline materials constructed from metal ions and multitopic bridging ligands and have been widely studied in many fields such as gas storage and isolation, energy conversion and storage devices as well as electro-catalysts in water splitting [35,36,37,38]. However, most MOFs have poor electrical conductivity and a lack of stability in a strongly acidic or basic aqueous electrolyte [39,40]. However, MOFs are promising precursor candidates toward OER catalysts based on their high surface area and porosity [41]. For example, Ni-MOFs decompose in an alkaline medium (e.g., 1 mol L^{-1} KOH) into a mixture of α/β-$Ni(OH)_2$, α/β-NiOOH, and γ-NiOOH, which appear to be mainly responsible for the observed OER activity [42,43,44,45,46]. MOFs are often mixed with conductive carbon materials such as graphene [47], graphene oxide (GO) [48], reduced graphene oxide (RGO) [49], or carbon nanotubes (CNT) [50,51] to overcome the low electrical conductivity of MOFs. These carbon materials not only elevate the electrical conductivity of MOFs, but also give a homogenous dispersion to enhance the electrocatalytic performances of the formed catalyst under highly alkaline conditions. Ketjenblack carbon (KB) has a high surface area (-1300 m^2 g^{-1}), low cost, excellent charge-transport properties, and superior chemical stability. Currently, there are various reports on MOFs and nanoparticles with Ketjenblack carbon that were used as catalysts for OER [41,44,52,53,54,55]. For instance, Sondermann et al. [44] showed that the bimetallic Ni_{10}Co-BTC/KB composite gave an overpotential of 344 mV at a current density of 10 mA cm^{-2} and a Tafel slope of 47 mV dec^{-1} in 1 mol L^{-1} KOH. Öztürk et al. [55] reported Ni(Fe)-MOF-74 with KB for OER in an alkaline media. The required overpotential to achieve a current density of 10 mA cm^{-2} was only 274 mV and the Tafel slope was 40 mV dec^{-1}.

In this work, we report on the synthesis of the Ni-based pillared MOF [Ni_2(BDC)$_2$DABCO], (BDC = 1,4 benzenedicarbonxylate, DABCO = 1,4 diazabicyclo [2.2.2] octane), known as NiDMOF [56] as well as with the addition of iron as Ni(Fe) DMOF and as a composite with a modified Ketjenblack carbon (mKB) through a facile one-pot synthesis method (Scheme 1). NiDMOF consists of $\{Ni_2(O_2C\text{-})_4\}$ paddle-wheel clusters connected by BDC linkers to form two dimensional layers that are pillared by DABCO ligands to a three-dimensional structure (Figure S1, Supplementary Materials). The pretreatment of KB with nitric acid adds hydrophilic oxygen-containing groups that are beneficial to improving the wettability and interfacial area, promoting electrolyte accessibility. Good wettability strengthens the ion-accessible surface area, promotes charge transportation between the electrolyte and electrode, and enables effective electrical integration to minimize the ohmic losses, leading to the improved OER activity [57,58]. The Ni(Fe)DMOF/mKB materials were employed as precursors to OER electrocatalysts in a KOH electrolyte. Stability tests of the MOF in KOH and post-mortem analyses showed the transformation of the precursor into α/β-$Ni(OH)_2$ and β/γ-NiOOH (Scheme 1), albeit with the retention of the MOF morphology.

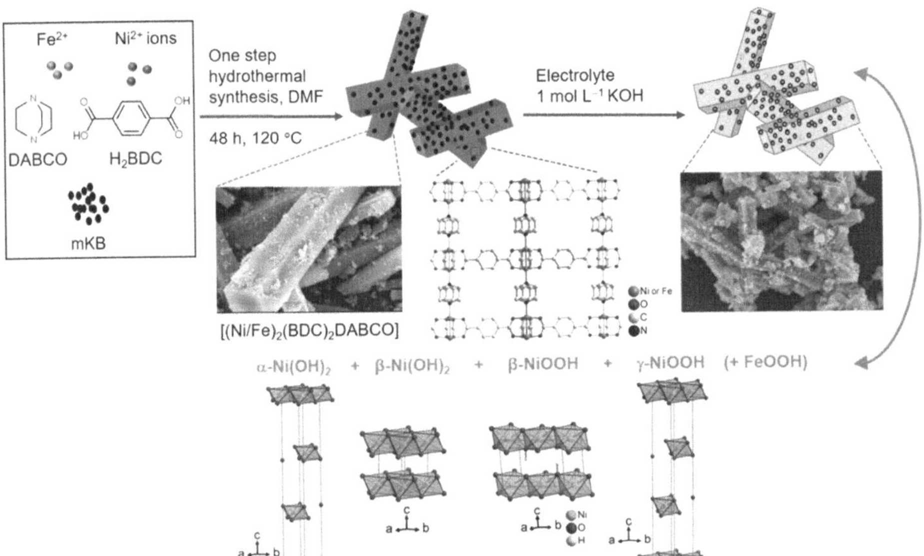

Scheme 1.: Schematic illustration of the process used for the synthesis of the NiDMOF/mKB and Ni(Fe)DMOF/mKB composites and the transformation of the MOF into metal oxide-hydroxides with the retention of the MOF morphology. For the nickel (oxy)hydroxides, the unit cells of the crystal structures are given in polyhedral mode.

2. RESULTS AND DISCUSSION

2.1. Characterization

Scheme 1 illustrates the procedure to prepare Ni(Fe)DMOF and its mKB composite. Nickel nitrate, together with iron acetate, benzene-1,4-dicarboxylic acid (H_2BDC), and 1,4 diazabicyclo[2.2.2]octane (DABCO), afford the products [Ni_2(B-

DC)$_2$DABCO] and [(Ni/Fe)$_2$(BDC)$_2$DABCO], abbreviated as NiDMOF and Ni(Fe)DMOF, respectively. In the presence of mKB, the composites NiDMOF/mKBx with different mKB fractions x = 7, 14, 22, and 34 wt.%. or Ni(Fe)DMOF)/mKB14 were obtained by using a one-step solvothermal reaction at 120 °C in DMF for 48 h. The synthesis of Ni(Fe)DMOF was carried out with a targeted molar Ni to Fe ratio of 32:1. A molar Ni to Fe ratio of 30:1 was achieved in the synthesized Ni(Fe)DMOF, as determined from atomic absorption spectroscopy (AAS) and SEM-EDX (Tables S2 and S3, Figure S9, Supplementary Materials). The weight fractions of mKB in the MOF/mKBx composites were about 7, 14, 22, and 34 wt.%, calculated from the MOF content, which was derived from the nickel amount by the atomic absorption spectroscopy (AAS) data (Table S2, Supplementary Materials).

As seen in the literature, the electric conductivity of the samples increases significantly by adding mKB (or another carbon material), thereby improving the excellent OER activities and cycling stability [59,60,61,62,63,64]. However, since mKB has almost no electrochemical activity, increasing the fraction of mKB would lead to lower energy density [65]. Only a small amount of mKB is needed to enhance the rate of OER performance as the electrons can also move due to incoherent jumps between neighboring locations. Furthermore, we will see later that the MOF is only the precursor. The active species are better conducting Ni–Fe oxide-hydroxides, for which the mKB then serves as a contact to the glassy carbon or nickel foam electrode. The experimental powder X-ray diffractogram (PXRD) of the MOF product was positively matched to the reported structure of [Ni$_2$(BDC)$_2$DABCO]·(DMF)$_4$·(H$_2$O)$_{1.5}$ (NiDMOF) (Figure 1a and Scheme S1) [56]. The PXRD also demonstrated that the presence of mKB influenced the crystallinity of the NiDMOF structure only at high mKB content in NiDMOF/mKB34, where peak broadening occurs (Figure 1 and Figure S2, Supplementary Materials). This is due to the formation of a larger amount of smaller MOF crystallites, also evident from the scanning electron microscopy images (Figure S7, Supplementary Materials). In Figure 1, the main PXRD reflection of Ni(Fe)DMOF/mKB14 shifted to lower degrees compared with other counterparts, indicating an enlarged structural unit and the existence of a tensile strain. The crystal tensile strain will help to adsorb OH^- reactants to improve the OER performance [66,67]. The functional groups –OH, C–O, and –COOH of mKB can serve as crystallization points, leading to more crystal seeds, which subsequently do not grow into large crystals [68,69,70]. Furthermore, the ratio of the 001/100 reflex intensities increases in the NiDMOF/mKBx composites with mKB content, indicating that in the presence of mKB, the DMOF crystallites may become more oriented along the [001] direction of the NiDMOF structure (Figure S1, Supplementary Materials).

The electrochemical performance is strongly correlated to the surface area, porosity and pore size distribution of electrode materials [71]. To explore these properties, nitrogen sorption measurements were performed at 77 K and gave the expected Type I isotherms [72] for microporous NiDMOF and Ni(Fe)DMOF (Figure 2a and Figure S6a). The specific BET surface area and total pore volume of NiDMOF was 2104 m^2 g^{-1} and 0.82 cm^3 g^{-1}, respectively, in good agreement with the literature values (2050 m^2 g^{-1}, 0.80 cm^3 g^{-1}) (Table 1) [73].

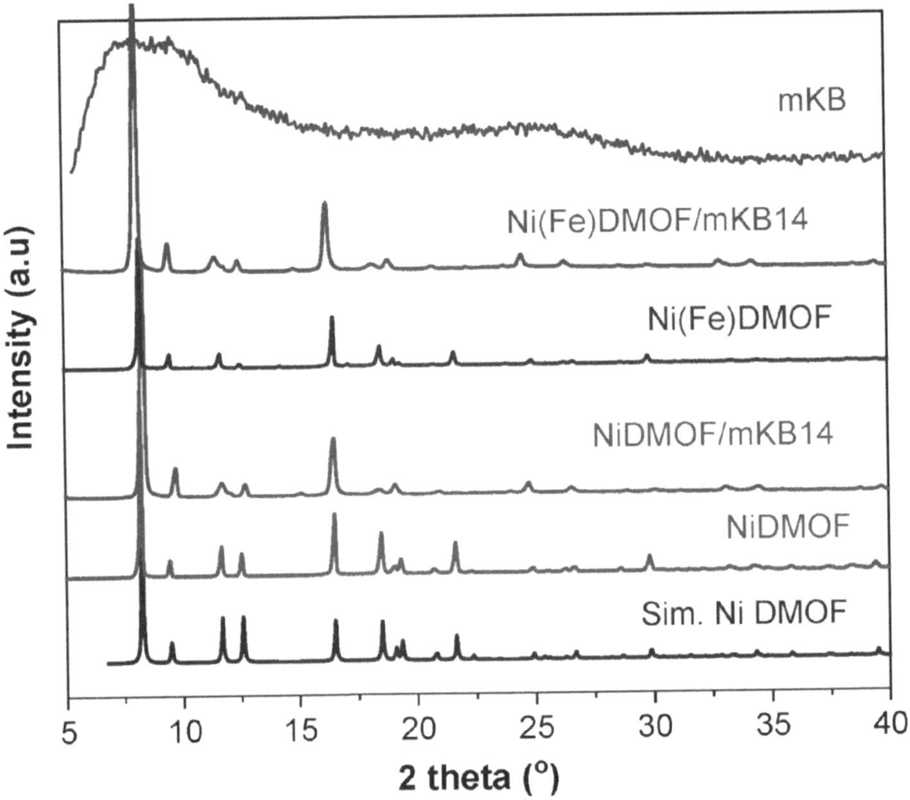

Figure 1.: *The PXRD patterns of NiDMOF, NiDMOF/mKB14, Ni(Fe)DMOF, Ni(Fe)DMOF/mKB14, and mKB. The simulated PXRD pattern of NiDMOF was obtained from CCDC no. 802892. The first five reflexes at 2θ = 8.2°, 9.4°, 11.7°, 12.4°, and 16.6° corresponded to the 100, 001, 110, 101 and 200 planes, respectively.*

Ni(Fe)DMOF also showed a similar high BET surface area and total pore volume with 1942 $m^2\ g^{-1}$ and 0.80 $cm^3\ g^{-1}$, respectively. The BET surface area and total pore volume of modified Ketjenblack (mKB) were 1234 $m^2\ g^{-1}$ and 1.50 $cm^3\ g^{-1}$, identical to the original Ketjenblack carbon (KB) (Table 1 and Figure S5a,b, Supplementary Materials). For the DMOF/mKBx composites, the desorption branch additionally displayed a Type H4 hysteresis [72], as seen in the isotherm of the mesoporous mKB carbon itself (Figure 2a and Figure S5c,d). The experimental surface area and pore volume parameters of the composites NiDMOF/mKBx (x = 7, 14, 22, 34 wt.%) and Ni(Fe)DMOF/mKB14 were somewhat lower than those estimated from the mass-weighted values of the neat MOF and mKB components (Table 1). This reduction in the BET value of composites was also observed for similar work and indicates that mutual pore blocking effects occur [55]. The NiDMOF can grow into the mKB pores or on the mKB surface, and therefore, access to the mKB and NiDMOF pores becomes restricted. The SEM images for NiDMOF/mKBx with a higher amount of mKB (x ≥ 22 wt.% mKB) illustrate the surface coverage of the NiDMOF rods with mKB particles (Figure S7, Supplementary Materials). At the same time, the combination of mKB with MOFs can increase the total (micro-meso)pore volume in the composites above the estimated value. The composite NiDMOF/mKB14 exhibited an experimental total pore volume of 1.18 $cm^3\ g^{-1}$, considerably higher than the estimate of 0.92 $cm^3\ g^{-1}$ for a physical mixture of 86 wt.% NiDMOF and 14 wt.% mKB. This pore volume of 1.18 $cm^3\ g^{-1}$ was also the maximum among the prepared DMOF/mKB composites.

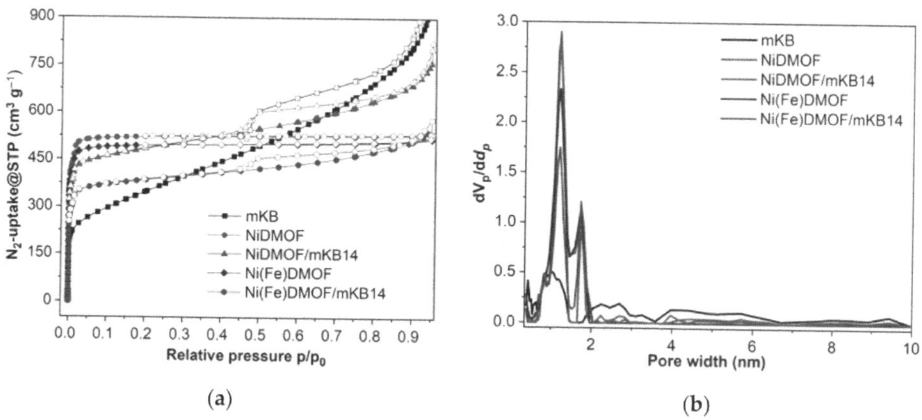

Figure 2.: *(a) Nitrogen sorption isotherms at 77 K (filled symbols adsorption; empty symbols desorption) and (b) the pore size distribution of NiDMOF, NiDMOF/mKB14, Ni(Fe)DMOF, Ni(Fe) DMOF/mKB14, and mKB.*

Table 1.: *The BET surface area and total pore volume of the NiDMOF and DMOF/mKBx composites.*

Material	BET Surface Area [a] (Estimated) [c] ($m^2\ g^{-1}$)	Total Pore Volume [b] (Estimated) [c] ($cm^3\ g^{-1}$)
NiDMOF-Literature [73]	2050	0.80
NiDMOF	2104	0.82
NiDMOF/mKB7	1897 (2043)	0.86 (0.87)
NiDMOF/mKB14	1773 (1982)	1.18 (0.92)
NiDMOF/mKB22	1106 (1913)	0.70 (0.97)
NiDMOF/mKB34	1087 (1808)	0.85 (1.05)

Material	BET Surface Area [a] (Estimated) [c] ($m^2\ g^{-1}$)	Total Pore Volume [b] (Estimated) [c] ($cm^3\ g^{-1}$)
Ni(Fe)DMOF	1942	0.80
Ni(Fe)DMOF/mKB14	1486 (1843)	0.86 (0.90)
KB	1300	1.76
mKB	1234	1.50

[a] BET surface area calculated at $0.002 < p/p_0 < 0.15$ from the N_2 sorption isotherm 77 K with a standard deviation ± 20 m^2-g^{-1}. [b] The total pore volumes were determined at $p/p_0 = 0.95$ from the adsorption branch for pores ≤ 25 nm. [c] Estimated BET surface area as the sum of the mass-weighted surface areas of NiDMOF (2104 m^2-g^{-1}) and mKB (1234 m^2-g^{-1}) calculated from the following formula:\

$$\text{BET (estimated)} = \frac{\text{wt.\% NiDMOF}}{100} \times 2104\ m^2 \cdot g^{-1} + \frac{\text{wt.\% mKB}}{100} \times 1234\ m^2 \cdot g^{-1} \quad (1)$$

The estimated total pore volume was calculated accordingly. NiDMOF/mKB7 stands for 93 wt.% NiDMOF and 7 wt.% mKB.

The synthesized NiDMOF and Ni(Fe)DMOF samples exhibited micropores with widths of about 1.2 and 1.7 nm (Figure 2b and Figure S6d). In the NiDMOF/mKBx and Ni(Fe)DMOF/mKB14 composites, there were additional mesopores above 2 nm from the mKB component. The increased pore volume beyond the estimate could be traced to interparticle mesopores above 5 nm (Table 1 and Figure S6a). Open microporous channels can provide active sites for the electrochemical reaction and open mesoporous channels will improve the diffusion rate of electrolyte ions [74].

Scanning electron microscopy (SEM) of the MOFs and the composites showed rod or block microcrystals of the NiDMOF with an increasing amount of fine grains of mKB from NiDMOF/mKB7 to NiDMOF/mKB34 (Figure 3 and Figure S6, Supplementary Materials). The observed NiDMOF morphology was in a good agreement with the reported one [75]. The element mapping from the SEM-energy-dispersive X-ray spectroscopy (SEM-EDX) (Figure 3c and Figure S8, Supplementary Materials) demonstrated that in Ni(Fe)DMOF, the iron is homogeneously distributed in the MOF microcrystals.

The X-ray photoelectron spectroscopy (XPS) survey spectrum of Ni(Fe)DMOF/mKB14 confirmed the presence of all the elements (Ni, Fe, C, N, and O) of the synthesized material (Figure S10, Supplementary Materials for XPS of NiDMOF and Ni(Fe)DMOF). The high-resolution spectrum of Ni 2p (Figure 4a) showed two characteristic peaks at 855.7 and 873.2 eV, which were ascribed to $Ni^{2+}\ 2p_{3/2}$ and $Ni^{2+}\ 2p_{1/2}$, respectively, and two expected satellite peaks were located at 860.7 and

879 eV [25,76,77,78]. A spin-orbit coupling energy difference between the $2p_{3/2}$ and $2p_{1/2}$ binding energy of 17.6 eV supports the assignment of the +2 oxidation state [25,79]. Furthermore, in the Fe 2p spectrum, the peaks at 710 and 723 eV (Figure 4b) confirmed the +2 oxidation state of Fe [80,81]. The $2p^{3/2}$ spectrum range was 710 to 720 eV including a satellite peak at 715 eV, while the $2p_{1/2}$ spectrum range was from 721 to 735 eV with a satellite peak at 733 eV. It should be noted that the Fe 2p spectral background had a contribution from a Ni_{LMM} Auger peak [82]. More detailed information on the high-resolution XPS of C 1s, O 1s, N 1s, and Fe 3p of Ni(Fe)DMOF/mKB14, Ni 2p and Fe 2p of Ni(Fe)DMOF, Ni 2p of NiDMOF is given in the Supplementary Materials (Figures S11 and S12).

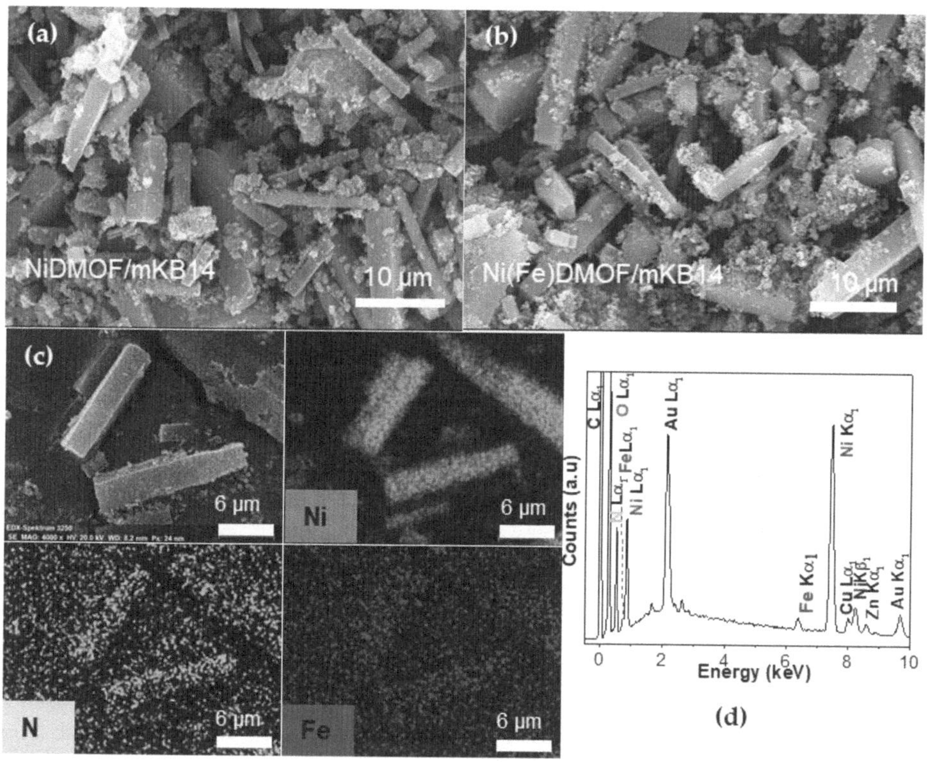

Figure 3.: SEM images of (a) NiDMOF/mKB14, (b) Ni(Fe)DMOF/mKB14, (c) EDX scanning element mapping for nickel, nitrogen and iron for Ni(Fe)DMOF/mKB14, and (d) the EDX spectrum with the composition for the selected mapping area of Ni(Fe)DMOF/mKB14.

2.2. Electrochemical Properties

The OER performance of different samples was evaluated by using a glassy carbon rotating disk electrode (GC-RDE) under alkaline conditions (1 mol L^{-1} KOH). OER polarization curves were collected from the linear sweep voltammetry (LSV) measurements at a sweep rate of 5 mV s^{-1}.

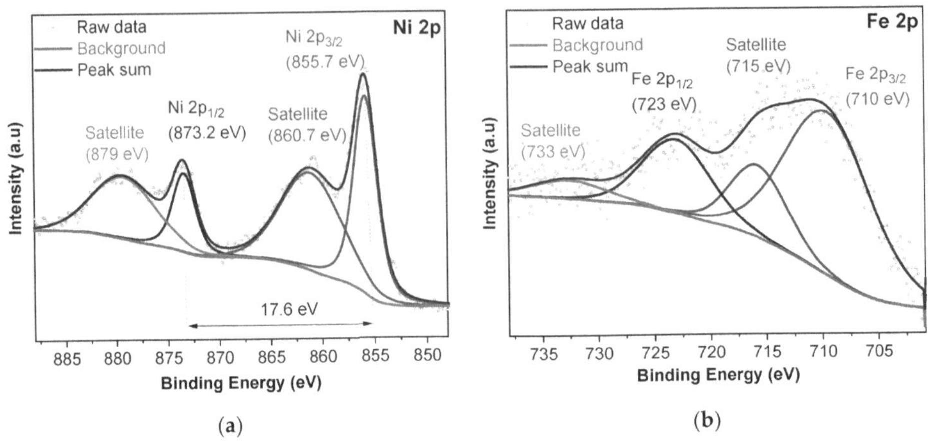

***Figure 4.**: High-resolution XPS spectra of (a) Ni 2p and (b) Fe 2p of Ni(Fe)DMOF/mKB14.*

As shown in the initial LSV curves (Figure 5a), the NiDMOF/mKBx composites were more efficient for OER activity than pristine NiDMOF and mKB alone at a current density of 10 mA cm^{-2}. Thereby, it is worth noting that the OER activity of the NiDMOF sample is already as good as the commercial RuO_2 benchmark sample. Among all NiDMOF/mKBx composites (x = 7, 14, 22, 34 wt.% mKB), NiDMOF/mKB14 exhibited the best electrocatalytic activity with the smallest overpotential of 294 mV (vs. RHE at 1.23 V) to achieve a current density of 10 mA cm^{-2}, which was much lower than the overpotential of mKB (375 mV), NiDMOF (315 mV), NiDMOF/mKB7 (308 mV), NiDMOF/mKB22 (302 mV), and NiDMOF/mKB34 (304 mV), and competes with the performance of the benchmark material RuO_2 (317 mV). Thereby, the overpotential of RuO_2 is in good accordance with the literature [83]. A low overpotential means a lower demand of energy for the oxygen evolution reaction. The kinetics on the GCE toward NiDMOF and its DMOF/mKBx composites were described on the basis of the Tafel equation. The OER kinetic parameters of the samples were analyzed by the Tafel equation ($\eta = a + b \log(j)$), which is used to determine the reaction mechanism and the kinetics [84]. The Tafel slope indicates how much one has to increase the overpotential to increase the reaction rate by a factor of ten. It describes the influence of the overpotential on the steady-state current density and is an important parameter for the evaluation of OER kinetics. The Tafel slope was calculated from corresponding LSV plots to obtain a quantitative idea about the electrocatalytic performance. Krasil'shchikov's OER mechanism is one of the established mechanisms, which is described by Reactions (3)–(6) with the corresponding Tafel slopes b [44,85,86].

$$M + OH^- \leftrightharpoons MOH + e^-, \ b = 120 \text{ mV dec}^{-1} \tag{2}$$

$$MOH + OH^- \leftrightharpoons MO^- + H_2O, \ b = 60 \text{ mV dec}^{-1} \tag{3}$$

$$MO^- \rightarrow MO + e^-, \ b = 45 \text{ mV dec}^{-1} \tag{4}$$

$$2MO \rightarrow 2M + O_2, \ b = 19 \text{ mV dec}^{-1} \tag{5}$$

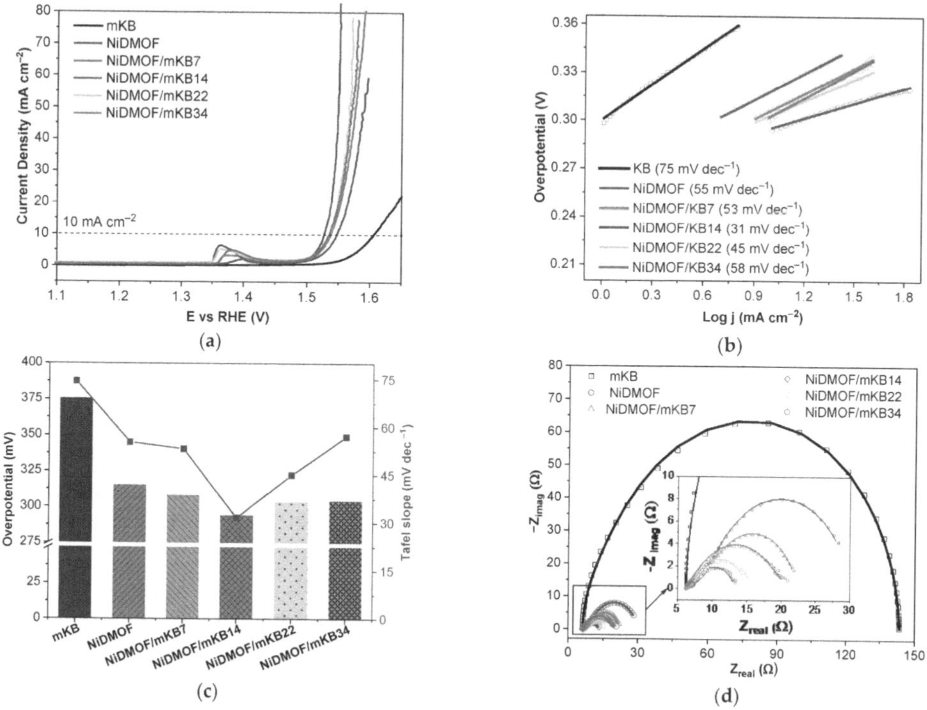

Figure 5.: (a) Polarization curves (LSV) operated at a scan rate of 5 mV s^{-1} with the iR correction. (b) Tafel plots, (c) corresponding overpotentials (columns) and Tafel slopes (red dots) at 10 mA cm^{-2}, (d) raw data of Nyquist plots at 1.5 V vs. RHE (points) and fitting to an equivalent circuit model (Figure S16, Supplementary Materials) (solid line) from the EIS test of mKB, NiDMOF, NiDMOF/mKB7, NiDMOF/mKB14, NiDMOF/mKB22, and NiDMOF/mKB34.

The value of the Tafel slope in NiDMOF (55 mV dec^{-1}) is in between Reactions (3) and (4), and indicates that the deprotonation of a metal hydroxide (3) and the oxidation of a metal oxide species (4) could occur together as rate-determining steps [86,87]. The corresponding Tafel curves derived from Figure 5b show the Tafel slopes for NiDMOF/mKBx composites (7, 14, 22, 34 wt.% mKB) with 53, 32, 45, and 51 mV dec^{-1}, respectively. Thus, integrating mKB with NiDMOF can enhance the kinetics of the catalyst. For the 14 and 22 wt.% composites, there was also a change in the rate-determining step toward Reaction (4) for 22 wt.% (slope of 45 mV dec^{-1}) and toward Reaction (5) of the evolution of O_2 for 14 wt.% (slope of 32 mV dec^{-1}). The NiDMOF/mKB14 composite had the smallest Tafel slope (32 mV dec^{-1}), confirming that the interaction between NiDMOF and 14 wt.% of mKB in the composite gives the most proficient material.

It is acknowledged that due to the surface structure reconstruction of Ni-based catalysts, an activation occurs during the water oxidation process [30,44]. In order to understand the electrocatalytic behavior and the activation process, 100 cyclic voltammetry (CV) scans were applied. With mKB present in the composites, the oxidation peak at 1.35–1.45 (V vs. RHE) became more noticeable and shifted into a positive direction, indicating the synergetic effect of mKB and NiDMOF on the oxidation of Ni^{+2} to Ni^{+3} (Figure S13, Supplementary Materials). As seen in Figure S14, the CV curves of all samples after 100 CVs showed that NiDMOF/mKB14 provided the most active catalyst, while mKB alone exhibited worse OER activity due to carbon corrosion in alkaline conditions [88]. The mKB additive led to the electrical conductivity between the active Ni sites and the GCE, and thereby enhanced the electrochemical activity [89,90]. However, an excessive amount of mKB lowered the carbon dispersion due to agglomeration from the $\pi-\pi$ interaction between the carbon particles, which would lead to lower energy density for OER [65,91,92].

Furthermore, electrochemical impedance spectroscopy (EIS) was carried out to assess the kinetics of the electrode reaction. The corresponding Nyquist plots (Figure 5d) obtained in a frequency range from 0.01 to 100 kHz with an AC potential amplitude of 5 mV at 1.5 V (RHE) were fitted to the simplified equivalent circuit by Doyle et al. for OER (Figure S16, Supplementary Materials) [30,93,94,95]. The kinetics of the interfacial charge transfer reaction corresponded to the resistive components R_s and R_p in terms of the electrocatalytic capabilities of the oxide layer. In particular, the polarization resistance (R_p) is considered as a total charge transfer resistance across multiple steps of OER, while R_s is related to the rate of production of one or more surface intermediates [30,93,94,95]. More detailed information concerning the circuit parameters is given in Figure S16. From the shape of the semicircles in the Nyquist plots, it can be seen that NiDMOF and the NiDMOF/mKBx composites follow a similar trend. Smaller semicircles are displayed by the NiDMOF/mKBx composites, which reflect smaller resistances at high frequency compared with both the pure NiDMOF and mKB, which illustrates that the kinetic performances of the neat MOF materials were improved by the integration of mKB. The low polarization resistance R_p (4Ω) and resistance adsorption components of the reaction intermediate R_s (2 Ω) for NiDMOF/mKB14 indicate a superior charge transfer rate and the easier formation of active species for OER, respectively, which contribute to its highest catalytic activity among the composites (Table S4, Supplementary Materials). Thus, mKB lowers the overall resistances, which is considered to be a key factor in enhancing the electrochemical performance of MOF-based composites.

It has been intensively studied that Fe can be advantageously incorporated into Ni-based catalysts. With a small amount of Fe, the catalytic activity toward OER can be significantly improved [26,28,30,31,33,44,55]. In order to investigate the effect of Fe on the electrocatalytic NiDMOF performance, a bimetallic Ni(Fe) DMOF and the Ni(Fe)DMOF/mKB14 composite were prepared and measured under the same conditions. The presence of a small amount of Fe at a Ni:Fe ratio of ~30:1 can significantly enhance the OER performance. The Ni(Fe)DMOF without mKB already exhibited an enhanced OER performance with a much lower overpotential of 301 mV at the current density of 10 mA cm^{-2}, which was much smaller compared to the commercial RuO_2 benchmark (317 mV) and pristine NiDMOF (315 mV). The composite of Ni(Fe)DMOF with 14 wt.% mKB further enhanced the OER activity, giving the smallest overpotential of only 279 mV among the materials reported here (Figure 6a,c). These results strongly demonstrate that Fe plays a key role in improving the OER activities of pure NiDMOF and its composites. As shown in Figure 6b,c, the Tafel slope calculated from the corresponding LSV curve of bimetallic Ni(Fe)DMOF was 40 mV dec^{-1}, which is in agreement with a thin-film of Ni–Fe oxide [96] and outperformed the monometallic NiDMOF (55 mV dec^{-1}). Ni(Fe)DMOF/mKB14 presented the lowest Tafel slope (25 mV dec^{-1}) compared to RuO_2 (56 mV dec^{-1}) and all DMOFs and composites investigated here (Table S4, Supplementary Materials), reflecting the fastest kinetics. The results indicate a change in the mechanism where the rate-determining step becomes the evolution of O_2 (Reaction (5)). Furthermore, the Ni(Fe)DMOF/mKB14 composite stands out as one of the best in the recently reported most advanced Ni-based electrocatalysts with both its low overpotential (279 mV) and Tafel slope (25 mV dec^{-1}) (Table S4, Supplementary Materials). A lower overpotential was reported for Ni(Fe)-MOF-74/KB (48 wt.% KB) (274 mV) [55], $Ni(Fe)(OH)_2$/KB (47 wt.% KB) (265 mV) [55], and Fe-doped HXP@NC800 (266 mV) [97] (all at 10 mA cm^{-2}), albeit with larger Tafel slopes of 40, 55, and 49 mV dec^{-1}, respectively. From Figure S15, it can be seen that even after 1000 CVs, the overpotential of the catalysts only slightly changed from 301 mV to 318 mV for the Ni(Fe)DMOF precursor and from 279 mV to 285 mV for Ni(Fe)DMOF/mKB14 (at 10 mA cm^{-2}), indicating their long-term activity. Furthermore, electrochemical impedance spectrometry (EIS) measurements were carried out to understand the charge-transfer kinetics during the OER process among the different samples. As shown by the Nyquist plots (Figure 6d) and fitted data of the equivalent circuit model in Figure S16, the resistances of the charge-transfer and adsorption components of the reaction intermediates of Ni(Fe)DMOF/mKB14 (R_p 4 Ω and R_s 2 Ω) was as low as NiDMOF/mKB14 (R_p 4 Ω and R_s 2 Ω), that is, lower than that of RuO_2 (R_p 23 Ω and R_s 4 Ω), NiDMOF (R_p 21 Ω and R_s 7 Ω), and Ni(Fe)DMOF (R_p 6 Ω and R_s 4 Ω) (Table S4, Supplementary Materials). The polarization resistance (R_p) of Ni(Fe)DMOF/KB14 was much smaller than that of $Ni_{32}Fe$ oxide (17.1 Ω) [30], IrO_2 (~5 Ω) [86], $NiCo_2O_4$ (~8 Ω) [86], $Ni_{1.7}Co_{1.3}O_4$ (~6 Ω) [98], H-$SCF_{0.55}$ (~58 Ω) [99] (all data fitted with the same equivalent circuit model). The catalyst from the Ni(Fe)DMOF/mKB14 precursor presented the smallest charge-transfer resistance, which indicates its faster electron transport kinetics and intrinsic excellent electrical conductivity, well in accordance with its lowest overpotential. This result is in good agreement with the identified higher catalytic activity of Ni(Fe)-MOFs compared to analogous Ni-MOFs [30,55,100,101]. In particular, the presence of iron promotes the oxidation of nickel from +2 to +3, the latter seen as the active state of fast reaction kinetics and enhanced conductivity.

Loading a catalyst on a porous Ni foam (NF) electrode can further improve the reaction rate of OER [102]. Because of its metallic conductivity and 3D macroporous structure, NF provides a large surface area and facilitates mass transport during OER [30]. The catalyst ink was loaded by the drop-casting method on the surface of NF with a good distribution of the ink layer on the scaffold of the NF. The sample Ni(Fe)DMOF/mKB14, which showed the best OER activity on the glassy carbon RDE, was deposited on NF (1 cm^2) (Figure S17, Supplementary Materials). The LSV curves in Figure 7a showed that the activity of Ni(Fe)DMOF/mKB14@NF was much higher than the pure NF substrate, reaching current densities of 10 and 50 mA cm^{-2} with overpotentials of 247 and 291 mV, respectively.

A practical current density of 400 mA cm^{-2} could be delivered for Ni(Fe)DMOF/ mKB14@NF at a low overpotential of 381 mV. This performance was better than the benchmark RuO_2@NF as an OER catalyst with overpotentials of 278 and 340 mV at current densities of 10 and 50 mA cm^{-2}, respectively (Figure 7a), the latter being in good accordance with the literature [103]. The pure NF electrode showed a much lower OER activity with an overpotential of 370 mV at 10 mA cm^{-2}, similar to the literature [103].

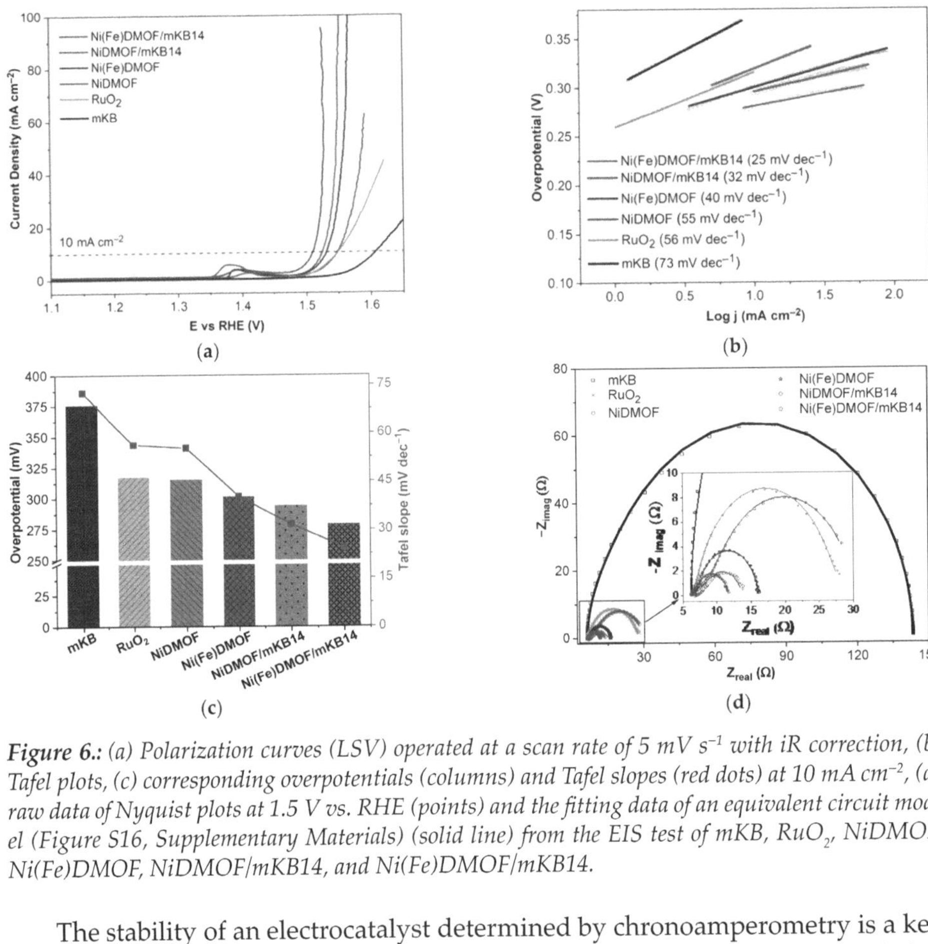

Figure 6.: (a) Polarization curves (LSV) operated at a scan rate of 5 mV s^{-1} with iR correction, (b) Tafel plots, (c) corresponding overpotentials (columns) and Tafel slopes (red dots) at 10 mA cm^{-2}, (d) raw data of Nyquist plots at 1.5 V vs. RHE (points) and the fitting data of an equivalent circuit model (Figure S16, Supplementary Materials) (solid line) from the EIS test of mKB, RuO_2, NiDMOF, Ni(Fe)DMOF, NiDMOF/mKB14, and Ni(Fe)DMOF/mKB14.

The stability of an electrocatalyst determined by chronoamperometry is a key parameter to evaluate the practical application of a material. A long-term stability measurement was carried out by applying a potential to reach a constant current density of 50 mA cm^{-2} for 30 h (Figure 7b).

The applied potential could be gradually decreased during the first 2.5 h, indicating that the Ni–Fe catalyst was activated under the anodic potential in 1 mol L^{-1} KOH electrolyte. After this time, the potential remained nearly unchanged during the rest of the 30 h measurement. In contrast, the bare NF electrode showed an increasing potential over time for delivering a current density of 50 mA

cm^{-2} due to the oxidation of nickel, which occurs at the surface of the foam but can also reach its subsurface during electrolysis under a high current density [30,104].

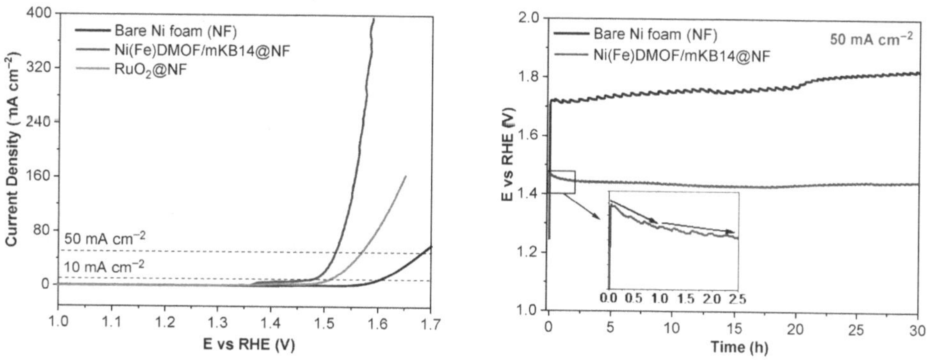

(a) (b)

Figure 7.: (a) Linear sweep voltammetry (LSV) curves at a scan rate of 5 mV s^{-1} with iR correction. (b) Chronopotentiometric curves for 30 h at a current density of 50 mA cm^{-2} of the NiFe-DMOF/ mKB14 loaded on nickel foam (NF) and bare NF for comparison in a 1.0 mol L^{-1} KOH solution.

Pure NF also necessitates a much higher potential at the current density of 50 mA cm^{-2} than Ni(Fe)DMOF/mKB14@NF during the 30 h of the chronoamperometry test. The catalyst derived from Ni(Fe)DMOF/mKB14 does not only have a robust stability, but also an outstanding activity to deliver a high current density, which demonstrates that the Ni(Fe)DMOF/mKB14 precursor has the potential to serve as a good OER catalyst for practical applications.

In order to understand the transformation of the precursors Ni(Fe)DMOF and Ni(Fe)DMOF/mKB14 in the alkaline electrolyte (1 mol L^{-1} KOH), we reacted macroscopic amounts of these samples in 1 mol L^{-1} KOH, followed by PXRD, FTIR, SEM, and N_2 sorption measurements to mimic post-mortem experiments of the minuscule electrode materials. The derived materials were collected by filtration after soaking Ni(Fe)DMOF and Ni(Fe)DMOF/mKB14 in 1 mol L^{-1} KOH electrolyte for 24 h and were dried for at least 12 h at 120 °C under vacuum (<10^{-2} mbar). The PXRD pattern of the derived-Ni(Fe)DMOF/mKB14 (Figure 8a) showed the disappearance of the crystalline MOF and suggests the formation of α/β–$Ni(OH)_2$, β–NiOOH, and γ`–NiOOH from the reflections assigned in Figure 8a based on the patterns of nickel(II) hydroxides and nickel(III) oxide-hydroxides. This implies the transformation from Ni(Fe)DMOF/mKB14 to nickel and iron oxide-hydroxides under a strong alkaline environment (1 mol L^{-1} KOH) [105], which has also been observed by other researchers [106,107,108,109,110]. The FTIR of the derived-materials (Figure 8b and Table S1) showed two broad bands at ~3400 and ~3600 cm^{-1}, which corresponded to the stretching vibrations of the adsorbed water molecules and to the O–H stretching vibrations, as in $Ni(OH)_2$. At the same time, bands due to the asymmetric vibration $\nu_{asym}COO^-$ and the bending vibrations of the adsorbed water molecules of the δO–H hydroxyl groups of the BDC phenyl ring at ~1600 and ~1570 cm^{-1}, along with $\nu_{sym}COO^-$ at ~1350 cm^{-1}, decreased in intensity. New strong bands at 516 and 450 cm^{-1} were attributed to the Ni–OH bending vibrations in $Ni(OH)_2$ and the oxide-hydroxides [111]. Interestingly, the morphology of Ni(Fe) DMOF/mKB14, as seen with SEM (Figure 8c), did not change profoundly. The rod shape of Ni(Fe)DMOF was largely retained in the alkaline treatment. Accordingly, N_2 physisorption measurements at 77 K revealed a residual BET surface area and micro-mesoporosity in the derived-Ni(Fe)DMOF and derived-Ni(Fe)DMOF/mKB14 (Figure 8d). The derived-Ni(Fe)DMOF displayed a Type IV isotherm with H2(b) hysteresis (associated with pore blocking) [72] and a BET surface area of 222 $m^2\ g^{-1}$. The adsorption isotherm of the derived-Ni(Fe)DMOF/mKB14 lacked the final saturation plateau of a Type IV isotherm and appeared like a Type II isotherm with an H_3 hysteresis, which would be given by a largely macroporous adsorbent. The pore size distribution curves showed mostly mesopores (>2 nm) for the derived-Ni(Fe)DMOF and the coexistence of micropores and mesopores for the derived-Ni(Fe)DMOF/mKB14 with micropores and small mesopores due to the mKB part (the macropore size is not given by N_2 sorption). The BET surface area of the derived-Ni(Fe)DMOF/mKB14 was 352 $m^2\ g^{-1}$ with the increase over derived-Ni(Fe)DMOF coming from the mKB portion. Overall, a hierarchical porous nature of derived-Ni(Fe)DMOF/mKB14 was evidenced by its isotherm and hysteresis shape and the analysis of the pore size distribution. The decomposition of the Ni-MOF precursor and transformation to $Ni(OH)_2$/NiOOH was also observed in other work [112,113,114]. The OER activity of the Ni(Fe)DMOF precursor is due to the in situ formation of α/β-$Ni(OH)_2$/FeOOH, followed by β/γ–NiOOH under the oxidizing anodic potential, which favors the kinetics of OER [43,44,115,116,117]. Furthermore, the degradation of the micro-mesoporosity of Ni(Fe)DMOF in an alkaline environment provides a large number of accessible active Ni (and promoter Fe) sites in the still porous hydroxides, which show better electrochemical activities and longer cycle numbers than crystalline and dense metal oxide catalysts [42,43,118,119,120,121]. In addition, a suitable amount of mKB (14 wt.%) in the composites provides electrical conductivity with its small particle size, mixes well with the Ni active sites, and avoids the carbon corrosion effect of carbon materials [59,60,61,62,63,64,65,66].

3. MATERIALS AND METHODS

3.1. Materials

Nickel(II) nitrate hexahydrate ($Ni(NO_3)_2\ 6H_2O$, Merck, Darmstadt, Germany), iron(II) acetate ($Fe(OAc)_2$, 99.99%, Sigma-Aldrich, St. Loius, MO, USA), ruthenium(IV) oxide (RuO_2, 99.9%, Sigma-Aldrich, St. Loius, MO, USA), benzene-1,4-dicarboxylic acid (H_2BDC, 98%, Alfa Aesar, Karlsruhe, Deutschland), 1,4-diazabicyclo[2.2.2]octane (DABCO, 98%, Sigma-Aldrich, St. Loius, MO, USA), perfluorinated resin solution containing Nafion™ 1100W (5 wt.% in lower aliphatic alcohols and water, Sigma-Aldrich, St. Loius, MO, USA), potassium hydroxide solution (KOH, 1 mol L^{-1}, Carl Roth, Karlsruhe, Germany), nitric acid (HNO_3, 65%, Sigma-Aldrich, Darmstadt, Germany), N,N-dimethylformamide (DMF, 99.99 %, Fisher, Schwerte, Germany), Ketjenblack EC 600 JD (AkzoNobel, Amsterdam, The Netherlands), and methanol (MeOH, 99.98 %, Sigma-Aldrich, Darmstadt, Germany) were purchased and used without further purification. Nickel foam (NF) was obtained from Racemat BV (Dodewaard, The Netherlands), cleaned with 1 mol L^{-1} HCl solution in an ultrasound bath for 5 min to remove the surface nickel oxide layer of the NF and then rinsed with Millipore water (residual conductivity 18.2 MΩ·cm).

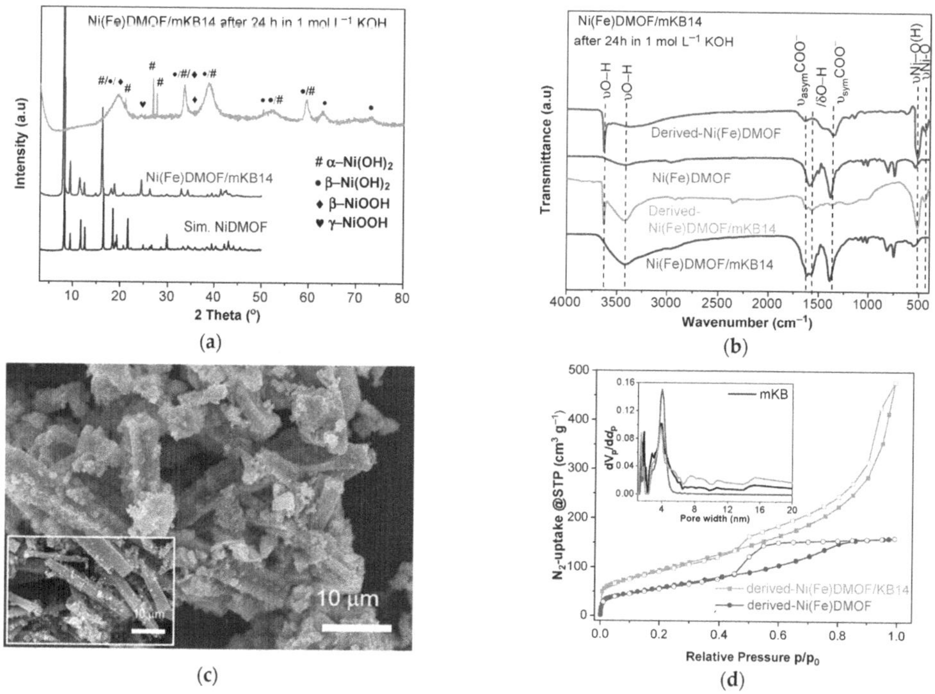

Figure 8.: (a) PXRD patterns of the experimental Ni(Fe)DMOF/mKB14 and derived-Ni(Fe)DMOF/mKB14 after 24 h in 1 mol L^{-1} KOH and simulated NiDMOF (CCDC Nr. 802892). Reflections from β–Ni(OH)$_2$ (•, ICDD: 14-0117), α–Ni(OH)$_2$ (#, ICDD: 38-0715), β–NiOOH (■, ICDD: 06-0141), and γ–NiOOH (♦, ICDD: 06-0075). (b) FTIR of Ni(Fe)DMOF and Ni(Fe)DMOF/mKB14 and its derived products in 1 mol L^{-1} KOH after 24 h. (c) SEM images of the derived-Ni(Fe)DMOF/mKB14 material (large image) and neat Ni(Fe)DMOF/mKB14 (small image). (d) Nitrogen sorption isotherms at 77 K (filled symbols adsorption, empty symbols desorption isotherm) and pore size distribution (small graph) of derived-Ni(Fe)DMOF and derived-Ni(Fe)DMOF/mKB14 after 24 h in 1 mol L^{-1} KOH.

3.2. Synthesis of Modified Ketjenblack Carbon (mKB)

The amount of 10 g KB was dispersed into 1 L of HNO_3 (20 wt.%) to form a homogeneous black dispersion that was heated to 80 °C under continuous stirring. After 3 h, the product was separated by centrifugation (10,000 rpm, 10 min) and washed with deionized water (500 mL) until the pH value was 7. Yield: 9.58 mg. This modified Ketjenblack carbon was denoted as mKB. Elemental analysis for Ketjenblack carbon: C 97.91 (O 2.91)% and modified Ketjenblack carbon: C 87.93,

H 0.79 (O 11.28)%. IR (ATR, cm^{-1}); 3435 (νOH), 1740 (νC=O), 1550 (ν_δOCO$^-$), and 1181 (ν_sC–OO).

3.3. Synthesis of NiDMOF

NiDMOF was synthesized by a solvothermal reaction according to the literature with a slight modification [75]. $Ni(NO_3)_2{\cdot}6H_2O$ (183.12 mg, 0.63 mmol), H_2BDC (104.66 mg, 0.63 mmol), and DABCO (36.33 mg, 0.32 mmol) were added into 16 mL of N,N-dimethylformamide (DMF). The resulting slurry was stirred overnight at room temperature to obtain a homogeneous dispersion. The clear solution was collected and transferred into a Pyrex tube and then heated at 120 °C for 48 h. After cooling to room temperature, the green crystalline powder was separated by centrifugation (10,000 rpm, 10 min) and washed three times with DMF (20 mL each). The washing process was repeated three more times with MeOH (20 mL each) and the product collected by centrifugation (10,000 rpm, 10 min). The product was activated at 120 °C under vacuum (<10^{-2} mbar) for at least 12 h and finally stored under nitrogen.

Yield: 200 mg (57% based on H_2BDC). Elemental analysis for $[Ni_2(C_8H_4O_4)_2(C_6H_{12}N_2)]{\cdot}(H_2O)_{0.5}$ (MW 566.79 g mol^{-1}), calculated C 46.62, H 3.73, N 4.94, Ni 20.71 %; $[Ni_2(C_8H_4O_4)_2(C_6H_{12}N_2)]{\cdot}H_2O$ (MW 575.81 g mol^{-1}), calculated C 45.89, H 3.85, N 4.87, Ni 20.39%; found C 45.26, H 3.61, N 4.94, Ni 20.82%.

3.4. Synthesis of Ni(Fe)DMOF

The bimetallic Ni(Fe)DMOF was synthesized by a similar procedure to the monometallic NiDMOF. The amount of $Ni(NO_3)_2{\cdot}6H_2O$ (177.65 mg, 0.61 mmol), H_2BDC (104.66 mg, 0.63 mmol), and DABCO (36.33 mg, 0.32 mmol) were dispersed into 11 mL of DMF. The resulting slurry was stirred overnight at room temperature to obtain a homogeneous dispersion. To this solution, 0.02 mmol of $Fe(OAc)_2$ (3.32 mg) in 5 mL of DMF was added dropwise (for an intended molar Ni:Fe ratio of 32:1). After sonicating for 30 min, the mixture was sealed into a Pyrex tube and heated at 120 °C for 48 h. After cooling to room temperature, the palm-leaf crystalline powder was separated by centrifugation (10,000 rpm, 10 min) and washed three times with DMF (20 mL each). The washing process was repeated three more times with MeOH (20 mL each) and the product collected by centrifugation (10,000 rpm, 10 min). The product was activated at 120 °C under vacuum (<10^{-2} mbar) for at least 12 h and stored under nitrogen.

Yield: 191 mg (54% based on H_2BDC). Elemental analysis for $[Ni_{1.94}Fe_{0.06}(C_8H_4O_4)_2(C_6H_{12}N_2)]\,(H_2O)_{0.5}$ (MW 566.61 g mol^{-1}), calculated: C 46.63, H 3.74, N 4.94, Ni 20.10, Fe 0.59%; $[Ni_{1.94}Fe_{0.06}(C_8H_4O_4)_2(C_6H_{12}N_2)]{\cdot}(H_2O)$ (MW 573.88 g mol^{-1}), calculated: C 45.90, H 3.85, N 4.87, Ni 19.78, Fe 0.58%; found C 46.35, H 3.57, N 5.01, Ni 20.85, Fe 0.66%.

3.5. Synthesis of NiDMOF/mKBx and Ni(Fe)DMOF/mKB14

The composites NiDMOF / mKBx and Ni(Fe)DMOF / mKB14 were synthesized according to the above procedures described for NiDMOF and Ni(Fe)DMOF, except that the modified Ketjenblack carbon (mKB) powder (12, 24, 40 and 66 mg) was dispersed in 3 mL of DMF by sonication for 30 min in advance and then added into the metal salt/H_2BDC/DABCO/DMF solution. In order to determine the MOF and mKB content in the final composite, the metal wt.% in the NiDMOF/mKBx composites was determined by flame atomic absorption spectroscopy (AAS). From these metal wt.% of the AAS results, the mass fractions of the MOF were calculated from which the mKB wt.% in the composites was taken as the difference to 100% (Table S2, Supplementary Materials).

The NiDMOF/mKBx composites had mKB contents of x = 7, 14, 22, and 34 wt.%. For Ni(Fe)DMOF, the mKB content in the composite was 14 wt.%. NiDMOF is a green crystalline powder, but its mKB composites had an increasing darker color with increasing mKB content.

Yields: 61% NiDMOF/mKB7, 65% NiDMOF/mKB14, 70% NiDMOF/mKB22, 76% NiDMOF/mKB34, 62% Ni(Fe)DMOF/mKB14, based on H_2BDC.

3.6. Materials Characterization

Powder X-ray diffraction (PXRD) data were collected with a Rigaku Miniflex 600 powder diffractometer (Rigaku, Tokyo, Japan) using a low background silicon sample holder and Cu-Kα irradiation (λ = 1.54184 Å). The measurements were conducted over a 2θ = 2–100° range with a scan speed of 1.5 deg min^{-1} (600 W, 40 kV, 15 mA). The diffractograms were analyzed using the software Match 3.1.0.

The Fourier transform infrared (FTIR) spectra of all the samples were recorded on a Bruker FTIR Tensor 37 spectrometer (Bruker AXS, Karlsruhe, Germany) in attenuated total reflection (ATR) mode with a diamond crystal or as KBr pellets in the range of 400–4000 cm^{-1}.

^1H MNR measurements were performed with a Bruker Advance III-300 (Bruker, Karlsruhe, Germany) operating at 300 MHz. Before dissolution under digestion for the solution NMR measurement, the MOF samples were activated at 150 °C for at least 20 h under vacuum (<10^{-2} mbar) in order to remove the residual solvent molecules. Then, the amount of 5 mg of each activated sample was dissolved with decomposition in 0.8 mL DMSO-d_6 and 50 µL D_2SO_4.

Nitrogen sorption isotherms were obtained with a Belsorp MAXII (Microtrac MRB, Haan, Germany) high precision gas/vapor adsorption measurement instrument at 77 K and evaluated with the BELMaster MAXII software (version 7.3.2.0). Prior to the measurement, the materials were activated by degassing under vacuum (10^{-2} mbar) at 120 °C for 12 h. Brunauer–Emmett–Teller (BET) surface areas were determined from the nitrogen adsorption isotherms and the pore size distributions were derived by non-local density functional theory (NLDFT) calculations based on the "nitrogen at 77 K with slit pores" method. The total pore volumes were calculated from the adsorbed volume at $p/p_0 = 0.95$.

Elemental analysis (CHNS) was carried out using an Elementar Analysensysteme vario MICRO cube instrument (Elementar Analysensysteme, Langenselbold, Germany). The samples were dried at 150 °C under a vacuum (<10^{-2} mbar) for at least 20 h prior before the measurement.

Flame atomic absorption spectroscopy (AAS) was conducted with a PinAAcle 900T from PerkinElmer (Perkin Elmer LAS GmbH, Rodgau-Jügesheim, Germany).

Exactly weighted samples (15–20 mg) were heated and stirred with concentrated hydrochloric acid overnight. The solution was carefully filtered and diluted with Millipore water to a volume of 25 mL and again by a factor of 1:50 for the AAS measurements.

Scanning electron microscopy (SEM) images were collected with a Jeol JSM-65 10 LV QSEM (Jeol, Akishima, Japan) advanced electron microscope with a LaB_6 cathode at 20 kV equipped with a Bruker Xflash 410 (Bruker AXS, Karlsruhe, Germany) silicon drift detector for energy-dispersive X-ray spectrometric (EDX) elemental composition analysis. The signals of Cu, Zn, and Au in the EDX spectra originated from the sample holder and from the sputtering of the sample with gold prior to the investigation.

X-ray photoelectron spectroscopy (XPS) measurements were made using an ULVAC-PHI VersaProbe II microfocus X-ray photoelectron spectrometer (UL-VAC-PHI, Chigasaki, Japan) equipped with a polychromatic aluminum Kα X-ray source (1486.8 eV). Experimental XP spectra were fitted by the CasaXPS, version 2.3.19PR1.0 from Casa Software Ltd. Program (Casa Software Ltd., Teignmouth, UK). A standard Shirley baseline with no offset was used for the background correction. Binding energies were calibrated to the carbon 1s orbital with a binding energy of 284.8 eV. In the case of the Fe 2p spectra, an additional correction was necessary due to the presence of a Ni-LMM Auger peak.

3.7. Electrochemical Measurements

Electrocatalytic measurements were carried out on an Interface 1010E potentiostat from Gamry Instruments with an RRDE-3A station from ALS Japan using a three-electrode configuration. The system consists of a coiled platinum wire as a counter electrode and a reversible hydrogen electrode (RHE) from Gaskatel (Kassel, Deutschland) as a reference electrode. Working electrodes were fabricated by depositing the prepared materials on a rotating disc electrode (RDE, 5 mm diameter, 0.196 cm^2 area), here, a rotating disc glassy carbon electrode (GCE, GC-RDE). The amount of 2.5 mg of MOF or MOF/mKB was dispersed in methanol (0.5 mL) with 25 μL of Nafion (around 5% in a mixture of water and lower aliphatic alcohols) as the binding agent. The suspension was sonicated for 30 min to form a homogeneous ink. Then, 10 μL of the prepared ink was dropped onto the rotating disc electrode (loading 0.25 mg cm^{-2}) and dried at room temperature. As an electrolyte, 1 mol L^{-1} KOH solution (pH: 14) was used and nitrogen was purged for 10 min through the cell to remove oxygen before the test. The working electrode was kept rotating at a rate of 1600 rpm during the measurements. The electrode was pre-cycled for at least 20 cycles at a sweep rate of 100 mV cm^{-1} between 1.0 and 1.7 V until reaching the steady state. All linear scanning voltammetry (LSV) curves were collected by sweeping the potential from 0.8 to 1.7 V vs RHE with a scan rate of 5 mV s^{-1}. Cyclic voltammetry (CV) measurements were carried out in the potential range between 1.0 and 1.7 V vs. RHE with a scan rate of 100 mV s^{-1}. iR compensation was completely applied to the recorded data afterward, as recommended in the literature [122]. In all measurements, R_u (uncompensated electrolyte resistance) was measured by electrochemical impedance spectroscopy (EIS) at the open circuit potential (OCP) and the iR_u drop was compensated automatically by the R_u value using the software Gamry Echem Analyst., Version 7.8.2.

The overpotential η was derived from the standard potential E vs. RHE using the formula:

$$\eta = ERHE - 1.23\ V \tag{6}$$

The Tafel slope b was obtained by fitting the linear portion of the Tafel plots, which were derived from the LSV curves according to the Tafel equation:

$$\eta = a + b \times \log(j). \tag{7}$$

where a is a constant and j represents the current density. Electrochemical impedance spectroscopy (EIS) data were recorded with the frequency range of 0.1–10,000 Hz at the potential of 1.5 V. To check the long-term stability of the best catalyst material at a constant current by chronopotentiometry, a Ni(Fe)DMOF/mKB@NF electrode was fabricated by drop-casting the catalyst ink on the surface of pre-treated nickel foam and drying at room temperature. The loading was around 1 mg cm^{-2} by weighing the electrode before and after material deposition. The chronopotentiometry at 50 mA cm^{-2} was carried out with 1 cm^2 of commercial nickel foam (NF) for 30 h in 1 mol L^{-1} KOH electrolyte.

4. CONCLUSIONS

In summary, we successfully fabricated a series of novel DMOF/mKB composites containing a bimetallic nickel-iron-based pillared MOF (DMOF) and modified Ketjenblack carbon using a simple one-step hydrothermal procedure, which can be directly used as efficient electrocatalysts in OER. The appropriate mass weight % of mKB in the composite DMOF/mKB is important for the fabrication of the optimum catalyst. Benefitting from enhanced conductivity, remarkably high surface areas, and the high intrinsic catalytic activity of mKB and the integration of Ni and Fe active sites from the MOF precursors, the Ni(Fe)DMOF/mKB14 with 14 wt.% mKB exhibited superior OER performance, which only needed a low overpotential of 279 mV and a Tafel slope of 25 mV dec^{-1} to reach the current density of 10 mA cm^{-2} in 1 mol L^{-1} of KOH electrolyte, outperforming the benchmark RuO_2 catalyst and many state-of-the art MOF-based and Ni-Fe-based OER catalysts. For the practical application, the working electrode was fabricated by depositing optimized Ni(Fe)DMOF/mKB14 electrocatalysts on commercial nickel foam, where the current density at 10 and 50 mA cm^{-2} was realized at 247 and 291 mV, respectively, and the activity was maintained for 30 h of applied bias. The superior OER performance is associated with the transformation of Ni(Fe) DMOF into highly functionalized α/β-$Ni(OH)_2$/β/γ-NiOOH and FeOOH inherited from the DMOF structure. The presented DMOF/mKB material preparation strategy is applicable to the development of high-performance electrode materials for other advanced energy storage systems.

Author Contributions

Conceptualization, C.J. and T.H.Y.B.; Methodology, T.H.Y.B.; Validation, T.H.Y.B.; Formal analysis, T.H.Y.B.; Investigation, T.H.Y.B., L.R. and R.O.; Resources, C.J.; Data curation, T.H.Y.B.; Writing—original draft preparation, T.H.Y.B.; Writing—review and editing, C.J.; Visualization, T.H.Y.B.; Supervision, C.J.; Project administration, C.J.; Funding acquisition, C.J. All authors have read and agreed to the published version of the manuscript.

Acknowledgments

The authors also thank the Center for Molecular and Structural Analytics at Heinrich Heine University (CeMSA@HHU) for recording the NMR-spectrometric data. The authors thank Birgit Tommes for carrying out the FTIR experiments, Annette Ricken for the AAS measurements.

REFERENCES

1. Lianos, P. Review of recent trends in photoelectrocatalytic conversion of solar energy to electricity and hydrogen. Appl. Catal. B 2017, 210, 235–254.
2. Chu, S.; Cui, Y.; Liu, N. Opportunities and challenges for a sustainable energy future. Nat. Mater. 2017, 16, 16–22.
3. Wang, J.; Zhang, H.; Wang, X. Recent Methods for the Synthesis of Noble-Metal-Free Hydrogen-Evolution Electrocatalysts: From Nanoscale to Sub-nanoscale. Small Methods 2017, 1, 1700118.
4. Shrivastav, V.; Sundriyal, S.; Goel, P.; Kaur, H.; Tuteja, S.K.; Vikrant, K.; Kim, K.-H.; Tiwari, U.K.; Deep, A. Metal-organic frameworks (MOFs) and their composites as electrodes for lithium battery applications: Novel means for alternative energy storage. Coord. Chem. Rev. 2019, 393, 48–78.
5. Lee, Y.; Suntivich, J.; May, K.J.; Perry, E.E.; Shao-Horn, Y. Synthesis and Activities of Rutile IrO2 and RuO2 Nanoparticles for Oxygen Evolution in Acid and Alkaline Solutions. J. Phys. Chem. Lett. 2012, 3, 399–404.
6. Pi, Y.; Zhang, N.; Guo, S.; Guo, J.; Huang, X. Ultrathin Laminar Ir Superstructure as Highly Efficient Oxygen Evolution Electrocatalyst in Broad pH Range. Nano Lett. 2016, 16, 4424–4430.
7. Hu, C.; Zhang, L.; Gong, J. Recent progress made in the mechanism comprehension and design of electrocatalysts for alkaline water splitting. Energy Environ. Sci. 2019, 12, 2620–2645.
8. Miao, X.; Wu, L.; Lin, Y.; Yuan, X.; Zhao, J.; Yan, W.; Zhou, S.; Shi, L. The Role of Oxygen Vacancies in Water Oxidation for Perovskite Cobalt Oxide Electrocatalysts: Are More Better? Chem. Commun. 2019, 55, 1442.
9. Li, X.; Hao, X.; Wang, Z.; Abudula, A.; Guan, G. In-situ intercalation of NiFe LDH materials: An efficient approach to improve electrocatalytic activity and stability for water splitting. J. Power Sources 2017, 347, 193–200.

10. Han, L.; Dong, S.; Wang, E. Transition-Metal (Co, Ni, and Fe)-Based Electrocatalysts for the Water Oxidation Reaction. Adv. Mater. 2016, 28, 9266–9291.
11. Yu, M.; Budiyanto, E.; Tüysüz, H. Principles of water electrolysis and recent progress in cobalt-, nickel-, and iron-based oxides for the oxygen evolution reaction. Angew. Chem. Int. Ed. 2022, 61, e202103824.
12. Vij, V.; Sultan, S.; Harzandi, A.M.; Meena, A.; Tiwari, J.N.; Lee, W.-G.; Yoon, T.; Kim, K.S. Nickel-based electrocatalysts for energy-related applications: Oxygen reduction, oxygen evolution, and hydrogen evolution reactions. ACS Catal. 2017, 7, 7196–7225.
13. Hall, D.S.; Lockwood, D.J.; Bock, C.; MacDougall, B.R. Nickel hydroxides and related materials: A review of their structures, synthesis and properties. Proc. Math. Phys. Eng. Sci. 2015, 471, 20140792.
14. Chen, Y.; Rui, K.; Zhu, J.; Dou, S.X.; Sun, W. Recent Progress on Nickel-Based Oxide/(Oxy) Hydroxide Electrocatalysts for the Oxygen Evolution Reaction. Eur. J. Chem 2019, 25, 703–713.
15. Beall, C.E.; Fabbri, E.; Schmidt, T.J. Perovskite Oxide Based Electrodes for the Oxygen Reduction and Evolution Reactions: The Underlying Mechanism. ACS Catal. 2021, 11, 3094–3114.
16. Tareen, A.K.; Priyanga, G.S.; Khan, K.; Pervaiz, E.; Thomas, T.; Yang, M. Nickel-Based Transition Metal Nitride Electrocatalysts for the Oxygen Evolution Reaction. ChemSusChem 2019, 12, 3941–3954.
17. Wang, C.; Wu, Y.; Zhou, Z.; Wang, J.; Pei, S.; Liu, S. Electrodeposited amorphous nickel–iron phosphide and sulfide derived films for electrocatalytic oxygen evolution. Int. J. Hydrog. Energy 2022, 47, 40849–40859.
18. Zhao, C.-X.; Liu, J.-N.; Wang, C.; Wang, J.; Song, L.; Li, B.-Q.; Zhang, Q. An anionic regulation mechanism for the structural reconstruction of sulfide electrocatalysts under oxygen evolution conditions. Energy Environ. Sci. 2022, 15, 3257–3264.
19. Wan, K.; Luo, J.; Zhou, C.; Zhang, T.; Arbiol, J.; Lu, X.; Mao, B.-W.; Zhang, X.; Fransaer, J. Hierarchical Porous Ni3S4 with Enriched High-Valence Ni Sites as a Robust Electrocatalyst for Efficient Oxygen Evolution Reaction. Adv. Funct. Mater. 2019, 29, 1900315.
20. Wan, K.; Luo, J.; Zhang, X.; Subramanian, P.; Fransaer, J. Sulfur-modified nickel selenide as an efficient electrocatalyst for the oxygen evolution reaction. J. Energy Chem. 2021, 62, 198–203.
21. Wang, T.; Liu, X.; Yan, Z.; Teng, Y.; Li, R.; Zhang, J.; Peng, T. Facile Preparation Process of NiCoP–NiCoSe2 Nano-Bilayer Films for Oxygen Evolution Reaction with High Efficiency and Long Duration. ACS Sustain. Chem. Eng. 2020, 8, 1240–1251.
22. Saha, S.; Ganguli, A.K. FeCoNi Alloy as Noble Metal-Free Electrocatalyst for Oxygen Evolution Reaction (OER). ChemistrySelect 2017, 2, 1630–1636.
23. Landon, J.; Demeter, E.; İnoğlu, N.; Keturakis, C.; Wachs, I.E.; Vasić, R.; Frenkel, A.I.; Kitchin, J.R. Spectroscopic Characterization of Mixed Fe–Ni Oxide Electrocatalysts for the Oxygen Evolution Reaction in Alkaline Electrolytes. ACS Catal. 2012, 2, 1793–1801.

24. Subbaraman, R.; Tripkovic, D.; Chang, K.-C.; Strmcnik, D.; Paulikas, A.P.; Hirunsit, P.; Chan, M.; Greeley, J.; Stamenkovic, V.; Markovic, N.M. Trends in activity for the water electrolyser reactions on 3d M (Ni,Co,Fe,Mn) hydr(oxy)oxide catalysts. Nat. Mater. 2012, 11, 550–557.

25. Stern, L.-A.; Hu, X. Enhanced oxygen evolution activity by NiOx and Ni(OH)2 nanoparticles. Faraday Discuss. 2014, 176, 363–379.

26. Corrigan, D.A. The Catalysis of the Oxygen Evolution Reaction by Iron Impurities in Thin Film Nickel Oxide Electrodes. J. Electrochem. Soc. 1987, 134, 377–384.

27. Louie, M.W.; Bell, A.T. An Investigation of Thin-Film Ni–Fe Oxide Catalysts for the Electrochemical Evolution of Oxygen. J. Am. Chem. Soc. 2013, 135, 12329–12337.

28. Qi, J.; Zhang, W.; Xiang, R.; Liu, K.; Wang, H.-Y.; Chen, M.; Han, Y.; Cao, R. Porous Nickel–Iron Oxide as a Highly Efficient Electrocatalyst for Oxygen Evolution Reaction. Adv. Sci. 2015, 2, 1500199.

29. Stevens, M.B.; Trang, C.D.M.; Enman, L.J.; Deng, J.; Boettcher, S.W. Reactive Fe-Sites in Ni/Fe (Oxy)hydroxide Are Responsible for Exceptional Oxygen Electrocatalysis Activity. J. Electrochem. Soc. 2017, 139, 11361–11364.

30. Yu, M.; Moon, G.; Bill, E.; Tüysüz, H. Optimizing Ni–Fe Oxide Electrocatalysts for Oxygen Evolution Reaction by Using Hard Templating as a Toolbox. ACS Appl. Energy Mater. 2019, 2, 1199–1209.

31. Fominykh, K.; Chernev, P.; Zaharieva, I.; Sicklinger, J.; Stefanic, G.; Döblinger, M.; Müller, A.; Pokharel, A.; Böcklein, S.; Scheu, C.; et al. Iron-Doped Nickel Oxide Nanocrystals as Highly Efficient Electrocatalysts for Alkaline Water Splitting. ACS Nano 2015, 9, 5180–5188.

32. Burke, M.S.; Enman, L.J.; Batchellor, A.S.; Zou, S.; Boettcher, S.W. Oxygen Evolution Reaction Electrocatalysis on Transition Metal Oxides and (Oxy)hydroxides: Activity Trends and Design Principles. Chem. Mater. 2015, 27, 7549–7558.

33. Burke, M.S.; Kast, M.G.; Trotochaud, L.; Smith, A.M.; Boettcher, S.W. Cobalt–Iron (Oxy) hydroxide Oxygen Evolution Electrocatalysts: The Role of Structure and Composition on Activity, Stability, and Mechanism. J. Am. Chem. Soc. 2015, 137, 3638–3648.

34. Ray, C.; Lee, S.C.; Jin, B.; Kundu, A.; Park, J.H.; Jun, S.C. Stacked Porous Iron-Doped Nickel Cobalt Phosphide Nanoparticle: An Efficient and Stable Water Splitting Electrocatalyst. ACS Sustain. Chem. Eng. 2018, 6, 6146–6156.

35. Li, B.; Wen, H.-M.; Zhou, W.; Chen, B. Porous metal–organic frameworks for gas storage and separation: What, how, and why? J. Phys. Chem. Lett. 2014, 5, 3468–3479.

36. Ma, S.; Zhou, H.-C. Gas storage in porous metal–organic frameworks for clean energy applications. Chem. Commun. 2010, 46, 44–53.

37. Yaghi, O.M.; Li, H. Hydrothermal Synthesis of a Metal-Organic Framework Containing Large Rectangular Channels. J. Am. Chem. Soc. 1995, 117, 10401–10402.

38. Yaghi, O.M.; Li, G.; Li, H. Selective binding and removal of guests in a microporous metal–organic framework. Nature 1995, 378, 703–706.

39. Scheurle, P.I.; Mähringer, A.; Jakowetz, A.C.; Hosseini, P.; Richter, A.F.; Wittstock, G.; Medina, D.D.; Bein, T. A highly crystalline anthracene-based MOF-74 series featuring electrical conductivity and luminescence. Nanoscale 2019, 11, 20949–20955.

40. Allendorf, M.D.; Schwartzberg, A.; Stavila, V.; Talin, A.A. A roadmap to implementing metal–organic frameworks in electronic devices: Challenges and critical directions. Chem. Eur. J. 2011, 17, 11372–11388.
41. Wu, H.B.; Lou, X.W. Metal-Organic Frameworks and Their Derived Materials for Electrochemical Energy Storage and Conversion: Promises and Challenges. Sci. Adv. 2017, 3, eaap9252.
42. Zheng, W.; Lee, L.Y.S. Metal–organic frameworks for electrocatalysis: Catalyst or precatalyst? ACS Energy Lett. 2021, 6, 2838–2843.
43. Singh, B.; Yadav, A.; Indra, A. Realizing electrochemical transformation of a metal–organic framework precatalyst into a metal hydroxide–oxy (hydroxide) active catalyst during alkaline water oxidation. J. Mater. Chem. A 2022, 10, 3843–3868.
44. Sondermann, L.; Jiang, W.; Shviro, M.; Spieß, A.; Woschko, D.; Rademacher, L.; Janiak, C. Nickel-Based Metal-Organic Frameworks as Electrocatalysts for the Oxygen Evolution Reaction (OER). Molecules 2022, 27, 1241.
45. Wang, J.; Kim, S.-J.; Liu, J.; Gao, Y.; Choi, S.; Han, J.; Shin, H.; Jo, S.; Kim, J.; Ciucci, F. Redirecting dynamic surface restructuring of a layered transition metal oxide catalyst for superior water oxidation. Nat. Catal. 2021, 4, 212–222.
46. Gong, L.; Yang, H.; Wang, H.; Qi, R.; Wang, J.; Chen, S.; You, B.; Dong, Z.; Liu, H.; Xia, B.Y. Corrosion formation and phase transformation of nickel-iron hydroxide nanosheets array for efficient water oxidation. Nano Res. 2021, 14, 4528–4533.
47. Hassan, M.H.; Soliman, A.B.; Elmehelmey, W.A.; Abugable, A.A.; Karakalos, S.G.; Elbahri, M.; Hassanien, A.; Alkordi, M.H. A Ni-loaded, metal–organic framework–graphene composite as a precursor for in situ electrochemical deposition of a highly active and durable water oxidation nanocatalyst. Chem. Commun. 2019, 55, 31–34.
48. Meng, J.; Zhou, Y.; Chi, H.; Li, K.; Wan, J.; Hu, Z. Bimetallic porphyrin mof anchored onto rgo nanosheets as a highly efficient 2d electrocatalyst for oxygen evolution reaction in alkaline conditions. ChemistrySelect 2019, 4, 8661–8670.
49. Zheng, X.; Cao, Y.; Liu, D.; Cai, M.; Ding, J.; Liu, X.; Wang, J.; Hu, W.; Zhong, C. Bimetallic metal–organic-framework/reduced graphene oxide composites as bifunctional electrocatalysts for rechargeable Zn–air batteries. ACS Appl. Mater. Interfaces 2019, 11, 15662–15669.
50. Srinivas, K.; Chen, Y.; Wang, X.; Wang, B.; Karpuraranjith, M.; Wang, W.; Su, Z.; Zhang, W.; Yang, D. Constructing Ni/NiS heteronanoparticle-embedded metal–organic framework-derived nanosheets for enhanced water-splitting catalysis. ACS Sustain. Chem. Eng. 2021, 9, 1920–1931.
51. Sohrabi, S.; Dehghanpour, S.; Ghalkhani, M. A cobalt porphyrin-based metal organic framework/multi-walled carbon nanotube composite electrocatalyst for oxygen reduction and evolution reactions. J. Mater. Sci. 2018, 53, 3624–3639.
52. Wang, C.; Zhou, M.; Ma, Y.; Tan, H.; Wang, Y.; Li, Y. Hybridized Polyoxometalate-Based Metal–Organic Framework with Ketjenblack for the Nonenzymatic Detection of H2O2. Chem. Asian J. 2018, 13, 2054–2059.

53. Peng, H.; Yang, X.; Ma, Y.; Liu, J.; Wang, Y.; Tan, H.; Li, Y. Polyoxometalate-based metal–organic framework loaded with an ultra-low amount of Pt as an efficient electrocatalyst for hydrogen production. CrystEngComm 2018, 20, 5387–5394.

54. Huang, Z.-H.; Xie, N.-H.; Zhang, M.; Xu, B.-Q. Nonpyrolyzed Fe–N Coordination-Based Iron Triazolate Framework: An Efficient and Stable Electrocatalyst for Oxygen Reduction Reaction. ChemSusChem 2019, 12, 200–207.

55. Öztürk, S.; Moon, G.-H.; Spieß, A.; Budiyanto, E.; Roitsch, S.; Tüysüz, H.; Janiak, C. A Highly-Efficient Oxygen Evolution Electrocatalyst Derived from a Metal-Organic Framework and Ketjenblack Carbon Material. ChemPlusChem 2021, 86, 1106–1115.

56. Maniam, P.; Stock, N. Investigation of Porous Ni-Based Metal–Organic Frameworks Containing Paddle-Wheel Type Inorganic Building Units via High-Throughput Methods. Inorg. Chem. 2011, 50, 5085–5097.

57. Zhang, Q.; Li, T.; Liang, J.; Wang, N.; Kong, X.; Wang, J.; Qian, H.; Zhou, Y.; Liu, F.; Wei, C. Highly wettable and metallic NiFe-phosphate / phosphide catalyst synthesized by plasma for highly efficient oxygen evolution reaction. J. Mater. Chem. A 2018, 6, 7509–7516.

58. Qiu, Y.; Xin, L.; Li, W. Electrocatalytic oxygen evolution over supported small amorphous Ni–Fe nanoparticles in alkaline electrolyte. Langmuir 2014, 30, 7893–7901.

59. Yaqoob, L.; Noor, T.; Iqbal, N.; Nasir, H.; Zaman, N.; Talha, K. Electrochemical synergies of Fe–Ni bimetallic MOF CNTs catalyst for OER in water splitting. J. Alloys Compd. 2021, 850, 156583.

60. Abbasi, M.; Noor, T.; Iqbal, N.; Zaman, N. Electrocatalytic study of Cu/Ni MOF and its γ-C3N4 composites for methanol oxidation reaction. Int. J. Energy Res. 2022, 46, 13915–13930.

61. Yaqoob, L.; Noor, T.; Iqbal, N.; Nasir, H.; Zaman, N. Development of Nickel-BTC-MOF-Derived Nanocomposites with rGO Towards Electrocatalytic Oxidation of Methanol and Its Product Analysis. Catalysts 2019, 9, 856.

62. Fang, Y.; Li, X.; Li, F.; Lin, X.; Tian, M.; Long, X.; An, X.; Fu, Y.; Jin, J.; Ma, J. Self-assembly of cobalt-centered metal organic framework and multiwalled carbon nanotubes hybrids as a highly active and corrosion-resistant bifunctional oxygen catalyst. J. Power Sources 2016, 326, 50–59.

63. Yu, S.; Wu, Y.; Xue, Q.; Zhu, J.-J.; Zhou, Y. A novel multi-walled carbon nanotube-coupled CoNi MOF composite enhances the oxygen evolution reaction through synergistic effects. J. Mater. Chem. A 2022, 10, 4936–4943.

64. Jahan, M.; Liu, Z.; Loh, K.P. A Graphene oxide and copper-centered metal organic framework composite as a tri-functional catalyst for HER, OER, and ORR. Adv. Funct. Mater. 2013, 23, 5363–5372.

65. Wang, H.; Han, J.; Li, L.; Peng, F.; Zheng, F.; Huang, D.; Lai, F.; Hu, S.; Pan, Q.; Li, Q. Effects of oxidized Ketjen Black as conductive additives on electrochemical performance of the LiMn2O4@Al2O3 cathode in lithium-ion batteries. J. Alloys Compd. 2021, 860, 158482.

66. Guan, D.; Zhong, J.; Xu, H.; Huang, Y.-C.; Hu, Z.; Chen, B.; Zhang, Y.; Ni, M.; Xu, X.; Zhou, W.; et al. A universal chemical-induced tensile strain tuning strategy to boost oxygen-evolving electrocatalysis on perovskite oxides. Appl. Phys. Rev. 2022, 9, 011422.
67. Guan, D.; Shi, C.; Xu, H.; Gu, Y.; Zhong, J.; Sha, Y.; Hu, Z.; Ni, M.; Shao, Z. Simultaneously mastering operando strain and reconstruction effects via phase-segregation strategy for enhanced oxygen-evolving electrocatalysis. J. Energy Chem. 2023, 82, 572–580.
68. Li, D.-J.; Li, Q.-H.; Gu, Z.-G.; Zhang, J. A surface-mounted MOF thin film with oriented nanosheet arrays for enhancing the oxygen evolution reaction. J. Mater. Chem. A 2019, 7, 18519–18528.
69. Henke, S.; Schneemann, A.; Wütscher, A.; Fischer, R.A. Directing the Breathing Behavior of Pillared-Layered Metal–Organic Frameworks via a Systematic Library of Functionalized Linkers Bearing Flexible Substituents. J. Am. Chem. Soc. 2012, 134, 9464–9474.
70. Asgharnejad, L.; Abbasi, A.; Shakeri, A. Ni-based metal-organic framework/GO nanocomposites as selective adsorbent for CO2 over N2. Microporous Mesoporous Mater. 2018, 262, 227–234.
71. Qi, J.; Zhang, W.; Cao, R. Porous Materials as Highly Efficient Electrocatalysts for the Oxygen Evolution Reaction. ChemCatChem 2018, 10, 1206–1220.
72. Thommes, M.; Kaneko, K.; Neimark, A.V.; Olivier, J.P.; Rodriguez-Reinoso, F.; Rouquerol, J.; Sing, K.S.W. Physisorption of gases, with special reference to the evaluation of surface area and pore size distribution (IUPAC Technical Report). Pure Appl. Chem. 2015, 87, 1051–1069.
73. Wang, X.; Niu, Z.; Al-Enizi, A.M.; Nafady, A.; Wu, Y.; Aguila, B.; Verma, G.; Wojtas, L.; Chen, Y.-S.; Li, Z.; et al. Pore environment engineering in metal–organic frameworks for efficient ethane/ethylene separation. J. Mater. Chem. A 2019, 7, 13585–13590.
74. Jiang, X.; Deng, S.; Sun, L.; Liu, J.; Qi, N.; Chen, Z. Pillared nickel-based metal-organic frameworks as electrode material with high electrochemical performance. J. Electroanal. Chem. 2020, 879, 114802.
75. Hungerford, J.; Bhattacharyya, S.; Tumuluri, U.; Nair, S.; Wu, Z.; Walton, K.S. DMOF-1 as a representative MOF for SO2 adsorption in both humid and dry conditions. J. Phys. Chem. C 2018, 122, 23493–23500.
76. McIntyre, N.S.; Cook, M.G. X-ray photoelectron studies on some oxides and hydroxides of cobalt, nickel, and copper. Anal. Chem. 1975, 47, 2208–2213.
77. Jiang, H.; Guo, Y.; Wang, T.; Zhu, P.-L.; Yu, S.; Yu, Y.; Fu, X.-Z.; Sun, R.; Wong, C.-P. Electrochemical fabrication of Ni(OH)2/Ni 3D porous composite films as integrated capacitive electrodes. RSC Adv. 2015, 5, 12931–12936.
78. Lee, J.W.; Ahn, T.; Soundararajan, D.; Ko, J.M.; Kim, J.-D. Non-aqueous approach to the preparation of reduced graphene oxide/α-Ni(OH)2 hybrid composites and their high capacitance behavior. Chem. Commun. 2011, 47, 6305–6307.
79. Yan, J.; Fan, Z.; Sun, W.; Ning, G.; Wei, T.; Zhang, Q.; Zhang, R.; Zhi, L.; Wei, F. Advanced asymmetric supercapacitors based on Ni(OH)2/graphene and porous graphene electrodes with high energy density. Adv. Funct. Mater. 2012, 22, 2632–2641.

80. Yamashita, T.; Hayes, P. Analysis of XPS spectra of Fe2+ and Fe3+ ions in oxide materials. Appl. Surf. Sci. 2008, 254, 2441–2449.

81. Biesinger, M.C.; Payne, B.P.; Grosvenor, A.P.; Lau, L.W.M.; Gerson, A.R.; Smart, R.S.C. Resolving surface chemical states in XPS analysis of first row transition metals, oxides and hydroxides: Cr, Mn, Fe, Co and Ni. Appl. Surf. Sci. 2011, 257, 2717–2730.

82. Yin, L.I.; Yellin, E.; Adler, I. X-Ray Excited LMM Auger Spectra of Copper, Nickel, and Iron. J. Appl. Phys. 1971, 42, 3595–3600.

83. Zhou, C.; Chen, X.; Liu, S.; Han, Y.; Meng, H.; Jiang, Q.; Zhao, S.; Wei, F.; Sun, J.; Tan, T.; et al. Superdurable Bifunctional Oxygen Electrocatalyst for High-Performance Zinc–Air Batteries. J. Am. Chem. Soc. 2022, 144, 2694–2704.

84. Shinagawa, T.; Garcia-Esparza, A.T.; Takanabe, K. Insight on Tafel slopes from a microkinetic analysis of aqueous electrocatalysis for energy conversion. Sci. Rep. 2015, 5, 13801.

85. Matsumoto, Y.; Sato, E. Electrocatalytic properties of transition metal oxides for oxygen evolution reaction. Mater. Chem. Phys. 1986, 14, 397–426.

86. Li, G.; Anderson, L.; Chen, Y.; Pan, M.; Chuang, P.-Y.A. New insights into evaluating catalyst activity and stability for oxygen evolution reactions in alkaline media. Sustain. Energy Fuels 2018, 2, 237–251.

87. Alobaid, A.; Wang, C.; Adomaitis, R.A. Mechanism and Kinetics of HER and OER on NiFe LDH Films in an Alkaline Electrolyte. J. Electrochem. Soc. 2018, 165, J3395.

88. Möller, S.; Barwe, S.; Masa, J.; Wintrich, D.; Seisel, S.; Baltruschat, H.; Schuhmann, W. Online Monitoring of Electrochemical Carbon Corrosion in Alkaline Electrolytes by Differential Electrochemical Mass Spectrometry. Angew. Chem. Int. Ed. 2020, 59, 1585–1589.

89. Li, J.; Zhou, N.; Song, J.; Fu, L.; Yan, J.; Tang, Y.; Wang, H. Cu–MOF-Derived Cu/Cu2O Nanoparticles and CuNxCy Species to Boost Oxygen Reduction Activity of Ketjenblack Carbon in Al–Air Battery. ACS Sustain. Chem. Eng. 2018, 6, 413–421.

90. Tee, S.Y.; Win, K.Y.; Teo, W.S.; Koh, L.-D.; Liu, S.; Teng, C.P.; Han, M.-Y. Recent Progress in Energy-Driven Water Splitting. Adv. Sci. 2017, 4, 1600337.

91. Liu, Q.; Cong, H.; Deng, H. Deciphering the Spatial Arrangement of Metals and Correlation to Reactivity in Multivariate Metal–Organic Frameworks. J. Am. Chem. Soc. 2016, 138, 13822–13825.

92. Gao, M.-R.; Xu, Y.-F.; Jiang, J.; Yu, S.-H. Nanostructured metal chalcogenides: Synthesis, modification, and applications in energy conversion and storage devices. Chem. Soc. Rev. 2013, 42, 2986–3017.

93. Doyle, R.L.; Lyons, M.E.G. An electrochemical impedance study of the oxygen evolution reaction at hydrous iron oxide in base. Phys. Chem. Chem. Phys. 2013, 15, 5224–5237.

94. Doyle, R.L.; Lyons, M.E.G. Kinetics and Mechanistic Aspects of the Oxygen Evolution Reaction at Hydrous Iron Oxide Films in Base. J. Am. Chem. Soc. 2013, 160, H142.

95. Doyle, R.L.; Godwin, I.J.; Brandon, M.P.; Lyons, M.E.G. Redox and electrochemical water splitting catalytic properties of hydrated metal oxide modified electrodes. Phys. Chem. Chem. Phys. 2013, 15, 13737–13783.

96. Miller, E.L.; Rocheleau, R.E. Electrochemical behavior of reactively sputtered iron-doped nickel oxide. J. Electrochem. Soc. 1997, 144, 3072.

97. Lin, Y.; Wan, H.; Wu, D.; Chen, G.; Zhang, N.; Liu, X.; Li, J.; Cao, Y.; Qiu, G.; Ma, R. Metal–organic framework hexagonal nanoplates: Bottom-up synthesis, topotactic transformation, and efficient oxygen evolution reaction. J. Am. Chem. Soc. 2020, 142, 7317–7321.

98. Negahdar, L.; Zeng, F.; Palkovits, S.; Broicher, C.; Palkovits, R. Mechanistic Aspects of the Electrocatalytic Oxygen Evolution Reaction over Ni–Co Oxides. ChemElectroChem 2019, 6, 5588–5595.

99. Zhang, H.; Gao, Y.; Xu, H.; Guan, D.; Hu, Z.; Jing, C.; Sha, Y.; Gu, Y.; Huang, Y.-C.; Chang, Y.-C.; et al. Combined Corner-Sharing and Edge-Sharing Networks in Hybrid Nanocomposite with Unusual Lattice-Oxygen Activation for Efficient Water Oxidation. Adv. Funct. Mater. 2022, 32, 2207618.

100. Feng, J.-X.; Xu, H.; Dong, Y.-T.; Ye, S.-H.; Tong, Y.-X.; Li, G.-R. FeOOH/Co/FeOOH hybrid nanotube arrays as high-performance electrocatalysts for the oxygen evolution reaction. Angew. Chem. Int. Ed. 2016, 55, 3694–3698.

101. Lu, C.; Tranca, D.; Zhang, J.; Rodríguez Hernández, F.; Su, Y.; Zhuang, X.; Zhang, F.; Seifert, G.; Feng, X. Molybdenum carbide-embedded nitrogen-doped porous carbon nanosheets as electrocatalysts for water splitting in alkaline media. ACS Nano 2017, 11, 3933–3942.

102. Luo, J.; Im, J.-H.; Mayer, M.T.; Schreier, M.; Nazeeruddin, M.K.; Park, N.-G.; Tilley, S.D.; Fan, H.J.; Grätzel, M. Water photolysis at 12.3% efficiency via perovskite photovoltaics and Earth-abundant catalysts. Science 2014, 345, 1593–1596.

103. Dong, G.; Fang, M.; Zhang, J.; Wei, R.; Shu, L.; Liang, X.; Yip, S.; Wang, F.; Guan, L.; Zheng, Z.; et al. In situ formation of highly active Ni–Fe based oxygen-evolving electrocatalysts via simple reactive dip-coating. J. Mater. Chem. A 2017, 5, 11009–11015.

104. Peugeot, A.; Creissen, C.E.; Karapinar, D.; Tran, H.N.; Schreiber, M.; Fontecave, M. Benchmarking of oxygen evolution catalysts on porous nickel supports. Joule 2021, 5, 1281–1300.

105. Yuan, Y.; Xia, X.; Wu, J.; Yang, J.; Chen, Y.; Guo, S. Nickel foam-supported porous Ni(OH)2/NiOOH composite film as advanced pseudocapacitor material. Electrochim. Acta. 2011, 56, 2627–2632.

106. Lee, S.-Y.; Kim, I.-S.; Cho, H.-S.; Kim, C.-H.; Lee, Y.-K. Resolving Potential-Dependent Degradation of Electrodeposited Ni(OH)2 Catalysts in Alkaline Oxygen Evolution Reaction (OER): In Situ XANES Studies. Appl. Catal. B 2021, 284, 119729.

107. Klaus, S.; Cai, Y.; Louie, M.W.; Trotochaud, L.; Bell, A.T. Effects of Fe electrolyte impurities on Ni (OH)2/NiOOH structure and oxygen evolution activity. Open J. Phys. Chem. C 2015, 119, 7243–7254.

108. Nadesan, J.B.; Tseung, A.C. Oxygen evolution on nickel oxide electrodes. J. Electrochem. Soc. 1985, 132, 2957.

109. Fominykh, K.; Feckl, J.M.; Sicklinger, J.; Döblinger, M.; Böcklein, S.; Ziegler, J.; Peter, L.; Rathousky, J.; Scheidt, E.-W.; Bein, T. Ultrasmall dispersible crystalline nickel oxide nanoparticles as high-performance catalysts for electrochemical water splitting. Adv. Funct. Mater. 2014, 24, 3123–3129.

110. Ma, X.; Zheng, D.J.; Hou, S.; Mukherjee, S.; Khare, R.; Gao, G.; Ai, Q.; Garlyyev, B.; Li, W.; Koch, M.; et al. Structure–Activity Relationships in Ni- Carboxylate-Type Metal–Organic Frameworks' Metamorphosis for the Oxygen Evolution Reaction. ACS Catal. 2023, 13, 7587–7596.

111. Xu, P.; Han, X.J.; Zhang, B.; Lv, Z.S.; Liu, X.R. Characterization of an ultrafine β-nickel hydroxide from supersonic co-precipitation method. J. Alloys Compd. 2007, 436, 369–374.

112. Jiang, X.; Deng, S.; Liu, J.; Qi, N.; Chen, Z. Enhanced Electrochemical Performance of Bimetallic Doped Ni-Based Metal–Organic Frameworks by Redox Additives in an Alkaline Electrolyte. ACS Appl. Energy Mater. 2021, 4, 4610–4619.

113. Qu, C.; Jiao, Y.; Zhao, B.; Chen, D.; Zou, R.; Walton, K.S.; Liu, M. Nickel-based pillared MOFs for high-performance supercapacitors: Design, synthesis and stability study. Nano Energy 2016, 26, 66–73.

114. Jiao, Y.; Qu, C.; Zhao, B.; Liang, Z.; Chang, H.; Kumar, S.; Zou, R.; Liu, M.; Walton, K.S. High-Performance Electrodes for a Hybrid Supercapacitor Derived from a Metal–Organic Framework/Graphene Composite. ACS Appl. Energy Mater. 2019, 2, 5029–5038.

115. Anantharaj, S.; Karthik, P.E.; Kundu, S. Petal-like hierarchical array of ultrathin Ni(OH)2 nanosheets decorated with Ni(OH)2 nanoburls: A highly efficient OER electrocatalyst. Catal. Sci. Technol. 2017, 7, 882–893.

116. Gao, M.; Sheng, W.; Zhuang, Z.; Fang, Q.; Gu, S.; Jiang, J.; Yan, Y. Efficient Water Oxidation Using Nanostructured α-Nickel-Hydroxide as an Electrocatalyst. J. Am. Chem. Soc. 2014, 136, 7077–7084.

117. Smith, R.D.L.; Prévot, M.S.; Fagan, R.D.; Trudel, S.; Berlinguette, C.P. Water Oxidation Catalysis: Electrocatalytic Response to Metal Stoichiometry in Amorphous Metal Oxide Films Containing Iron, Cobalt, and Nickel. J. Am. Chem. Soc. 2013, 135, 11580–11586.

118. Tian, J.; Jiang, F.; Yuan, D.; Zhang, L.; Chen, Q.; Hong, M. Electric-field assisted in situ hydrolysis of bulk metal–organic frameworks (MOFs) into ultrathin metal oxyhydroxide nanosheets for efficient oxygen evolution. Angew. Chem. Int. Ed. 2020, 59, 13101–13108.

119. Singh, B.; Prakash, O.; Maiti, P.; Menezes, P.W.; Indra, A. Electrochemical transformation of Prussian blue analogues into ultrathin layered double hydroxide nanosheets for water splitting. Chem. Commun. 2020, 56, 15036–15039.

120. Kou, Z.; Li, X.; Zhang, L.; Zang, W.; Gao, X.; Wang, J. Dynamic surface chemistry of catalysts in oxygen evolution reaction. Small Sci. 2021, 1, 2100011.

121. Li, J.; Huang, W.; Wang, M.; Xi, S.; Meng, J.; Zhao, K.; Jin, J.; Xu, W.; Wang, Z.; Liu, X. Low-crystalline bimetallic metal–organic framework electrocatalysts with rich active sites for oxygen evolution. ACS Energy Lett. 2018, 4, 285–292.

122. Anantharaj, S.; Noda, S. iR drop correction in electrocatalysis: Everything one needs to know! J. Mater. Chem. A 2022, 10, 9348–9354.

123. Brandenburg, K. Diamond; Version 4.6, Crystal and Molecular Structure Visualization, Crystal Impact; K. Brandenburg & H. Putz Gbr: Bonn, Germany, 2009–2022.

124. Zhang, C.; Liang, Y.; Wang, Y.; He, Y.; Majeed, A.; Yang, Z.; Yao, S.; Shen, X.; Li, T.; Qin, S. Polyamidoamine dendrimer modified Ketjen Black mixed sulfur coated cathode for enhancing polysulfides adsorbability in Li-S batteries. Ionics 2021, 27, 2997–3005.

125. Xing, S.; Liang, J.; Brandt, P.; Schäfer, F.; Nuhnen, A.; Heinen, T.; Boldog, I.; Möllmer, J.; Lange, M.; Weingart, O. Capture and separation of SO2 traces in metal–organic frameworks via pre-synthetic pore environment tailoring by methyl groups. Angew. Chem. Int. Ed. 2021, 60, 17998–18005.

126. Wu, Y.; Song, X.; Xu, S.; Chen, Y.; Oderinde, O.; Gao, L.; Wei, R.; Xiao, G. Chemical fixation of CO2 into cyclic carbonates catalyzed by bimetal mixed MOFs: The role of the interaction between Co and Zn. Dalton Trans. 2020, 49, 312–321.

127. Grosvenor, A.P.; Kobe, B.A.; Biesinger, M.C.; McIntyre, N.S. Investigation of multiplet splitting of Fe 2p XPS spectra and bonding in iron compounds. Surf. Interface Anal. 2004, 36, 1564–1574.

128. Paolella, A.; Bertoni, G.; Hovington, P.; Feng, Z.; Flacau, R.; Prato, M.; Colombo, M.; Marras, S.; Manna, L.; Turner, S. Cation exchange mediated elimination of the Fe-antisites in the hydrothermal synthesis of LiFePO4. Nano Energy 2015, 16, 256–267.

129. Pang, W.; Shao, B.; Tan, X.-Q.; Tang, C.; Zhang, Z.; Huang, J. Exfoliation of metal–organic frameworks into efficient single-layer metal–organic nanosheet electrocatalysts by the synergistic action of host–guest interactions and sonication. Nanoscale 2020, 12, 3623–3629.

130. Zhao, X.; Ding, X.; Xia, Y.; Jiao, X.; Chen, D. Coupling-Effect-Induced Acceleration of Electron Transfer for α-Ni(OH)2 with Enhanced Oxygen Evolution Reaction Activity. ACS Appl. Nano Mater. 2018, 1, 1476–1483.

131. Biesinger, M.C. Accessing the robustness of adventitious carbon for charge referencing (correction) purposes in XPS analysis: Insights from a multi-user facility data review. Appl. Surf. Sci. 2022, 597, 153681.

5 SYNTHESIS OF SILICA-BASED MATERIALS USING BIO-RESIDUES THROUGH THE SOL-GEL TECHNIQUE

Karine Zanotti [1], Katerine Igal [2], María Belen Colombo Migliorero [2], Vânia Gomes Zuin [1,3,4] and Patricia Graciela Vázquez [2]

[1]*Department of Chemistry, Federal University of São Carlos, São Carlos 13565-905, SP, Brazil*

[2]*Centro de Investigación y Desarrollo en Ciencias Aplicadas "Dr. Jorge J. Ronco" CINDECA, CONICET-CIC-UNLP, Facultad de Ciencias Exactas, Universidad Nacional de La Plata, La Plata 1900, Argentina*

[3]*Institute of Sustainable Chemistry, Leuphana University Lüneburg, 21335 Lüneburg, Germany*

[4]*Green Chemistry Centre of Excellence, University of York, York YO10 5DD, UK*

ABSTRACT

This study focused on the use of citrus bio-waste and obtention of silica-based materials through the sol-gel technique for promoting a greener and more sustainable catalysis. The sol-gel method is a versatile synthesis route characterized by the low temperatures the materials are synthesized in, which allows the incorporation of organic components. This method is carried out by acid or alkali hydrolysis combined with bio-waste, such as orange and lemon peels, generated as co-products in the food processing industry. The main objective was to obtain silica-based materials from the precursor TEOS with different catalysts—acetic, citric and hydro-chloric acids and ammonium hydroxide—adding different percentages of lemon and orange peels in order to find the influence of bio-waste on acids/alkali precursor hydrolysis. This was to partially replace these catalysts for orange or lemon peels. The solids obtained were characterized with different techniques, such as SEM, FT-IR, potentiometric titration and XRD. SEM images were compared with pure silica obtained to contrast the morphology of the acidic and alkali hydrolysis. However, until now, few attempts have been made to highlight the renewability of reagents used in the synthesis or to incorporate bio-based catalytic processes on larger scales.

Keywords: *sol-gel silica-based materials; acid/alkali hydrolysis; bio-waste; citrus peels*

1. INTRODUCTION

Based on the main notions of the circular economy and the bioeconomy [1,2], the concept of using agricultural waste is discussed. The common objective is to minimize the generation of waste from economic or urban activities related to agriculture. The circular economy turns out to be attractive for understanding the sustainable challenges we face in terms of social, economic and environmental aspects. However, in the case of bio-waste, taking advantage of the immense amount of agricultural waste generated, the processes could become a complex and difficult operation [3]. This is especially true for the development of new materials from citrus fruit bio-waste.

Citrus fruits are believed to have originated from the warm southern slopes of the Himalayas in north-eastern India and northern Myanmar [4]. At present, citrus crops are among the most cultivated fruits in the tropical and subtropical regions of the world. Regarding the bio-waste issue, approximately one-third of citrus fruits are utilized for processing, which produces around 50–60% of organic waste [5].

With the growing importance that organic farming has acquired, both in developed and developing countries, there are no studies of these socio-economic realities [6]. Organic agriculture [7,8] has remained a haven against the invasion of agrochemicals and industrialization in the food supply chain. It represents less than 1% of global agriculture, specifically reported as 0.98% by Willer and Lernoud [9]. On the contrary, agriculture is the fastest-growing food sector in the world [10]. South America has a strong presence in the agriculture sector: Argentina (3,191,255 ha), Uruguay (930,965 ha) and Brazil (705,233 ha) [9,10,11]. Approximately 40% of the oranges produced globally are used in the production of different commercial products [12]. Due to processing, large amounts of waste products (bio-waste), such as peels from the juice industry, are generated and accumulated [13].

Considering this waste accumulation, different investigations propose the use of waste and bio-waste in order to obtain new materials, especially in sol-gel synthesis to obtain silica-based materials [13,14,15,16,17,18,19]. The sol-gel process can be described as the formation of an oxidic network through the polycondensation reaction of a molecular precursor in a liquid. The term sol refers to a dispersion of colloidal particles in a liquid medium where the stability of the colloids is due to both Brownian motion and small particle size. The term gel refers to a three-dimensional network of a solid phase interwoven with an entrapped and immobilized continuous liquid phase [18,19,20]. In our previous research, the sol-gel method was used as a versatile synthesis route characterized by the low temperatures at which inorganic materials and organic/inorganic hybrids were synthesized [21,22].

In this research, two main issues were addressed at the same time: the use of citric bio-waste and the obtention of materials using the sol-gel technique—looking for their use in heterogeneous catalysis as a bi-functional support system. Silica has been identified as an ideal support material for its strong hydrophilicity, acknowledged biocompatibility, shape and chemistry surface [23]. Two different strategies have been explored to combine bio-waste with the synthesis of silica-based materials. The first step was to obtain silica through acidic or alkaline hydrolysis by using acetic, citric, hydrochloric acids and an alkali (ammonium hydroxide). In the second step, lemon or orange peels were added to the obtained mixtures. The objective of this mixed synthesis was to find the influence of organic bio-waste on acids/alkalis, to partially replace them with organic orange or lemon peels, which could presumable provide citric acid, among other compounds. For this, different amounts of bio-waste were used in the various syntheses [24,25].

As a precursor of silica, tetraethyl orthosilicate (TEOS) was used, and its hydrolysis and condensation were studied, using absolute ethanol as a green solvent due to its renewability [26,27]. This solvent plays an important role in the synthesis of silica-based materials, both in the case of the reaction with TEOS, once they act as a homogenizer agent for substances with different solubilities that participate in the condensation reaction, and with lemon and orange peels, extracting compounds of interest from them [28,29,30]. All synthesized solids were characterized in-depth by integrating different physical-chemical techniques, such as scanning electron microscopy (SEM) and x-ray diffraction (XRD). The chemical compositions of the solids were investigated through Fourier transform infrared spectroscopy (FT-IR) and the acidic properties by n-butylamine potentiometric titration.

With the physicochemical properties discussed in this research, silica-based materials synthesized with bio-residues could have the potential to be used as supports for heterogeneous reactions (whose analysis is reserved for future research).

2. MATERIALS AND METHODS

The starting point to specifying the proposed objectives includes having the necessary material and carrying out an adequate design of the experiments. For the design of these steps, two levels of complexity were selected, when required. This means that some of the operating variables remain fixed. Thus, reliable results can be obtained with the fewest possible experiences [21,22].

Before starting with the sol-gel synthesis, the oranges and lemons were peeled (20 of each of them), and the peels were cut into small pieces and put on the stove for 3 days, at 70 °C. Once calcined, the peels become brittle, and it is easier to grind them to a small size, so they were ground in a mortar (Scheme 1). This size allows uniformity between the liquid gel and ground peels mixture. It should be noted that in 2020 a doctoral thesis was carried out with citrus peels without calcining (lemon, orange and mandarin) to obtain silica-based materials [31].

The sol-gel synthesis was carried out in a chamber with a nitrogen-controlled atmosphere at room temperature. First, a portion of the solvent, absolute ethanol (CH_3CH_2OH, Carlo Erba), and the corresponding catalyst for acid hydrolysis were placed into a beaker: acetic acid glacial (CH_3COOH, HPLC grade, J. T. Baker), citric acid ($C_6H_8O_7$ anhydrous for synthesis, Merck) or hydrochloric acid (HCl, 37%, Merck). The same was done in the alkaline hydrolysis using ammonium hydroxide (NH_4OH, Anedra analytical reagent). Then, the TEOS ($Si(OC_2H_5)_4$, Aldrich 98%) was added, followed by the last portion of solvent and distilled water, respectively.

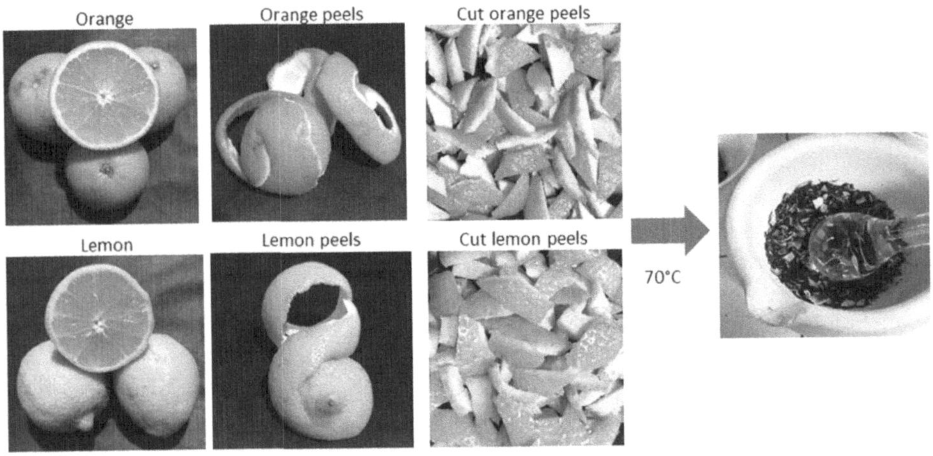

Scheme 1.: *Treatment of lemon and orange peels.*

Each mixture, outside the nitrogen-controlled atmosphere, was placed on a magnetic stirrer at 500 rpm; the respective amounts of orange or lemon peels were added, maintaining stirring for 2 h. The mixture was left for 3 weeks at room temperature; then, the gel formed was dried at 100 °C for 1 h. The details of each synthesis are shown in Table 1, Table 2, Table 3 and Table 4 (considering the total volume of ethanol added). Each table details the ratios of the compounds and the amounts used of lemon and orange peels (25 and 50%) in relation to the pure silica mass—considering as 100% the 5 g of pure silica obtained with fixed volumes of 17 mL of TEOS and 21.8 mL and, as appropriate, 5 mL of CH_3COOH; 0.43 mL of $C_6H_8O_7$; 7.5 mL of HCl (so proton concentration is the same in all cases) or 1.8 mL of NH_4OH. The amounts of all the reagents were varied in order to maintain the relationship between the TEOS precursor, the acid and alkaline catalysts and the solvent, with which the same silica-based solids should be generated, but adding citrus peels.

Table 1.: *Conditions employed in silica synthesis using acetic acid as a catalyst.*

Sample	Precursor	Solvent	Acid	Water	Bio-Waste
KZ1	TEOS	EtOH	CH_3COOH	H_2O	Pure silica
	17.0 mL	21.8 mL	5.0 mL	5.0 mL	5.0 g
KZ2	TEOS	EtOH	CH_3COOH	H_2O	50% orange peel
	8.5 ml	10.9 mL	2.5 mL	2.5 mL	2.5 g
KZ3	TEOS	EtOH	CH_3COOH	H_2O	25% orange peel
	12.8 mL	16.3 mL	3.75 mL	3.75 mL	1.25 g
KZ4	TEOS	EtOH	CH_3COOH	H_2O	50% lemon peel
	8.5 mL	10.9 mL	2.5 mL	2.5 mL	2.5 g

Sample	Precursor	Solvent	Acid	Water	Bio-Waste
KZ5	TEOS	EtOH	CH_3COOH	H_2O	25% lemon peel
	12.8 mL	16.3 mL	3.75 mL	3.75 mL	1.25 g

Table 2.: *Conditions employed in silica synthesis using citric acid as a catalyst.*

Sample	Precursor	Solvent	Acid	Water	Bio-Waste
KZ6	TEOS	EtOH	$C_6H_8O_7$	H_2O	Pure silica
	17.0 mL	21.8 mL	0.43 mL	5.0 mL	5.0 g
KZ7	TEOS	EtOH	$C_6H_8O_7$	H_2O	50% orange peel
	8.5 ml	10.9 mL	0.20 mL	2.5 mL	2.5 g
KZ8	TEOS	EtOH	$C_6H_8O_7$	H_2O	25% orange peel
	12.8 mL	16.3 mL	0.30 mL	3.75 mL	1.25 g
KZ9	TEOS	EtOH	$C_6H_8O_7$	H_2O	50% lemon peel
	8.5 mL	10.9 mL	0.20 mL	2.5 mL	2.5 g
KZ10	TEOS	EtOH	$C_6H_8O_7$	H_2O	25% lemon peel
	12.8 mL	16.3 mL	0.30 mL	3.75 mL	1.25 g

Table 3.: *Conditions employed in silica synthesis using hydrochloric acid as a catalyst.*

Sample	Precursor	Solvent	Acid	Water	Bio-Waste
KZ11	TEOS	EtOH	HCl	H_2O	Pure silica
	17.0 mL	21.8 ml	7.5 mL	5.0 mL	5.0 g
KZ12	TEOS	EtOH	HCl	H_2O	50% orange peel
	8.5 ml	10.9 mL	3.8 mL	2.5 mL	2.5 g
KZ13	TEOS	EtOH	HCl	H_2O	25% orange peel
	12.8 mL	16.3 mL	5.6 mL	3.75 mL	1.25 g
KZ14	TEOS	EtOH	HCl	H_2O	50% lemon peel
	8.5 mL	10.9 mL	3.8 mL	2.5 mL	2.5 g
KZ15	TEOS	EtOH	HCl	H_2O	25% lemon peel
	12.8 mL	16.3 mL	5.6 mL	3.75 mL	1.25 g

Table 4.: *Conditions employed in silica synthesis using ammonium hydroxide as a catalyst.*

Sample	Precursor	Solvent	Alkali	Water	Bio-Waste
KZ16	TEOS	EtOH	NH_4OH	H_2O	Pure silica
	17.0 mL	21.8 ml	11.8 mL	5.0 mL	5.0 g
KZ17	TEOS	EtOH	NH_4OH	H_2O	50% orange peel
	8.5 ml	10.9 mL	5.9 mL	2.5 mL	2.5 g
KZ18	TEOS	EtOH	NH_4OH	H_2O	25% orange peel
	12.8 mL	16.3 mL	8.8 mL	3.75 mL	1.25 g
KZ19	TEOS	EtOH	NH_4OH	H_2O	50% lemon peel
	8.5 mL	10.9 mL	5.9 mL	2.5 mL	2.5 g
KZ20	TEOS	EtOH	NH_4OH	H_2O	25% lemon peel
	12.8 mL	16.3 mL	8.8 mL	3.75 mL	1.25 g

The evaluation of the acidic properties was carried out by potentiometric titration with *n*-butylamine using a Metrohm 794 Basic Titrino titrator (Switzerland) with a double-junction electrode. For that, 0.025 mL / min of an *n*-butylamine solution in acetonitrile (0.025 N) was added to 0.025 g of sample, previously suspended in 45 mL of acetonitrile and keeping the stirring time constant (540 s) before adding the first drop and a waiting time of the drop of 60 s, while stirring constantly. FT-IR spectrum was obtained using Bruker IFS 66 equipment (Germany). Pellets of the sample in KBr, at room temperature, were measured in a range between 500 and 4000 cm^{-1}. SEM was applied to obtain different micrographs of the solids using the JEOL equipment, JSM-6390LV (Japan), with a voltage of 20 kV. Samples were supported on graphite and metalized with a sputtered gold film. Finally, XRD patterns were recorded by means of a PANalytical X'Pert Pro 3373 / 00 device with the following conditions: Cu Kα radiation (λ = 1.5417 Å); Ni filter; 20 mA and 40 kV in the high voltage source; scanning angle (2θ) from 5° to 40°, scanning rate 2° (2θ) / min.

3. RESULTS AND DISCUSSION

3.1. Digital Photographs

Among the basic principles of the sol-gel technique [32] is that it is an alternative to obtain non-metallic inorganic materials, such as glass and ceramics. This consists of the hydrolysis and condensation, originating from alkoxide precursors to form a polymeric network in the vitreous state, typically exhibiting a macroporous structure [33]. The gel is formed by a two-stage reaction: (1) hydrolysis and (2) polycondensation:

(1) The hydrolysis for gel formation is carried out using different alcohols as the solvent. This reaction is produced by the nucleophilic attack of the oxygen present in the water on the metal atom, forming silanols groups [34]. The solvent is the chemical compound that represents the largest proportion of the process.

(2) The condensation begins as soon as the alkoxide groups are hydrolyzed. The concentrations of -OH and/or -OR end groups depend on the hydrolysis and condensation mechanisms. They are determined by the nature of the catalyst used [35]. Brinker and Scherer [26] studied the effect of pH on the morphology of the siliceous structures. They found that the change in the pH of the solution affects the rates of hydrolysis and condensation, respectively. Figure 1 shows the silica synthesized with acid hydrolysis using acetic acid. The first photograph (pure silica) shows the silica obtained without bio-waste, with its glassy structure, coming from a gel that slowly dried at room temperature.

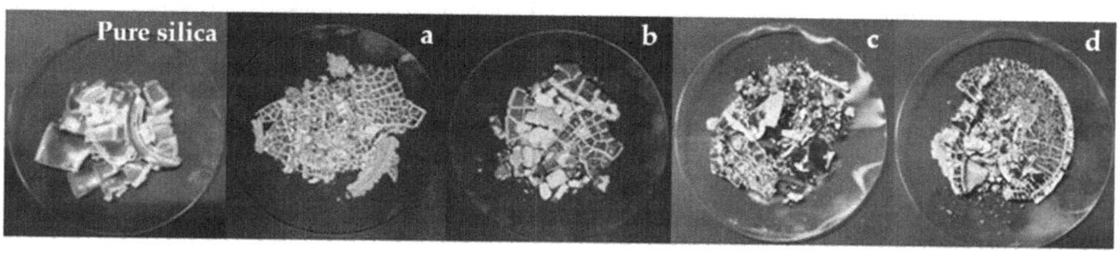

***Figure 1.**: Silica gel obtained with CH_3COOH: (a) 50% orange peel; (b) 25% orange peel; (c) 50% lemon peel; (d) 25% lemon peel.*

Milea et al. [36] found that the sol-gel technique allows the preparation of a new glass composition with superior properties determined by the specific properties of the gel [37,38]. In the following, Figure 1a,b correspond to the silica-based materials with orange peel aggregates, and Figure 1c,d correspond to silica-based materials with lemon peels. It is important to note that in all cases, the added peels remained in the siliceous structure, mainly on the surface of the dry gel, due to condensation; this occurred as the gel dried. If Figure 1a and c are compared, it can be distinguished that with 50% orange or lemon peel, the silica-based solid turns yellowish and has a surface structure similar to pure silica. This layer of siliceous material is brighter and has a vitreous consistency. However, in the case of 25% orange and lemon peel added to the gel mixture, when drying, they are less shiny, although the surface is similar to 50% added bio-residues.

In the case of the 50% lemon or orange peel, the silica-based solid obtained forms two phases: one given by the compounds extracted by the solvents (ethanol and water) from the peels, varying the pH [25], and the other phase by the acid used in the synthesis. In the case of 25% peels, the effect is the same, although the pH is lower [28,39]. At this point, it is interesting to compare the gelation when the hydrolysis has already been carried out. Gelation is physically manifested by a drastic increase in the viscosity of the solution [40]. These changes in viscosity occur without causing the generation of chemical transformations or endothermic or exothermic changes [41]. The wet gel can be strengthened by aging and by the syneresis mechanism of Ostwald maturation [42]. The syneresis is characterized by the contraction of the gel network, produced by the expulsion or extraction of liquid. The gel that is a homogeneous substance becomes a segregation of solid components. Smaller particles have a higher solubility, which leads to their precipitation and the formation of larger ones [39]. During drying, the effect of citrus peels on the surface of the solids obtained indicates that the silicon skeleton becomes increasingly rigid. When the surface tension cannot deform the network, the gel body becomes too stiff. At this point, the likelihood of it breaking is higher [37,43], as can be seen in the photographs.

3.2. Potentiometric Titration Characterization

The acid properties of all the synthesized solids were studied through potentiometric titration with n-butylamine, which allows knowing the number of acid sites and their acid strength. To interpret the results obtained, the initial electrode potential (Ei) indicates the maximum acid strength of the surface sites, and the values of m eq/g solid, in which the plateau is reached, indicate the total number of acid sites. The acid strength of the surface sites can be classified according to the following ranges: very strong sites Ei > 100 mV, strong sites 0 < Ei < 100 mV, weak sites −100 < Ei < 0 mV and very weak sites Ei < −100 mV, respectively. It is important to clarify that this technique only indicates the mass acidity trend of the solids obtained [44]. The initial electrode potentials are shown in Table 5 and their respective potentiometric curves in Figure 2.

Comparing the potentiometric curves of the synthesis with acid hydrolysis (Figure 2a–c)—samples that use acetic, citric and hydrochloric acid—the presence of free electrons in the acid gave a higher acidity (Figure 2c). Comparing the use of 50 and 25% orange peels in acid hydrolysis versus pure silica (with acetic acid), KZ1 gives an Ei of 128 mV for pure silica versus 110 mV for 50% (KZ2) and 132 mV for 25% (KZ3) (Table 5). Citric and hydrochloric acid showed similar behavior to acetic acid when silica-based solid synthesized with orange and lemon peels were analyzed (Figure 2b,c). For basic hydrolysis, the use of NH4OH generated the lowest value in both cases of 50 and 25% of the bio-waste. In the case of lemon peel, for both 50 and 25%, the Ei values (Table 5) and the potentiometric curves (Figure 2d) followed the same trend as those obtained for orange peels.

Among the constituents of lemon, approximately 70% correspond to citric acid, while folic and ascorbic acid are the predominant constituents in orange [31]. The incorporation of bio-waste in the structure of silica prepared by acid hydrolysis maintains the same Ei as the initial one, or decrease for all cases, since the acidity of the acids used as catalysts was greater than that of the bio-waste (compare Ei of Table 5 in the KZ11-Z13 series for orange and in the KZ11-KZ15 series for lemon). Although lemon was more effective in maintaining the high number of acidic sites for the samples synthesized with hydrochloric acid (compare the two series in Figure 2c), this trend did not hold for the samples synthesized with acetic or citric acid. Furthermore, at this point, one can see that none of the bio-residues contributed new strong acid sites to the final material because the acidity of pure silica was "diluted" by the incorporation of orange or lemon peels.

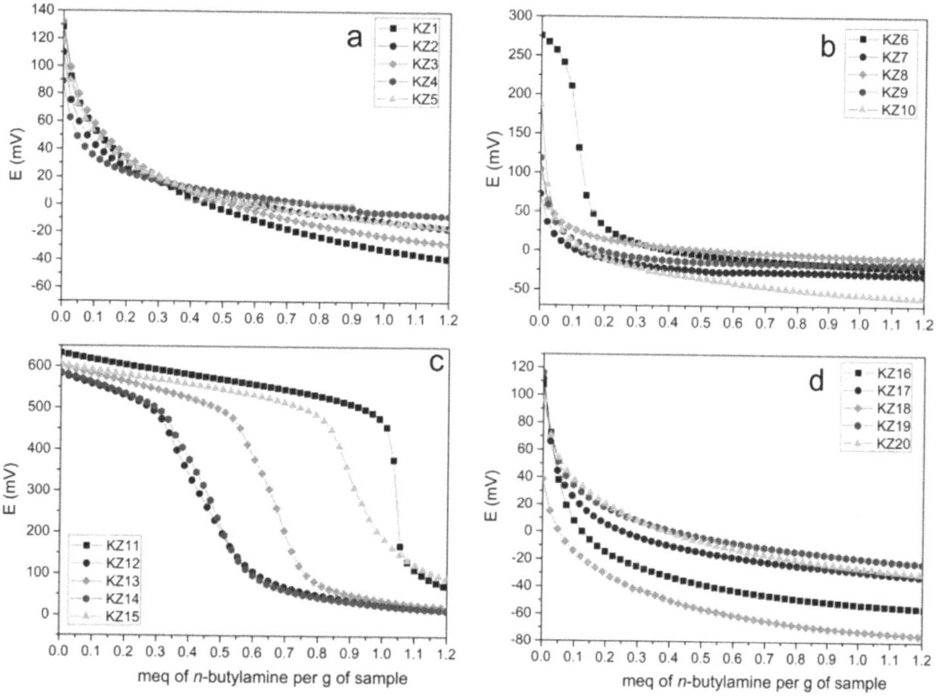

Figure 2.: *Potentiometric curves of the synthesized solids: (a) Acid hydrolysis: acetic acid (KZ1, KZ2, KZ3, KZ4, KZ5); (b) citric acid (KZ6, KZ7, KZ8, KZ9, KZ10); (c) hydrochloric acid (KZ11, KZ12, KZ13, KZ14, KZ15); (d) alkali hydrolysis: ammonium hydroxide (KZ16, KZ17, KZ18, KZ19, KZ20).*

Table 5.: *Initial electrode potential (Ei) values of the silica-based solid synthesized.*

Sample	Ei (mV)	Sample	Ei (mV)
KZ1	128	KZ11	632
KZ2	110	KZ12	582
KZ3	132	KZ13	604
KZ4	89	KZ14	585
KZ5	117	KZ15	604
KZ6	275	KZ16	107
KZ7	72	KZ17	112
KZ8	104	KZ18	38
KZ9	118	KZ19	116
KZ10	187	KZ20	96

When ammonium hydroxide was used, the acidic sites of the bio-waste were neutralized by the ammonia. Therefore, when the bio-waste amount was only

25%, the acidity of the bio-waste was neutralized and, hence, the obtained Ei of KZ18 (38 mV) and KZ20 (96 mV) was lower than pure silica KZ16 (107 mV). On the other hand, when the bio-waste amount was 50%, the acidity of the obtained silica slightly increased in comparison with the pure silica, as can be seen in the Ei values of KZ17 (112 mV) and KZ19 (116 mV), which can be understood as a contribution of the bio-residue to the final material acidity, slightly increasing it.

3.3. SEM Characterization

The pH of the catalyst determines the texture of the solid product, as can be seen in Figure 3. The morphology caused by acid hydrolysis (KZ1, KZ6 and KZ11) is laminar and typical of sol-gel silica. Curran and Stiegman [45] studied the morphology of siliceous structures at very high acidic concentrations; below the isoelectric point, dry xerogels become mesoporous. This is due to the protonation of silanols (-SiOH) to produce (-$SiOH_2^+$), which are leaving groups and increase the rate of condensation. First, hydrated monomers obtained from the hydrolysis of precursors undergo condensation reactions, forming silicate polymer chains and then small nuclei. These grow through silica monomer and polymer bonding, forming primary particles, usually 5–7 nm in mean diameter. Then, primary particles aggregate, producing larger SiO_2 particles that grow up to a stationary critical size [46]. In contrast, alkaline hydrolysis (KZ16) gives clusters of spheres because it is slower hydrolysis (formation of branched polymers or spherical aggregates). As expected, the use of only one alkoxide combined with appropriately chosen amounts of reagents allows for deep control over the nucleation and growth processes, leading to spherical and highly monodisperse nanoparticles [47].

Regarding the SEM micrographs, Figure 4 shows the samples prepared with HCl as a catalyst. It was possible to observe a similarity to the characteristic laminar format of silica regardless of the amounts of orange or lemon that were added to the synthesis. In this case, hydrolysis with the strong acid can form, on the one hand, only silica and, on the other, leach the shells, as can be seen in the micrographs. Zarib and Abdullah [48] obtained a smooth surface and a rough surface together in the silicate particle. This was when rice husk was used as a bio-waste.

The NH_4OH sample, which can be seen in Figure 5, had a rounded shape and its particles formed clusters of variable sizes, which may be due to the action of orange and lemon aggregates that modify the pH of the synthesis. An agglomeration of nanoparticles of silica-based material and bio-waste was observed due to the three-dimensional hydrophilic networks on the surface of the solids [49]. Other researchers have synthesized silica with low surface microspheres from wheat husk ash with an alkaline and acid precipitation process [50].

3.4. XRD Characterization

The structure of the synthesized materials was determined using the XRD technique. This conventional technique has been increasingly used in similar studies for mixed matrices using bio-waste, such as the work developed by Worathanakul et al. [51] in which bagasse ash was used as a silica source for zeolite synthesis, another alternative source for the production of valuable materials and chemicals.

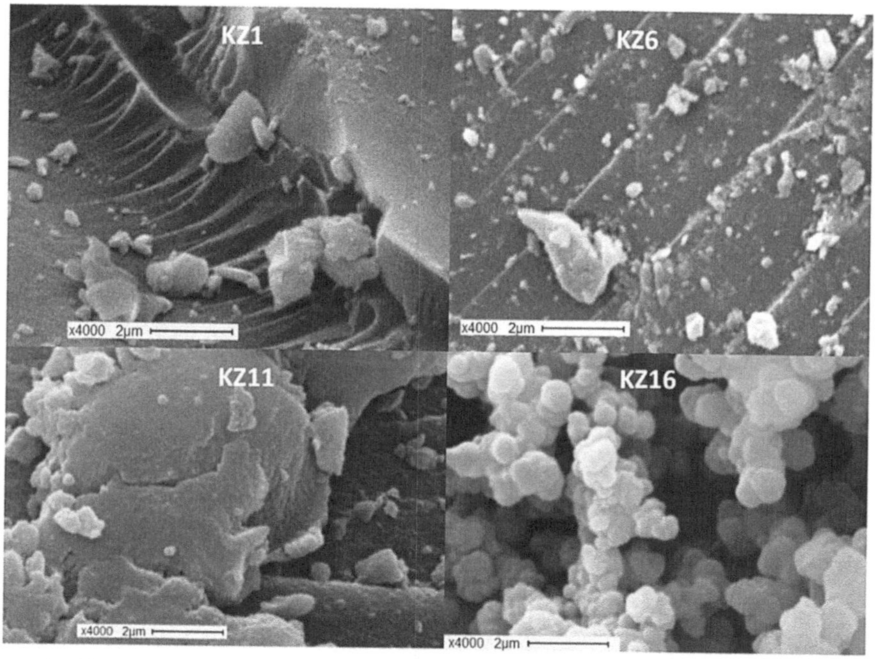

***Figure 3.:** SEM micrographs of synthesized silica-based solid by acid (KZ1, KZ6 and KZ11) and alkali (KZ16) hydrolysis. Magnification ×4000 (scale bar: 2 μm).*

***Figure 4.:** SEM micrographs of the silica-based solid synthesized using HCl with orange and lemon peels. Magnification ×1000 (scale bar: 10 μm).*

Figure 5.: SEM micrographs of the silica-based solid synthesized using NH_4OH with orange and lemon peels. Magnification ×1000 (scale bar: 10 μm).

Figure 6 shows the x-ray diffraction diagrams of the silica obtained using TEOS as a precursor employing the acids (acetic, citric and hydrochloric) and the alkali (ammonium hydroxide) as catalysts. The amorphous structure of silica xerogels was confirmed by the broad peaks in the interval around 15–30° 2θ, as reported by Czarnobaj [52], and by the band located around 23° 2θ, which is the typical diffraction of this type of silica [53,54].

Figure 7 presents the XRD diagrams of the synthesized silica-based solids with 50%, and 25% orange and lemon peels with acidic and alkali hydrolysis. The similarity of the diagrams with the silica without the addition of bio-waste indicates that the surface of the solids obtained was mostly silica-based solids.

3.5. FT_IR Characterization

FT-IR spectroscopy has been widely used for the characterization of materials synthesized by the sol-gel method [51,55]. This made it possible to perform an in-depth analysis to understand the relationships between the properties of the material—including its structure at the atomic level—and the IR spectrum. As the structure evolved in the silica xerogel, a three-dimensional network of interconnected tetrahedral units formed. FT-IR allows, depending on the width and position of the bands, the characterization of the material by interpreting its composition and structure. In this study, the region of 4000–500 cm^{-1} was used.

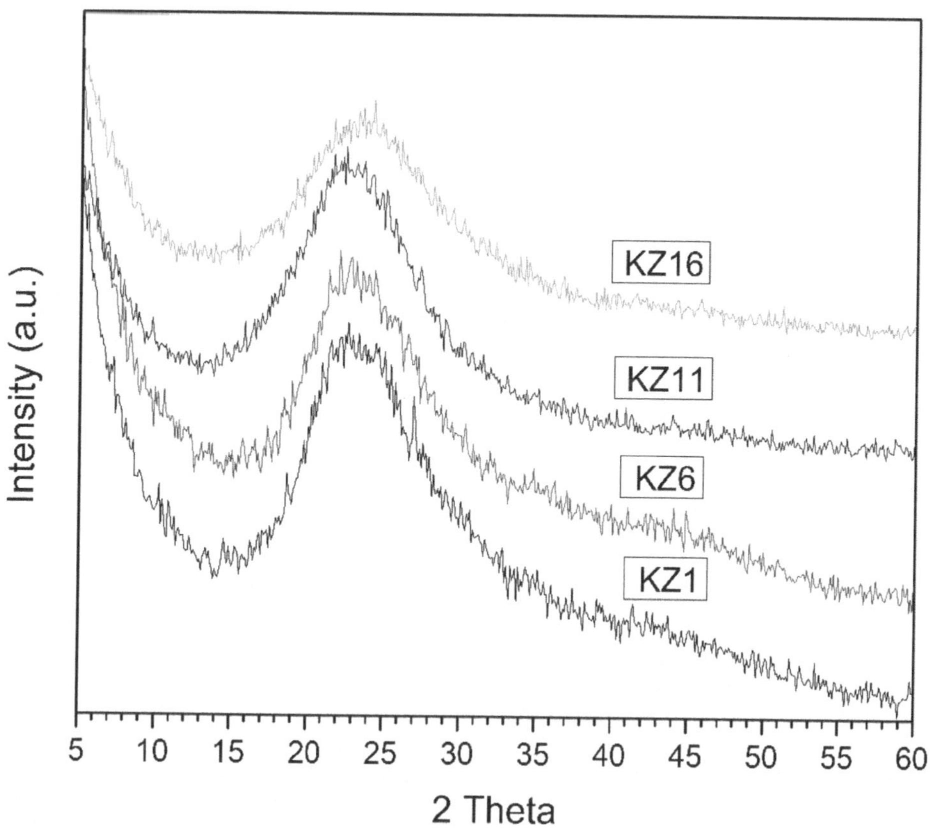

Figure 6.: *XRD diagram of silica-based solid synthesized.*

Figure 8 shows the complete FT-IR spectrum of the samples obtained by synthesis with TEOS as a precursor and acetic acid (KZ1), citric acid (KZ6) and hydrochloric acid (KZ11) as acid catalysts and ammonium hydroxide (KZ16) as the alkaline catalyst. Table 6 shows the characteristic absorption bands of silica obtained with acetic acid as a catalyst.

Figure 8 and Table 6 show bands at 500, 800, 950, 1080 and 1200 cm^{-1} (the five last bands), which are characteristic bands of silica, corresponding to vibrations of silicon-oxygen bonds. These bands can be classified by the movement (to roll, bend and stretch) of the oxygen atom related to the silicon atoms [56]. The symmetric and asymmetric stretching modes of the Si-O-Si bonds can be assigned to bands at 1166 and 1079 cm^{-1}, respectively. The vibration at 797 cm^{-1} is associated with the symmetric stretching of the Si-O bond or vibrational modes of structure rings. The bands centered at 948 and 1864 cm^{-1} were assigned to the vibration of the Si-O bonds (silanols) and are characteristic of silica xerogel [57]. The band located around 566 cm^{-1} was attributed to the deformation of cyclotetrasiloxanes (four-membered siloxane rings). This molecule can constitute a large fraction of the oligomeric species present in TEOS-derived systems because they are stable during the hydrolysis processes [58].

The broad band located at 3000 to 3700 cm^{-1} was assigned to the OH species due to stretching vibrations of hydrogen-bonded water molecules on the surface of silica [59], to silanols bound to molecular water and to free silanols on the surface of the gel [60,61]. The band located at 1639 cm^{-1} was assigned to the deformation of molecular water and resulted from the angular deformations of H-O-H bonds in H_2O. The description made for the KZ1 silica sample is common for the other three pure silica forms (KZ6, KZ11 and KZ16).

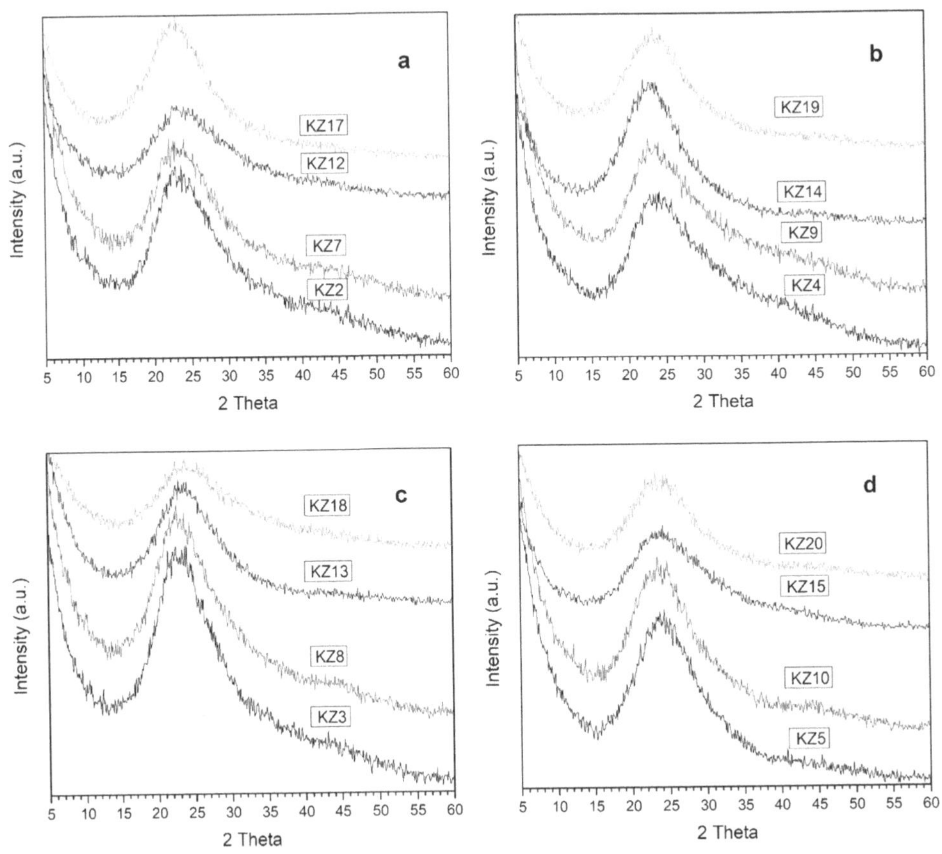

Figure 7.: *XRD diagram of all synthesized silica-based solids. (a) All solids synthesized with 50% orange peel. (b) All solids synthesized with 50% lemon peel. (c) All solids synthesized with 25% orange peel. (d) All solids synthesized with 25% lemon peel.*

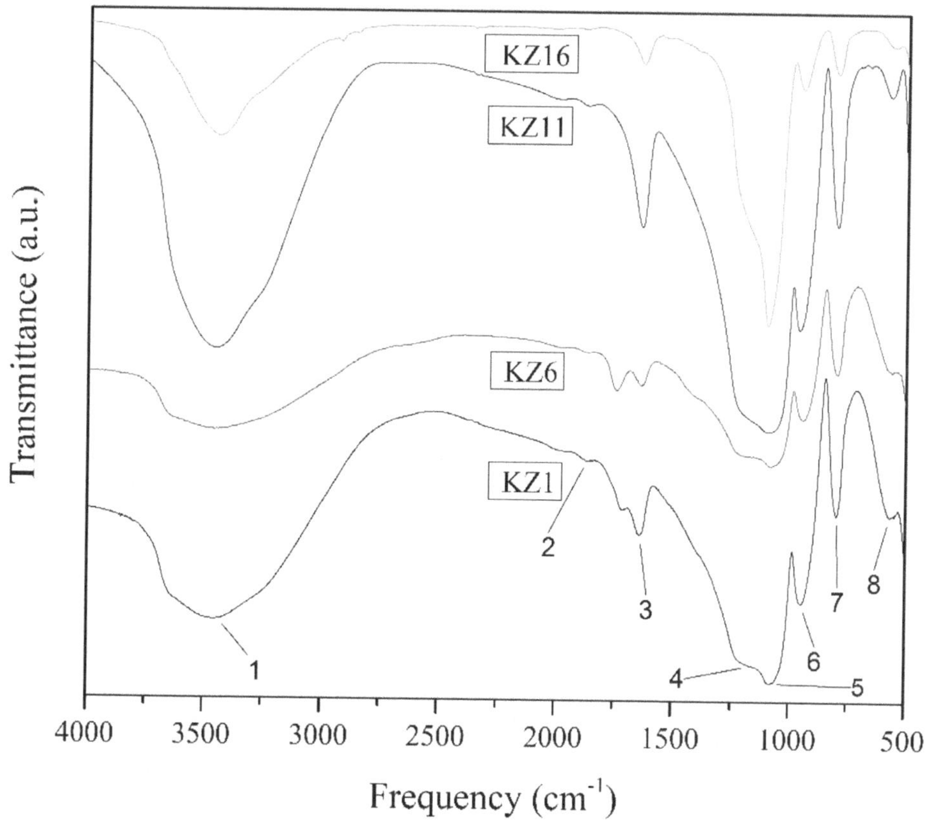

Figure 8.: *FT-IR spectrum of the pure silica obtained by synthesis with TEOS as a precursor and acetic acid (KZ1), citric acid (KZ6), hydrochloric acid (KZ11) and ammonium hydroxide (KZ16).*

Table 6.: *FT-IR of silica synthesized with acetic acid (KZ1).*

Bands	Wavenumber (cm^{-1})	Band Assignment	Functional Group
1	3437	ν OH [1]	O-H bonded
2	1864	νβ (Si-O) [2]	Si-OH
3	1639	δ H-O-H [3]	H-O-H
4	1166	νS (Si-O-Si) [4]	Si-O-Si
5	1079	νa (Si-O-Si) [5]	Si-O-Si
6	948	νβ (Si-O) [2]	Si-OH
7	797	νs (Si-O) [6]	Si-O-Si
8	566	ν (Si-O) [1]	SiO_2

[1] ν: vibration stretching; [2] νβ: symmetric vibration stretching; [3] δ: deformation vibration; [4] νS: symmetric deformation vibration; [5] νa: anti-symmetrical vibration stretching; [6] νs: symmetrical vibration.

Figure 9 shows the FT-IR spectra of the silica-based materials obtained with 50% and 25% orange and lemon peels with acid and alkaline hydrolysis. As previously mentioned, changes in the position and shape of the absorption bands can be related to structural changes, which can occur with different solvent compositions (such as water and ethanol), leading to a variation in the hydrolysis and condensation reactions.

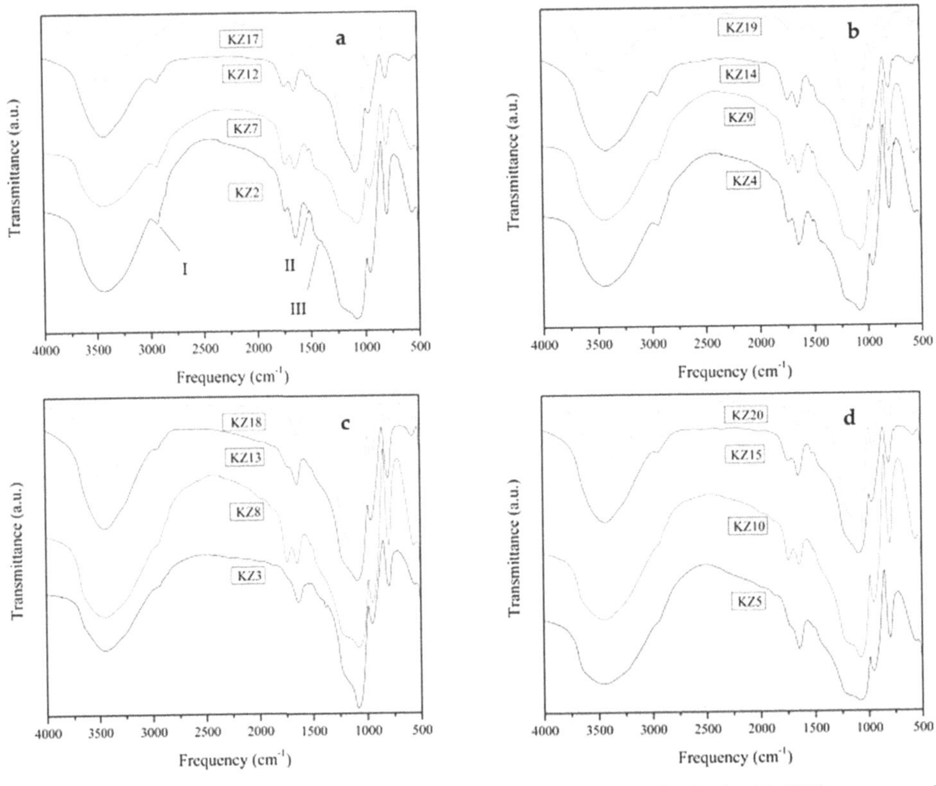

Figure 9.: *FT-IR spectrum of the synthesized silica. (a) All silica synthesized with 50% orange peel. (b) All silica synthesized with 50% lemon peel. (c) All silica synthesized with 25% orange peel. (d) All silica synthesized with 25% lemon peel.*

For Figure 9a, the band appearing at 2932 cm^{-1} (marked as I) can be assigned to the symmetric and asymmetric stretching of the C-H bonds in the residual groups CH_3-OH and CH_3-O-Si [60]. In this same figure, two absorption bands are observed at 1409 and 1518 cm^{-1} (marked as II and III, respectively) due to the CH_3 groups provided by the organic material (bio-waste); these bands are characteristic of a hydrophobic form of silica. The description made for the KZ2 sample is common for the other silica-based solids synthesized with 50% and 25% orange and lemon peels.

All the characterizations used and discussed above portray the first part of a broad study, in which we highlight the initial stage of understanding the morphology and behavior of silica supports synthesized using different amounts of bio-waste and acidic and basic catalysts in order to analyze the influence that such bio-residues have on the reaction to obtain the silica-based solid support. Based on this knowledge, and also based on previous studies on the same topic [24,25], orange and lemon bio-waste can provide acidity for the reaction with the precursor TEOS. There is an indication that it would be possible to partially replace the acid catalysts used in this synthesis by bio-waste, but not completely, as they subtly reduce the acidity of the mixed materials when compared to pure silica—since the acidity of the acids used is greater than the amount of acidic compounds that are extracted from the citrus peels in the mixed matrices.

4. CONCLUSIONS

In this study, we synthesized new materials using the sol-gel method, a simple and fast technique that allowed the inclusion of bio-waste into silica matrices, proposing a first stage for the development of new silica-based supports that can be used in heterogeneous catalysis. The analyses of the influence of the orange and lemon peel percentages in the acidity and morphology of siliceous solids revealed promising results, suggesting that the bio-waste used can provide acidity to partially replace the acid catalysts investigated in acid hydrolysis—since the values obtained using 25 and 50% of bio-waste are similar to those of pure silica—but it is not possible to replace them completely. In addition, the mixed materials also showed the ability to maintain the siliceous structure. This is an important step for future studies in order to replace, for example, strong inorganic acids with a renewable raw material (bio-waste), which has a lower cost and suitable morphology. Research based on sustainability aspects is increasingly important and necessary in the current model of production and consumption in which we live. The chemistry required for a circular model will materialize only through a new attitude toward research, chemical engineering and the design of new processes and products.

Author Contributions

Conceptualization, P.G.V. and V.G.Z.; conduct of the experiments and characterization techniques, K.Z., K.I. and M.B.C.M.; prepared and wrote the draft of the manuscript, P.G.V.; supervision and editing of the final version, P.G.V. All authors have read and agreed to the published version of the manuscript.

Acknowledgments

The authors would like to thank UNLP, CONICET, AUGM, CNPq and Capes for the financial support, at the service of FT-IR of CINDECA, Dra. R. Arreche, Dra. V. Palermo, Tco Qco L. Osiglio and G. Kürten for their collaboration in this study.

REFERENCES

1. Kümmerer, K. Sustainable Chemistry: A Future Guiding Principle. Angew. Chem. Int. Ed. 2017, 56, 16420–16421.

2. Kümmerer, K.; Clark, J.H.; Zuin, V.G. Rethinking Chemistry for a Circular Economy. Science 2020, 367, 369–370.
3. Zuin, V.G. Circularity in Green Chemical Products, Processes and Services: Innovative Routes Based on Integrated Eco-Design and Solution Systems. Curr. Opin. Green Sustain. Chem. 2016, 2, 40–44.
4. Hamilton-Smith, E. Terrestrial Ecoregions of the Indo-Pacific: A Conservation Assessment. Electron. Green J. 2002, 1.
5. Satari, B.; Karimi, K. Citrus Processing Wastes: Environmental Impacts, Recent Advances, and Future Perspectives in Total Valorization. Resour. Conserv. Recycl. 2018, 129, 153–167.
6. García, R.M.L. La agricultura ecológica en el quehacer científico. Tema incipiente en la geografía. An. Geogr. Univ. Complut. 1999, 19, 351.
7. DeLind, L.B. Transforming Organic Agriculture into Industrial Organic Products: Reconsidering National Organic Standards. Hum. Organ. 2000, 59, 198–208.
8. Planella, J.; Pagès Santacana, A.; Darnell i Vianya, M. Antropologia de l'Educació; Editorial UOC: Barcelona, Spain, 2007; ISBN 978-84-9788-576-8.
9. Willer, E.H.; Lernoud, J. The World of Organic Agriculture-Statistics and Emerging Trends 2019; Research Institute of Organic Agriculture (FiBL) and IFOAM—Organics International: Frick, Switzerland, 2019; 356p.
10. Willer, E.H.; Schlatter, B.; Trávní, J.; Kemper, L.; Lernoud, J. The World of Organic Agriculture Statistics and Emerging Trends 2020; Research Institute of Organic Agriculture (FiBL) and IFOAM—Organics International: Frick, Switzerland, 2020; 337p.
11. Paull, J.; Hennig, B. Atlas of Organics: Four Maps of the World of Organic Agriculture. J. Org. 2016, 3, 25–32.
12. Sarris, A. Medium-Term Prospects for Agricultural Commodities: Projections to the Year 2010; Food and Agriculture Organization of the United Nations, Ed.; Food and Agriculture Organization of the United Nations: Rome, Italy, 2003; ISBN 978-92-5-105077-4.
13. Sharma, K.; Mahato, N.; Cho, M.H.; Lee, Y.R. Converting Citrus Wastes into Value-Added Products: Economic and Environmently Friendly Approaches. Nutrition 2017, 34, 29–46.
14. Ananthi, A.; Ramesh, D. Preparation and Characterization of Silica Material from Rice Husk Ash-An Economically Viable Method. Chem. Mater. Res. 2016, 8, 1–7.
15. Shreya, M.K.; Indhumathi, C.; Rajarajeswari, G.R.; Ashokkumar, V.; Preethi, T. Facile Green Route Sol–Gel Synthesis of Nano-Titania Using Bio-Waste Materials as Templates. Clean Technol. Environ. Policy 2021, 23, 163–171.
16. Boffa, V.; Perrone, D.G.; Magnacca, G.; Montoneri, E. Role of a Waste-Derived Polymeric Biosurfactant in the Sol–Gel Synthesis of Nanocrystalline Titanium Dioxide. Ceram. Int. 2014, 40, 12161–12169.
17. Patel, K.G.; Shettigar, R.R.; Misra, N.M. Recent Advance in Silica Production Technologies from Agricultural Waste Stream–Review. J. Adv. Agric. Technol. 2017, 4, 274–279.

18. Srinath, P.; Azeem, P.A.; Reddy, K.V. Sol-Gel Synthesis of SiO2-CaO-Na2O Bio-Ceramics Using Bio-Waste. AIP Conf. Proc. 2020, 2265, 030042.
19. Silvestri, B.; Vitiello, G.; Luciani, G.; Calcagno, V.; Costantini, A.; Gallo, M.; Parisi, S.; Paladino, S.; Iacomino, M.; D'Errico, G.; et al. Probing the Eumelanin–Silica Interface in Chemically Engineered Bulk Hybrid Nanoparticles for Targeted Subcellular Antioxidant Protection. ACS Appl. Mater. Interfaces 2017, 9, 37615–37622.
20. Gonzalo-Juan, I.; Riedel, R. Ceramic Synthesis from Condensed Phases. Chem. Texts 2016, 2, 6.
21. Igal, K.; Vázquez, P. Antimicrobial Fabrics Impregnated with Ag Particles Included in Silica Matrices. In Waste in Textile and Leather Sectors; Körlü, A., Ed.; IntechOpen: London, UK, 2020; ISBN 978-1-78985-243-1.
22. Igal, K.; Arreche, R.A.; Sambeth, J.E.; Bellotti, N.; Vega-Baudrit, J.R.; Redondo-Gómez, C.; Vázquez, P.G. Antifungal Activity of Cotton Fabrics Finished Modified Silica-Silver-Carbon-Based Hybrid Nanoparticles. Text. Res. J. 2019, 89, 825–833.
23. Pota, G.; Venezia, V.; Vitiello, G.; Di Donato, P.; Mollo, V.; Costantini, A.; Avossa, J.; Nuzzo, A.; Piccolo, A.; Silvestri, B.; et al. Tuning Functional Behavior of Humic Acids through Interactions with Stöber Silica Nanoparticles. Polymers 2020, 12, 982.
24. Migliorero, M.B.C.; Palermo, V.; Vázquez, P.G.; Romanelli, G.P. Valorization of Citrus Waste: Use in Catalysis for the Oxidation of Sulfides. J. Renew. Mater. 2017, 5, 167–173.
25. Palermo, V.; Igal, K.; Colombo Migliorero, M.B.; Sathicq, A.G.; Quaranta, N.; Vazquez, P.G.; Romanelli, G.P. Valorization of Different Wastes and Their Use for the Design of Multifunctional Eco-Catalysts. Waste Biomass Valorization 2017, 8, 69–83.
26. Yilmaz, E.; Soylak, M. Functionalized Nanomaterials for Sample Preparation Methods. In Handbook of Nanomaterials in Analytical Chemistry; Elsevier: Amsterdam, The Netherlands, 2020; pp. 375–413. ISBN 978-0-12-816699-4.
27. Zanella, R. Metodologías Para La Síntesis de Nanopartículas Controlando Forma y Tamaño. Mundo Nano Rev. Interdiscip. En Nanociencia Nanotecnol. 2014, 5, 69–81.
28. Brinker, C.J.; Scherer, G.W. Sol.-Gel Science: The Physics and Chemistry of Sol.-Gel Processing; Academic Press, INC: San Diego, CA, USA, 2013; ISBN 978-0-08-057103-4.
29. Fidalgo, A.; Ilharco, L.M. Correlation between Physical Properties and Structure of Silica Xerogels. J. Non-Cryst. Solids 2004, 347, 128–137.
30. Jiménez-González, C.; Curzons, A.D.; Constable, D.J.C.; Cunningham, V.L. Expanding GSK's Solvent Selection Guide?Application of Life Cycle Assessment to Enhance Solvent Selections. Clean Technol. Environ. Policy 2004, 7, 42–50.
31. Cordero Castaño, A.F. Nuevos Materiales Utilizados Para la Obtención de Ecocatalizadores a Partir de Bio-Residuos. Ph.D. Thesis, Universidad Nacional de La Plata, La Plata, Argentina, 2020.
32. Schmidt, H. New Type of Non-Crystalline Solids between Inorganic and Organic Materials. J. Non-Cryst. Solids 1985, 73, 681–691.
33. Mendoza, E.; García, C. Recubrimientos por Sol-Gel Sobre Sustratos de Acero Inoxidable, Revisión del Estado del Arte. Dyna 2007, 74, 101–110.

34. Matlack, A.S. Introduction to Green Chemistry, 2nd ed.; CRC Press: Boca Raton, FL, USA, 2010; ISBN 978-1-4200-7811-4.

35. Byrne, F.P.; Jin, S.; Paggiola, G.; Petchey, T.H.M.; Clark, J.H.; Farmer, T.J.; Hunt, A.J.; Robert McElroy, C.; Sherwood, J. Tools and Techniques for Solvent Selection: Green Solvent Selection Guides. Sustain. Chem. Process. 2016, 4, 7.

36. Milea, C.A.; Bogatu, C.; Du, A. The Influence of Parameters in Silica Sol-Gel Process. Bull. Transilv. Univ. Brasov. Eng. Sci. Ser. I 2011, 4, 59.

37. Arreche, R. Inclusión de Ag en Materiales Basados en Sílice y Circonia, Sintetizados Por el Método Sol-Gel, Para Su aplicación Como Aditivos Antimicrobianos en Pinturas. Ph.D. Thesis, Universidad Nacional de La Plata, La Plata, Argentina, 2016.

38. Igal, K. Telas Antimicrobianas Impregnadas Con Partículas de Ag o Zn Reciclado Incluidas en Matrices Silíceas Modificadas. Ph.D. Thesis, Universidad Nacional de La Plata, La Plata, Argentina, 2019.

39. Meixner, D.L.; Dyer, P.N. Influence of Sol-Gel Synthesis Parameters on the Microstructure of Particulate Silica Xerogels. J. Sol.-Gel Sci. Technol. 1999, 14, 223–232.

40. de Gennes, P.-G. Scaling Concepts in Polymer Physics; Cornell University Press: New York, NY, USA, 1979; ISBN 978-0-8014-1203-5.

41. Kirkbir, F.; Murata, H.; Meyers, D.; Chaudhuri, S.R.; Sarkar, A. Drying and Sintering of Sol-Gel Derived Large SiO2 Monoliths. J. Sol.-Gel Sci. Technol. 1996, 6, 203–217.

42. Soleimani Dorcheh, A.; Abbasi, M.H. Silica Aerogel; Synthesis, Properties and Characterization. J. Mater. Process. Technol. 2008, 199, 10–26.

43. Carman, P.C. Constitution of Colloidal Silica. Trans. Faraday Soc. 1940, 36, 964–973.

44. Romanelli, G.; Vázquez, P.; Pizzio, L.; Quaranta, N.; Autino, J.; Blanco, M.; Cáceres, C. Phenol Tetrahydropyranylation Catalyzed by Silica-Alumina Supported Heteropolyacids with Keggin Structure. Appl. Catal. Gen. 2004, 261, 163–170.

45. Curran, M.D.; Stiegman, A.E. Morphology and Pore Structure of Silica Xerogels Made at Low PH. J. Non-Cryst. Solids 1999, 249, 62–68.

46. Zhao, S.; Xu, D.; Ma, H.; Sun, Z.; Guan, J. Controllable Preparation and Formation Mechanism of Monodispersed Silica Particles with Binary Sizes. J. Colloid. Interface Sci. 2012, 388, 40–46.

47. Tadanaga, K.; Morita, K.; Mori, K.; Tatsumisago, M. Synthesis of Monodispersed Silica Nanoparticles with High Concentration by the Stöber Process. J. Sol.-Gel Sci. Technol. 2013, 68, 341–345.

48. Zarib, N.S.M.; Abdullah, S. Effect of C6H8O7 Concentration on Silica Extraction of Rice Husk, Rice Husk Ash and Mixture of Rice Husk With Rice Husk Ash Via Acid Leaching Process. Int. J. Eng. 2018, 7, 190–195.

49. Saini, J.; Garg, V.K.; Gupta, R.K. Green Synthesized SiO2@OPW Nanocomposites for Enhanced Lead (II) Removal from Water. Arab. J. Chem. 2020, 13, 2496–2507.

50. Cui, J.; Sun, H.; Luo, Z.; Sun, J.; Wen, Z. Preparation of Low Surface Area SiO2 Microsphere from Wheat Husk Ash with a Facile Precipitation Process. Mater. Lett. 2015, 156, 42–45.

51. Worathanakul, P.; Trisuwan, D.; Phatruk, A.; Kongkachuichay, P. Effect of Sol–Gel Synthesis Parameters and Cu Loading on the Physicochemical Properties of a New SUZ-4 Zeolite. Colloids Surf. Physicochem. Eng. Asp. 2011, 377, 187–194.
52. Czarnobaj, K. Preparation and Characterization of Silica Xerogels as Carriers for Drugs. Drug Deliv. 2008, 15, 485–492.
53. Ng, E.-P.; Mohammad, A.G.S.; Rigolet, S.; Daou, T.J.; Mintova, S.; Ling, T.C. Micro-and Macroscopic Observations of the Nucleation Process and Crystal Growth of Nanosized Cs-Pollucite in an Organotemplate-Free Hydrosol. New J. Chem. 2019, 43, 17433–17440.
54. Pakizeh, M.; Omidkhah, M.R.; Zarringhalam, A. Synthesis and Characterization of New Silica Membranes Using Template–Sol–Gel Technology. Int. J. Hydrogen Energy 2007, 32, 1825–1836.
55. Hu, S.; Hsieh, Y.-L. Preparation of Activated Carbon and Silica Particles from Rice Straw. ACS Sustain. Chem. Eng. 2014, 2, 726–734.
56. Kirk, C.T. Quantitative Analysis of the Effect of Disorder-Induced Mode Coupling on Infrared Absorption in Silica. Phys. Rev. B 1988, 38, 1255–1273.
57. Wong, S.-F.; Deekomwong, K.; Wittayakun, J.; Ling, T.C.; Muraza, O.; Adam, F.; Ng, E.-P. Crystal Growth Study of K-F Nanozeolite and Its Catalytic Behavior in Aldol Condensation of Benzaldehyde and Heptanal Enhanced by Microwave Heating. Mater. Chem. Phys. 2017, 196, 295–301.
58. Araujo-Andrade, C.; Ortega-Zarzosa, G.; Selina, P.; Martinez, J.R.; Villegas-Aguirre, F.; Ruiz, F. Análisis de Las Reacciones de Hidrólisis y Condensación En Muestras de Sílica Xerogeles Usando Espectroscopía Infrarroja. Rev. Mex. Fis. 2000, 46, 593–597.
59. Orcel, G.; Phalippou, J.; Hench, L.L. Structural Changes of Silica Xerogels during Low Temperature Dehydration. J. Non-Cryst. Solids 1986, 88, 114–130.
60. Martınez, J.R.; Ruiz, F. Mapeo estructural de sílica xerogel utilizando espectroscopía infrarroja. Rev. Mex. Fis. 2002, 48, 142–149.
61. Duran, A.; Serna, C.; Fornes, V.; Fernandez Navarro, J.M. Structural Considerations about SiO2 Glasses Prepared by Sol-Gel. J. Non-Cryst. Solids 1986, 82, 69–77.

6 SYNTHESIS AND BEHAVIOR OF HEXAMETHYLENETETRAMINE-BASED IONIC LIQUIDS AS AN ACTIVE INGREDIENT IN LATENT CURING FORMULATIONS WITH ETHYLENE GLYCOL FOR DGEBA

Dawid Zielinski [1,2], Andrea Szpecht [1], Paulina Hinc [2] and Marcin Smiglak [1]

[1]*Poznan Science and Technology Park, Adam Mickiewicz University Foundation, 61-612 Poznań, Poland*

[2]*Faculty of Chemistry, Adam Mickiewicz University in Poznań, 61-614 Poznań, Poland*

ABSTRACT

The paper presents the preparation of new ionic liquids based on hexamethylenetetramine with bis(trifluoromethanesulfonyl)imide and dicyanamide anion, which were characterized in detail in terms of their purity (Ion Chromatography) and thermal properties (Differential Scanning Calorimetry), as well as stability. The obtained substances were used to develop curing systems with ethylene glycol, which were successfully tested for their application with bisphenol A diglycidyl ether molecule. In addition, the curing process and its relationship to the structure of the ionic liquid are characterized in detail. The research showed that hexamethylenetetramine-based new ionic liquids can be successfully designed using well-known and simple synthetic methods—the Delepine reaction. Moreover, attention was paid to their stability, related limitations, and the application of hexamethylenetetramine-based ionic liquids in epoxy-curing systems.

Keywords: *hexamethylenetetramine; ionic liquids; epoxy resin; ethylene glycol; curing agent; composites*

1. INTRODUCTION

It is commonly believed that ready-made composite materials are neutral to the environment and due to the lack of direct interactions at the molecular level, are not toxic [1]. This theorem is fulfilled for properly prepared composites, in which the network of connections between resins and the hardener are closed, well-known, predictable, and stable, and the resin itself does not contain solvents and other volatile organic compounds (VOCs) before the curing process [2].

The aim of this work was to provide new solution in the field of multifunctional initiating systems for the polymerization of epoxy resins. To be more precise, this solution is based directly on previously unknown ionic liquids, designed with well-known raw materials, reactions, and processes. The proposed cross-linking systems must be easy to prepare, stable and predictable in the context of their application and most importantly, well thought out. In addition, it is important to bring added value to the final products in the form of changing their mechanical properties and/or application potential through new functionalities. In the era of ecological awareness and social responsibility, an extremely important element of research is the use of substances for which the details of their environmental impact, toxicity and methods of obtainment are known. This is one of the ways to limit the emergence of new, unpredictable, and not fully tested substances in the industry, while still developing science and improving existing solutions, which this work was also intended to prove.

Ionic liquids (ILs) are chemical compounds with an ionic structure, consisting of a cation and an anion and having several specific properties. The cation of the ionic liquid is usually of organic origin and has a varied structure. On the other hand, anions of ionic liquids can be both organic and inorganic, and their nature significantly affects the properties of the entire compound. The most important feature of this group of compounds is undoubtedly an extremely wide range of applications, in which the following should be distinguished: (i) dissolution and processing of biomass [3,4,5], (ii) catalytic reactions [6,7,8], (iii) active ingredients in pharmaceutical preparations [9,10] or (iv) design of polymerizable monomers intended for modern material solutions [11,12,13]. The wide area of application of ionic liquids is due to their properties, among which the following should be mentioned: high thermal stability [14], high electrochemical window [15], low volatility [16] and flammability [17]. At the same time, it turns out that these compounds can successfully replace, among others, hardeners (as a latent curing agents) of epoxy resins [18,19,20,21] as well as a large group of property modifiers: (i) flame retardants [22,23,24], (ii) hydrophobizers [25,26] and (iii) compounds that increase the mechanical parameters of the composite [21,27,28,29,30].

However, despite so many literature reports and numerous confirming the unique applications of ionic liquids, there are still groups of these compounds that require further research. One of these understudied groups is the non-aromatic polycyclic amines with catalytic properties, which can potentially be used as structures of ionic liquid cations. The most important substances in this group include: (i) 1,4-diazabicyclo[2.2.2]octane (DABCO), (ii) 1,8-diazabicyclo[5.4.0]undec-7-ene (DBU), (iii) 1-azabicyclo[2.2.2]octane (ABCO), (iv) 1,5-diazabicyclo[4.3.0]non-5-ene (DBN) and hexamethylenetetramine (HMTA). The first two, DABCO and DBU, remain in the interest of many researchers, especially as new ionic liquids with catalytic properties [21]; DBN is not very popular due to its high similarity to DBU, while ABCO, because of its very high price, is practically not studied at all. In contrast, the simplest, best known and most easily available, as well as the cheapest polycyclic amine—HMTA, remains to this day the most unexplored source of ionic liquids, which may have a few unique properties, from catalytic, electrochemical and anti-corrosion to fuel applications. There are only a few reports of HMTA-based ionic liquids being obtained and used as components for the production of strong catalysts for obtaining hexahydroquinolines [31] and pyrido[2,3-d:5,6-d0]dipyrimidines [32]. In addition, it was confirmed that HMTA-ILs exhibit high ionic conductivity, which increases significantly with temperature [33]. A huge impact on the importance of HMTA in chemistry was the discovery of the Delepine reaction in 1895, which allows for a simple and easy synthesis of primary amines through the intermediate product, which is a quaternary ammonium derivative of HMTA. The key here is the fact that this intermediate product can be easily isolated and shows very high stability [34,35].

Another view is ethylene glycol (EG), which is a well-known, simple chemical compound that is constantly gaining new applications based on the excellent knowledge of the properties of this substance. Especially in combination with ILs there is an interesting set of reports about its applications as a microemulsifier [36,37], co-solvent [38,39,40] and a substance able to stabilize ionic liquids through a specific interaction [41].

All the literature reports cited above were an inspiration to conduct this research, the results of which will be presented in the following subsections.

2. RESULTS

2.1. Synthesis of Ionic Liquids

The research proposed a portfolio of new HMTA derivatives, differing in the length of the carbon chain attached to the nitrogen atom. Structures of all cations are shown in the Figure 1. As part of the research, alkyl chains with both an even and an odd number of carbon atoms were used to precisely determine the influence of the chain length on the properties of the obtained compounds. Moreover, two groups of anions were selected and combined with cation structures to obtain new ionic liquids. The bis(trifluoromethanesulfonyl)imide (TFSI) anion is highly hydrophobic [42], while the dicyanamide (DCA) anion is inherently more hydrophilic [43]. Such a correlation between anions allows for a thorough look at the nature of the compounds obtained and opens up more application possibilities and areas (mainly those indicated in the introduction) where new substances may prove value. To obtain the proposed substances, it was necessary to conduct quaternization reactions first, leading to the desired halogen derivatives, which were then subjected to metathesis reactions leading to the final products. Afterward, all obtained substances were subjected to a thorough analysis of physicochemical properties before they were subjected to application tests.

2.1.1. Quaternization

The first step of the Delepine reaction is the attachment of an alkylated chain to the nitrogen atom, followed by a quaternary ammonium salt (QAS) of HMTA formation. The reaction proceeds at room temperature in chloroform, from which the product is easily separated by simple filtration.

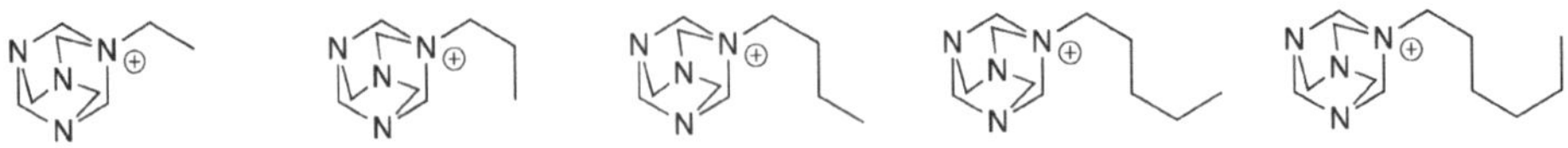

1 2 3 4 5

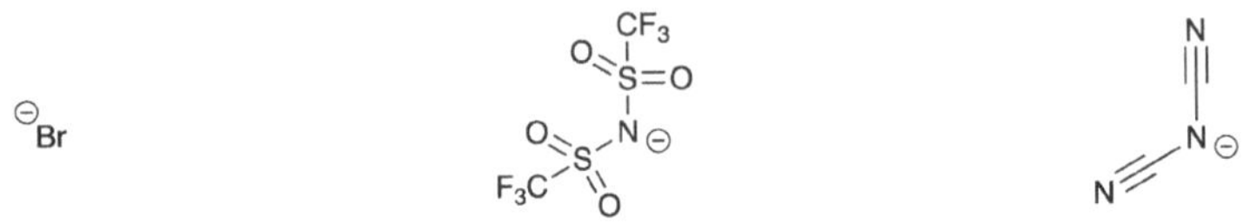

a b (TFSI) c (DCA)

Figure 1.: Chemical structures of the obtained compounds.

Thus, the process was highly efficient, and the products were recrystallized to obtain QASs of very high purity (Table 1). On the other hand, the obtained products do not show strong hygroscopicity in contrast to analogous derivatives for, e.g., DABCO [17]. Determination of the purity of the obtained substances was carried out using ion chromatography, considered to be the best method for determining the purity of ionic compounds [12,17,40,41,42]. In addition, the same technique was used to monitor the progress of the reaction. Through routine analysis of product increment and substrate depletion, the reaction time was precisely determined to enable high process yields to be obtained in the shortest possible time. As a result of the conducted experiments, five bromide derivatives were successfully obtained, reaching the efficiency exceeding 99% and the purity of the obtained substances at the average level of 99.2%. To fully identify the obtained compounds, they were subjected to NMR and mass spectrometry analysis. Another equally important value for ILs is their water content, especially for halides, which are a key step on the way for obtaining the desired further products. None of the salts obtained exceeded the level of 1% of water; moreover, compounds 1a, 2a and 5a, did not exceed the level of 0.5% water. The lowest value was recorded for the derivative having a propyl chain (2a) in its structure.

***Table 1.:** Parameters of the synthesis of HMTA derivatives.*

Compound	Yield [%]	Purity [%]	Water Content [wt%]
1a	99.6	99.6	0.47
2a	99.0	99.0	0.17
3a	99.1	99.1	0.68
4a	99.1	99.1	0.57
5a	99.0	99.1	0.42

2.1.2. Metathesis Reaction

However, ten new TFSI and DCA ionic liquids were selected for research. The bromide salts obtained earlier were precursors on the way to their preparation and required further processing. For this purpose, two different pathways of the metathesis reaction were used between the bromide anion in the HMTA quaternary salt and (i) the dicyandiamide anion derived from the freshly prepared (according previously published work [21]) silver salt, or the (ii) bis(trifluoromethanesulfonyl)imide anion derived from lithium salt. Each pathway was analyzed at room temperature, however, using different process conditions, as shown in Figure 2, and synthetic techniques, as described in detail in the synthesis section. In the case of hydrophobic ionic liquids, the average yield of the metathesis reaction was 98.4%, while the purity of the obtained compounds ranged from 99.1 to 99.8%, with the lowest values for ionic liquids 4b and 5b, with the longest alkyl chains. Similar results were recorded for the synthesis of dicyandiamide derivatives, for which the average yield of the reaction was 94.5%, with the lowest values of 87.9% for compound 2c and 93.9% for compound 4c, for compounds with an odd number of carbon atoms in the alkyl chain. On the other hand, the purity of all dicyandiamide derivatives exceeded 97.5%, reaching the highest value of 98.1% for compounds 1c and 3c containing ethyl and butyl alkyl chains, respectively.

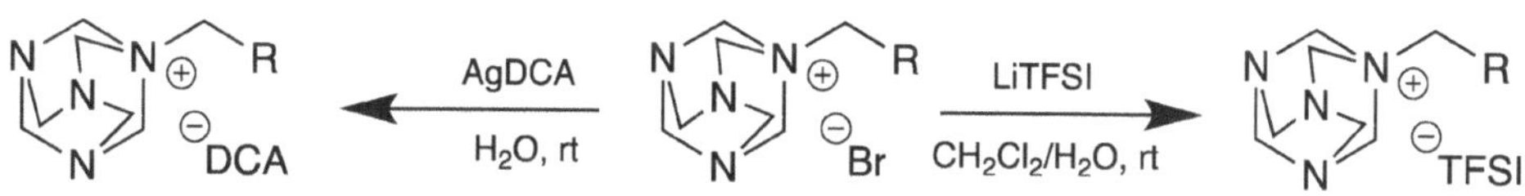

1c–5c 1a–5a 1b–5b

Figure 2.: *Scheme of metathesis reaction leading to HMTA derivatives with TFSI anion and DCA.*

2.1.3. Thermal Properties of the Compounds

Measurements of thermal properties of the obtained substances (thermograms from analyzes available in Supplementary Materials), show numerous differences between both individual groups of compounds (depending on the anion present in the IL structure) and depending on the length of the carbon chain, in particular the relationship between the even and odd number of carbon atoms in the alkyl chain (Table 2). Namely, for quaternary ammonium salts with a bromide anion, only an onset exothermic peak (T_{decomp}) was observed in the thermogram, characteristic of the process of partial decomposition of the salt with the release of a large portion of energy. Differences in the decomposition temperature were observed, depending on whether an alkyl chain with an even or an odd number of carbon atoms was present in the structure. Compounds 2a and 4a, which contain a propyl and a pentyl chain structure, have onset peak decomposition temperatures of 146.5 °C and 143.6 °C, respectively. Between this group of compounds and analogs containing even carbon chains in the structure, an average temperature difference of 14 °C was observable. The highest decomposition temperatures, 168.3 °C and 155.4 °C, respectively, were recorded for compounds 1a and 3a, i.e., those with an ethyl or butyl chain. On the contrary, all ionic liquids obtained in the metathesis reaction, both DCA and TFSI, are characterized by lower crystallization/decomposition peak temperatures (T_{decomp}) observed during the measurements. The lowest values were recorded for compounds 1b, 4b, 1c and 3c and amounted to the following: 119.3 °C, 121.6 °C, 120.3 °C and 101.6 °C, respectively. However, the highest values were for the derivatives 2b (126.8 °C), 5b (134.7 °C), 2c (135.3 °C) and 5c (126.8 °C). There was no clear relationship between the chain length and the parameters obtained. In addition, glass transition temperature (T_g) data were collected for ILs with the TFSI anion, and melting point temperatures (T_m) were collected for ILs with the DCA anion. For the glass transition temperature, there is a general shift towards lower temperatures with increasing chain length, resulting in the lowest temperature values for 4b and 5b being −29.1 °C and −19.9 °C, respectively. Similarly, the melting temperatures observed for the dicyanamides also show a shift towards higher temperatures with increasing chain length, except for compound 2c (57.4 °C), which decreases relative to a compound 1c (69.1 °C) with a shorter alkyl chain. All collected values are below 100 °C, reaching the highest value of 75.5 °C and 86.6 °C for 4c and 5c compounds, respectively.

Table 2.: *Thermal properties of the HMTA-based compounds obtained in this research.*

Compound	T_m [°C]	T_g [°C]	T_{decomp} [°C]
1a	-	-	168.3 [a]
2a	-	-	146.5 [a]
3a	-	-	155.4 [a]
4a	-	-	143.6 [a]
5a	-	-	152.4 [a]
1b	-	−5.0	119.3 [b]
2b	-	−16.6	126.8 [b]
3b	-	−17.1	126.3 [b]
4b	-	−29.1	121.6 [b]
5b	-	−19.9	134.7 [b]
1c	69.1	-	120.3 [b]
2c	57.4	-	135.3 [b]
3c	72.1	-	101.6 [b]
4c	75.5	-	124.0 [b]
5c	86.6	-	126.8 [b]

[a] Decomposition onset peak, [b] Start of exothermic decomposition during DSC measurement.

2.2. Curing Process

In the next stage of the research, a selected group of ten ILs was used to prepare ten curing mixtures intended for cross-linking the epoxy resin (Figure 3). The combinations were prepared by mixing the ionic liquid with ethylene glycol in a molar ratio of 1:2 in such a way as to ensure the highest possible curing activity in relation to the resin while maintaining high stability of the mixture itself. By experimenting with different ratios of IL to ethylene glycol, a system was selected in which there are two equivalents of EG for one equivalent of ionic liquid. Such a relationship between these substances resulted in the most stable mixtures, whose stability was validated by observing phase separation over a period of up to 60 days. Moreover, it was confirmed that ethylene glycol forms more stable systems with DCA-based ionic liquids, which successfully maintain their homogeneity beyond 60 days, while TFSI anion ionic liquids lose their homogeneity after about 2 weeks, requiring re-mixing. EG, due to its emulsifying properties, allowed us to prepare homogeneous curing systems with a high degree of compatibility with the resin, which in this case was the Bisphenol A diglycidyl ether (DGEBA, M = 340 g/mol) molecule. An important feature here is the fact that ethylene glycol creates much more stable emulsions in combination with more hydrophilic ionic liquids such as DCA ILs. It has been observed that TFSI causes emulsions to become heterogenous rather quickly. Specifically, to thoroughly study the curing process and compatibility, a system consisting of unmodified DGEBA molecules was used, thus eliminating the influence of possible modifiers and solvents that could be present in commercially available epoxy resins. Next, the prepared curing systems were mixed with a resin particle in a ratio that guaranteed full cross-linking (detailed data on sample preparation of resin systems are provided in the chapter Methods and in the reference [17]), degassed, and then cured in a vacuum oven with process temperature, developed based on data from simultaneously conducted tests using Differential Scanning Calorimetry (DSC). An important property of the curing mixtures is their initiating effect occurring only at elevated temperatures. At room temperature, they remain weakly active against the DGEBA molecule. In particular, this is a property observed with many ionic liquids used as curing initiators for epoxy resins and is fairly well explained [18,19,20,21].

2.2.1. Thermal Analysis

Subsequently, ten curing systems were used to prepare mixtures with DGEBA, which were subjected to tests of curing process nature. For this purpose, Differential Scanning Calorimetry was again used (thermograms from analyzes available in Supplementary Materials), and several parameters of the process (Table 3) were determined, such as polymerization start temperature (T_{onset}), temperature peak (T_{max}) and polymerization end temperature (T_{end}). At the same time, the amount of energy released (ΔH) during the process was monitored. In the case of curing systems based on ionic liquids with the TFSI anion, comparable results for individual parameters were obtained, which remained at a similar level of values, regardless of the type and length of the alkylated chain in the cation structure. The exception is the derivative with an ethyl chain (1b), for which a lower polymerization start temperature (109.4 °C) and a much higher process energy (84.4 J/g) were recorded.

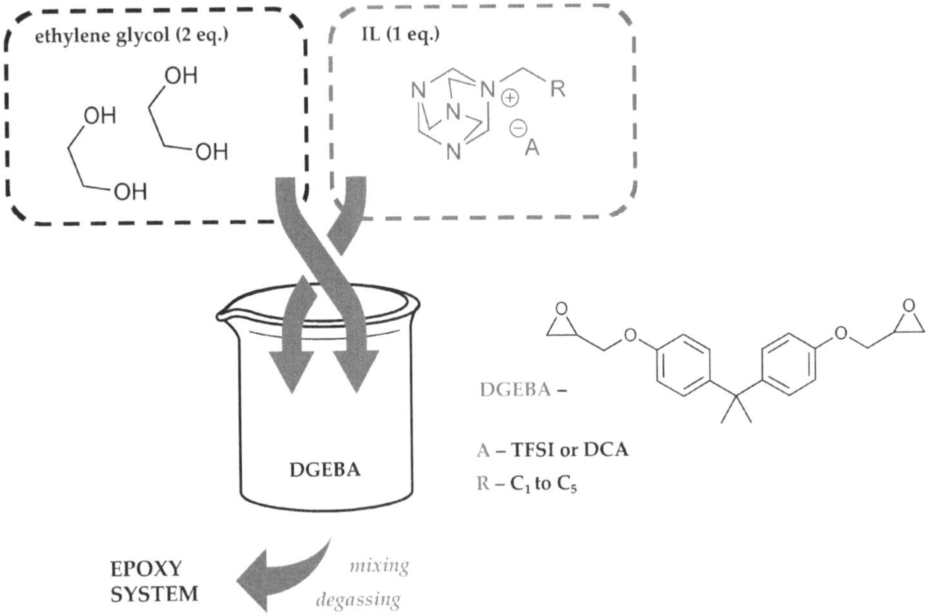

Figure 3.: *Scheme of the process of preparing resin systems with ionic liquids.*

At the same time, for the polymerization initiated by these curing systems, the appearance of two independent areas of temperature peaks in the form of two thermal processes occurring in the sample was observed. The highest temperature values of the first thermal peak, respectively 154.4 °C and 149.0 °C, were obtained for systems containing compounds 3b and 5b, however, these differences are insignificant. Similarly, in the case of the second thermal peak, the highest values were recorded for systems containing compounds 2b (224.4 °C) and 4b (229.5 °C). Simultaneously, with the increase in the length of the alkyl chain in the cation structure, the amount of energy released during the second thermal process occurring in the sample decreases. On the other hand, polymerization initiated by systems based on dicyanamide ionic liquids proceeds in the form of a single thermal transformation process. In this case, the relationship between the polymerization starting temperature and the energy released during the process, dependent on an odd or even number of carbon atoms in the alkyl chain, is clearly visible. Thus, with an odd number of carbon atoms in the cation structure, relatively lower values of the process start temperature were observed, while the amount of released energy decreases with the increase in the length of the odd alkyl chain. In contrast, if there is an even number of carbon atoms in the alkyl chain of the ionic liquid, the energy of thermal transformation increases with its length, reaching a maximum of 151.5 J/g for the system containing compound 5c, for which the lowest polymerization initiation temperature (101.3 °C) and polymerization peak temperature (151.8 °C) were also recorded. To summary, all characteristic temperature points of the cross-linking process are within the range of the decomposition temperature of ionic liquids.

Table 3.: *Thermal properties of the curing process.*

System	Anion	T_{onset} [°C]	T_{max} [°C]	T_{end} [°C]	ΔH [J/g]
1b/EG/DGEBA	TFSI	109.4	137.0	164.8	84.4
		200.7	223.5	236.9	24.3
2b/EG/DGEBA		122.2	148.1	201.5	26.5
		215.9	224.4	236.4	11.0
3b/EG/DGEBA		122.1	154.4	185.6	19.8
		210.3	218.9	239.5	11.7
4b/EG/DGEBA		119.1	147.0	205.7	29.3
		221.0	229.5	242.3	8.9
5b/EG/DGEBA		119.8	149.0	193.2	28.5
		209.3	221.7	223.0	7.2
1c/EG/DGEBA	DCA	140.8	170.7	224.0	49.5
2c/EG/DGEBA		123.6	164.7	213.2	81.6
3c/EG/DGEBA		140.0	163.9	188.1	49.8
4c/EG/DGEBA		129.5	165.1	199.1	23.0
5c/EG/DGEBA		101.3	151.8	210.6	151.5

2.2.2. Hardness of the Cured Resin

Afterward, the prepared samples of the cured DGEBA were tested for the basic assessment of their hardness/brittleness depending on the curing system used. Such a quick experiment made it possible to determine the relationship between the structure of the IL and the behavior of the hardened material in laboratory conditions. Furthermore, the samples were evaluated 48 h after hardening and after 7 and 14 days to determine the effect of aging of the samples on the basic parameter, which was hardness. For comparison, analogous samples cured with two amines commonly used as initiators of epoxy resin polymerization, triethylenetetramine (TETA, RM1) and isophorone diamine (IPDA, RM2), whose hardening properties have already been thoroughly investigated [44], were made. The samples are tabulated and related to the results obtained for the new curing systems. With the passage of time, the changes in the hardness parameters of the samples did not change beyond 2D, therefore the Table 4 summarizes the collected data as average values for all measurements carried out for a given sample. The average value of the hardness of the samples in which systems with TFSI ILs were used was 79D, while systems with DCA IL was 85D and all these values are higher than the reference samples. The influence of the type of anion in the ionic liquid on the obtained hardness results is visible. DCA IL-based systems show a relatively higher hardness of the obtained materials than TFSI IL-based systems. However, in the case of these DCA systems, composed of ILs with an odd number of carbon atoms in the alkyl chain, they were characterized by significant brittleness, which made measurement impossible. Systems containing TFSI ILs have a smaller increase in hardness in relation to reference samples and at the same time, slightly higher values were observed for compounds 2b (80D) and 4b (81D), in which their structures contain chains with an odd number of carbon atoms.

Table 4.: *Hardness measurements results.*

Compound	**Hardness [D]**
RM1/RM2	74/71
1b	78
2b	80
3b	76
4b	81
5b	n/a
1c	87
2c	n/a
3c	89
4c	n/a
5c	80

n/a—not available due to too high brittleness of the samples.

3. DISCUSSION AND CONCLUSIONS

3.1. Synthesis and ILs Stability

Hexamethylenetetramine is an unusual amine, obtained by the reaction of formaldehyde and ammonia. It does not melt, only sublimates, and is very sensitive to any changes in pH. Improperly stored or contaminated, it slowly decomposes to formaldehyde and ammonia. The decomposition phenomenon is probably caused by the hygroscopicity of the amine itself and by the presence of water, initiating a slow hydrolysis process that leads to the formation of formaldehyde and ammonia [45]. On the other hand, under the influence of temperature or during combustion (with access to oxygen), the HMTA decomposition pathways change. To illustrate, thermal decomposition mainly leads to nitrogen oxides, carbon dioxide, carbon monoxide and formaldehyde, while the combustion process generates carbon dioxide, nitrogen, nitrogen oxides and trace amounts of ammonia. Based on studies of the kinetics of HMTA decay conducted to understand the behavior of this molecule in the human body, the following conclusions can be drawn: (i) HMTA solution in water has a pH of 9.5, (ii) pH values lower than 9.5 initiate the decomposition process, (iii) no release of formaldehyde is observed up to pH 7.4, (iv) at a pH around 2.0, formaldehyde is released in higher quantities [46]. We assume that during the thermal curing of the resin with HMTA-based ionic liquids, formaldehyde is not released due to the high process temperature, and carbon oxides are released in its place. Nevertheless, there is a risk that residual unreacted ionic liquid in the cured resin structure will be able to release formaldehyde during storage. This is not the subject of the presented research, but in the case of further experiments, it may be advisable to carry out tests in the direction of determination of volatile organic compounds (VOCs) from the cured material.

In this work, several new chemical compounds are presented. From our point of view, the key aspect to discuss here is undoubtedly the very high purity of the salts obtained and the high efficiency of the reactions. This shows how useful the first step of the Delepine reaction is for the synthesis of new potential ionic liquids. However, this reaction is limited to the amine substrate—HMTA. While the halide salts obtained directly in it show high stability, signs of lower stability appear in the case of metathesis reaction products. The spatial structure of the quaternary HMTA cation, especially as a product of only a single chain attachment, is likely to be energetically highly unstable. Moreover, the liquid form of the compound, which allows the migration of molecules and the presence of, for example, anions derived from strong acids, additionally affects the stability of the compounds obtained due to high impact on pH. As HMTA quaternary ammonium salts, the products are highly sensitive to any changes in pH that may occur through spontaneous partial decomposition of the cation, which will shift the ionic balance. In addition, the nature of ionic liquids favors the absorption of even small amounts of water, which is absorbed in the compounds and facilitates the hydrolysis of the compound. In the case of HMTA, this leads directly to the release of ammonia, the smell of which is observable in the case of TFSI and DCA ionic liquids over time of their storage at room temperature. Undoubtedly, this phenomenon is influenced primarily by the presence of water, but also by the fact that along with the breakdown of the cation structure, the concentration of anions increases, which results in a decrease in pH. In the case of the synthesis of quaternary ammonium salts based on HMTA, it is very important to monitor the purity of the obtained compounds, e.g., using ion chromatography. In this way, it is possible to quickly detect the potential presence of ammonia or other ionic structures that could be formed by the decomposition of the compound and, consequently, affect the possibility of using a given substance. On the one hand, the synthesis of this type of derivatives is exceptionally simple and gives a very wide field for the introduction of many alkylating agents from all groups of this type of compounds. However, one should be always carefully analyze the stability of the obtained substances and consider the possibility of changes in pH, which, by initiating the degradation process of the compound, may sometimes make it impossible to use them in specific application areas. In the case of application presented in this paper, the phenomenon of slow decay does not have a significant impact, and sometimes it is in favor of our solution.

3.2. Role of Ethylene Glycol

Ionic liquids are compounds with very different rheological parameters. Depending on many factors, such as (i) the length of the alkylated chain used or (ii) the type of anion, they can be compounds of very low viscosity as well as those with a viscosity resembling glue [47]. In fact, in some application areas, it is necessary to introduce additional substances that are designed to adjust the ionic liquids to a specific application, not only by changing the rheological properties, but sometimes also by additional interactions with ILs [48].

Ethylene glycol was not accidentally used in this work, which was supposed to fulfill the following tasks: (i) to reduce the viscosity of long-chain ionic liquids and at the same time increase their compatibility in combination with the epoxy resin molecule, (ii) due to its emulsifying properties [36,37], to form stable curing mixtures, (iii) be a substance capable of absorbing ammonia [49] and formaldehyde or eventually carbon dioxide [50] that could be released from the IL over time of its storage and (iv) remain in the structure of the cured resin, thanks to its high boiling point exceeding the temperature of the cross-linking process. In addition, during the study, the effect of ethylene glycol on slowing down the pH change process was observed, which was systematically controlled before each application experiment, using simple strip tests. Moreover, no ammonia odor was observed in mixture samples stored for 60 days. We suppose that through emulsification and connection with the structure of the IL, the EG undergoes a specific incorporation into the structure of the cation through a network of hydrogen bonds. Thus, this stabilizes the quaternary structure of HMTA and slows down the process of its breakdown. However, this is only an assumption requiring much extended research and theoretical calculations that would not affect the application presented in the work.

3.3. Curing Process and the Mechanism

The aim of this work was to demonstrate the use of new ionic liquids as active ingredients in the curing of epoxy resins. For this purpose, studies of the curing process were carried out, drawing key conclusions: (i) HMTA-based ionic liquids can successfully initiate the curing of epoxies, (ii) the low stability of these compounds favors their use in this area, resulting in lower temperatures of the curing process, (iii) the choice of anion affects the curing process itself, in particular the temperature range of the process. The quaternary hexamethylenetetramine cation can disintegrate under the influence of temperature into a whole spectrum of different types of molecules. Most of them are highly reactive, energetically unstable molecules, and able to react quickly with the epoxide groups of the resin. In Figure 4 several potential transition structures that may be generated during the process of thermal decomposition of HMTA-ILs have been proposed. The presented structures were inspired by the Delepine reaction mechanism, but also by data on HMTA decay directions [40,41]. The process generates, among others, primary amines based on an alkyl chain that was attached to the generic cation structure. In addition, the process may also produce ammonia, which is easily reacted with the epoxy group to form a quaternary adduct, which is unstable and quickly initiates a chain reaction leading to full cross-linking of the resin structure.

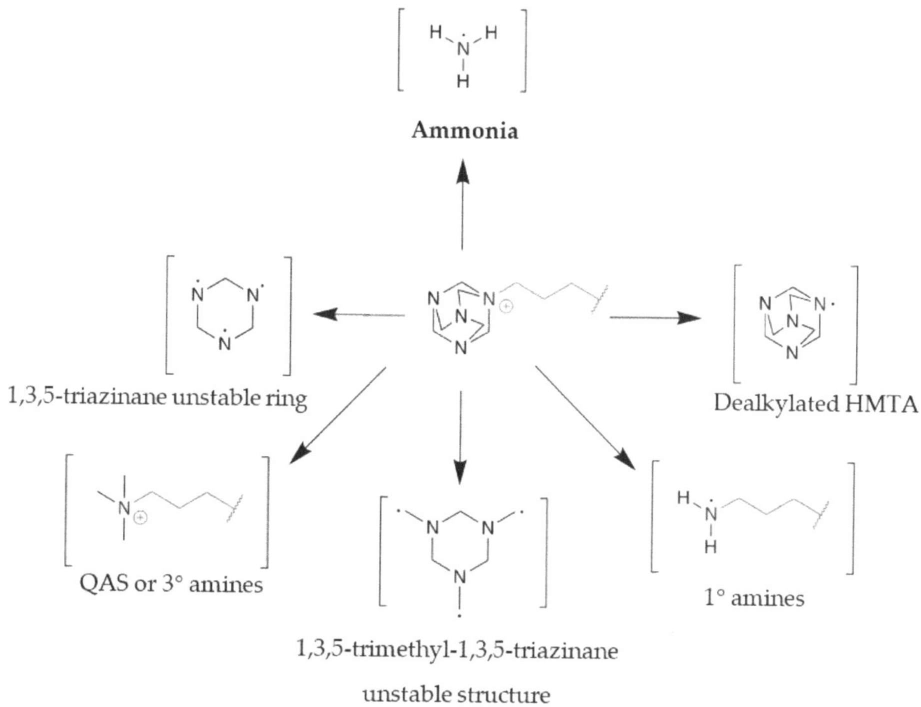

***Figure 4.**: Proposed pathways of thermal decomposition of HMTA-based ILs.*

Undoubtedly, in the case of curing the resin with ionic liquids based on HMTA, we are dealing with independently occurring processes of disintegration and simultaneous reaction with the epoxy groups present in the DGEBA molecule. In addition, the epoxy group can react with both basic (amines) and acidic compounds, which means that the anions present in the ionic liquid also react with the resin, leading to the formation of fully cross-linked structures. This phenomenon can be defined as cationic or anionic polymerization depending on which part of the ionic liquid is reacting with the epoxide at a given moment [21]. Both processes usually take place simultaneously, but if they occur at different temperatures, they are sometimes observable as separate polymerization peaks in thermal analysis. From the considerations regarding the ability to initiate the epoxy resin curing process, (i) due to their high thermal stability, and (ii) the lack of ionic liquid characteristics that would allow these compounds to be classified into this group, halide derivatives were excluded. So, the curing process of epoxy resin initiated by systems based on ionic liquids with TFSI anion is a process independent of the structure of the cation and its chain length. For all the tested compounds, the polymerization start, peak and end temperatures were similar, with the only exception being the derivative with an ethyl chain, which, probably due to its short chain, quickly joins the curing process and does not require the time necessary to generate appropriate, active transitional structures. What is more, this thesis becomes more likely with the information that a much greater amount of energy is released in the thermal process than in the case of analogous structures with longer carbon chains. In general, it can be concluded that the initiators based on HMTA–TFSI-ILs are characterized by a predictable and independent curing process. On the other hand, the polymerization reaction initiated by dicyanamides is characterized by a single thermal peak with a maximum temperature peak that depends on the chain length and generally decreases as it increases. More precisely, we assume that with the increase in the length of the alkyl chain in the cation structure, the thermal stability of the compounds decreases, which means that active molecules directly participating in the reaction with epoxides are generated faster. At the same time, we noticed differences in the polymerization start temperature depending on whether the alkyl chain has an even or odd number of carbon atoms. We suggest that the reason for this phenomenon may be that structures with an odd number of carbon atoms in the chain will decay faster, especially at the very beginning of the process, through an energetically lower favorable structure. Against the background of all the tested substances, the 5c derivative stands out the most, for which the polymerization start temperature is significantly lower than for the rest of the tested substances. However, the temperature peak for this substance is slightly lower than for the other representatives of this group, and the polymerization end temperature remains the same for all. We suggest that the reason for this may be the fact that the hexyl alkyl chain strains the cation structure significantly and the effect of ethylene glycol is not able to compensate for this phenomenon, so the molecule undergoes a slight disintegration, generating active initiators of resin polymerization, especially at the beginning after supplying temperature to the system, which can be seen by the recorded low polymerization start temperature and high process energy.

3.4. Hardness of the Materials

For additional characterization of the cured materials obtained, a simple measurement of the hardness of the resin after curing was chosen. Such a parameter allows one to quickly determine the effect of the tested substances on the final product and define certain regularities. The results showed that with the increase in the length of the carbon chain in the cation structure, the brittleness of the obtained material increases, which in some cases made it impossible to measure the hardness. We suggest that such a hardness test result correlates with the values of water absorbed by ionic liquids, and thus with the level of hygroscopicity of a particular ionic liquid. Moreover, as a rule, ionic liquids with TFSI anions are hydrophobic in nature, while dicyanamides are representatives of mainly hydrophilic compounds and it is for this group of compounds that information about significant brittleness has been reported more often. In addition, the brittle effect of the structure can sometimes also be caused by the presence of ethylene glycol.

In conclusion, it was proved during research that by using the first stage of the Delepine reaction, leading to the production of hexamethylenetetramine quaternary ammonium salts, it was possible to obtain a series of new ionic liquids that can be successfully used in epoxy resin curing mixtures operating only at elevated temperatures. What is more, the use of ethylene glycol as a component of such mixtures is presented, which simultaneously performs many functions important for the proposed area of application as well as for the obtained ionic liquids, especially in terms of their stability.

4. MATERIALS AND METHODS

4.1. Materials

All reagents were purchased from suppliers: (i) Merck (alkylating agents, hexamethylenetetramine, triethylenetetramine, isophorone diamine, lithium bis(trifluoromethylsulfonyl)imide, sodium dicyanamide, ethylene glycol, organic solvents, reagents necessary to perform analyzes), and (ii) abcr GmbH (bisphenol A diglycidyl ether, DGEBA, silver nitrate). The minimum purity for alkylating agents, amines (beyond triethylenetetramine—97%) and compounds used for metathesis is 99%, while all solvents had a minimum HPLC grade. DGEBA was 85% purity.

4.2. Synthesis

4.2.1. Bromides

To obtain products 1a–5a, a solution of hexamethylenetetramine (1.0 eq.) in chloroform was placed in a two-necked round-bottomed flask equipped with a reflux condenser with simultaneous stirring on magnetic stirrer. Then, corresponding alkyl bromide (1.0 eq.) was added dropwise at room temperature. After that, a white precipitate was observed immediately. After the time required to complete the reaction (controlled with IC analysis), a large amount of white precipitate was filtered through a sintered funnel, then washed with hexane (3 × 10.0 mL) and diethyl ether (3 × 10.0 mL). Products were dried under vacuum (24 h) and recrystallized from a mixture of methanol and diethyl ether to furnish pure compounds.

Compound 1a

Product form: white solid; yield: 99.6%; water content: 0.47 wt%; IC: $t_R = 4.67$ min (cation), 4.14 min (anion), Purity: 99.6%; ^{1}H NMR (400 MHz, D_2O): $\delta = 5.05$ (s, 6H), 4.69 (d, $J = 12.4$ Hz, 3H), 4.54 (d, $J = 12.5$ Hz, 3H), 2.98 (q, $J = 7.5$ Hz, 2H), 1.25 (t, $J = 7.4$ Hz, 3H); ^{13}C {^{1}H} NMR (100 MHz, D_2O): $\delta = 77.72$, 70.14, 52.66, 4.95; MS (ESI): m/z (%) = 169 ([M]$^+$, 100), 155 ([HMTA-CH$_2$]$^+$), 140 ([HMTA]$^+$).

Compound 2a

Product form: white solid; yield: 99.0%; water content: 0.17 wt%; IC: $t_R = 4.89$ min (cation), 4.17 min (anion), Purity: 99.0%; ^{1}H NMR (400 MHz, D_2O): $\delta = 5.04$ (s, 6H), 4.63 (d, $J = 14.9$ Hz, 3H), 4.52 (d, $J = 13.4$ Hz, 3H), 2.81 (m, 2H), 1.67 (ddd, $J =$ 15.3, 7.4, 5.3 Hz, 2H), 0.89 (t, $J = 7.3$ Hz, 3H); ^{13}C {^{1}H} NMR (100 MHz, D_2O): 77.65, 70.15, 58.32, 20.77, 13.63; MS (ESI): m/z (%) = 183 ([M]$^+$, 100), 155 ([HMTA-CH$_2$]$^+$), 140 ([HMTA]$^+$).

Compound 3a

Product form: white solid; yield: 99.1%; water content: 0.68 wt%; IC: t_R = 5.12 min (cation), 4.16 min (anion), Purity: 99.1%; ^{1}H NMR (400 MHz, D_2O): δ = 5.06 (s, 6H), 4.65 (d, *J* = 15.5 Hz, 3H), 4.54 (d, *J* = 13.0 HZ, 3H), 2.87 (m, 2H), 1.65 (ddd, *J* = 15.6, 12.6, 7.5 Hz, 2H), 1.32 (dd, *J* = 14.8, 7.4 Hz, 2H), 0.89 (t, *J* = 7.4 Hz, 3H); ^{13}C {^{1}H} NMR (100 MHz, D_2O): 78.13, 70.14, 60.45, 26.82, 19.61, 14.03; MS (ESI): *m/z* (%) = 197 ([M]$^+$, 100), 168 ([HMTA-CH_2-CH_2]$^+$, 155 ([HMTA-CH_2]$^+$), 140 ([HMTA]$^+$).

Compound 4a

Product form: white solid; yield: 99.1%; water content: 0.57 wt%; IC: t_R = 5.78 min (cation), 4.12 min (anion), Purity: 99.1%; ^{1}H NMR (400 MHz, D_2O): δ = 5.05 (s, 6H), 4.66 (d, *J* = 13.5 Hz, 3H), 4.55 (d, *J* = 13.1 Hz, 3H), 2.85 (m, 2H), 1.65 (dd, *J* = 15.4, 7.5 Hz, 2H), 1.31 (m, 4H), 0.83 (t, *J* = 7.1 Hz, 3H); ^{13}C {^{1}H} NMR (100 MHz, D_2O): 77.52, 70.93, 60.43, 31.57, 25.32, 19.70, 13.05; MS (ESI): *m/z* (%) = 211 ([M]$^+$, 100), 182 ([HMTA-CH_2-CH_2-CH_2]$^+$, 168 ([HMTA-CH_2-CH_2]$^+$, 155 ([HMTA-CH_2]$^+$), 140 ([HMTA]$^+$).

Compound 5a

Product form: white solid; yield: 99.0%; water content: 0.42 wt%; IC: t_R = 6.34 min (cation), 4.10 min (anion), Purity: 99.1%; ^{1}H NMR (400 MHz, D_2O): δ = 5.03 (s, 6H), 4.67 (d, *J* = 13.1 Hz, 3H), 4.51 (d, *J* = 12.7 Hz, 3H), 2.84 (dd, *J* = 11.9, 5.5 Hz, 2H), 1.63 (dd, *J* = 15.8, 7.8 Hz, 2H), 1.26 (m, 6H), 0.80 (t, *J* = 6.9 Hz, 3H); ^{13}C {^{1}H} NMR (100 MHz, D_2O): 78.12, 70.12, 57.32, 30.25, 25.47, 21.64, 19.10, 13.15; MS (ESI): *m/z* (%) = 225 ([M]$^+$, 100), 182 ([HMTA-CH_2-CH_2-CH_2]$^+$, 155 ([HMTA-CH_2]$^+$), 140 ([HMTA]$^+$).

4.2.2. Bis(trifluoromethane)sulfonimides

Compounds 1b–5b were obtained according to a slightly modified, previously reported procedure. Corresponding bromide (1a–5a, 1.0 eq.) was dissolved in deionized water (100.0 mL) and transferred into a separating funnel. Then, methylene chloride was added followed by an aqueous solution of lithium bis(trifluoromethylsulfonyl)imide (80%, 1.1 eq.). The obtained mixture was vigorously shook for a few minutes and then the separated organic layer was washed with deionized water (3 × 20.0 mL) and then deionized water with small portion of lithium bis(trifluoromethylsulfonyl)imide (80%, 0.1 eq., 1 × 20.0 mL). The washing procedure was finished in an absence of bromide anion in the water layer (silver nitrate test, and washed until no observable yellow silver bromide precipitation). To obtain pure product (1b–5b), methylene chloride from the organic layer was evaporated under reduced pressure and bis(trifluoromethylsulfonyl)imide ionic liquid was dried under vacuum (24 h).

Compound 1b

Product form: slightly yellow liquid; yield: 98.8%; water content: 0.31 wt%; IC: t_R = 4.57 min (cation), 18.58 min (anion), Purity: 99.8%.

Compound 2b

Product form: yellow liquid; yield: 97.7 %; water content: 0.23 wt%; IC: t_R = 4.94 min (cation), 18.71 min (anion), Purity: 99.6%.

Compound 3b

Product form: transparent liquid; yield: 99.1%; water content: 0.28 wt%; IC: t_R = 5.19 min (cation), 18.73 min (anion), Purity: 99.5%.

Compound 4b

Product form: yellow liquid; yield: 98.9%; water content: 0.49 wt%; IC: t_R = 5.81 min (cation), 18.69 min (anion), Purity: 99.1%

Compound 5b

Product form: yellow, viscous liquid; yield: 97.6%; water content: 0.38 wt%; IC: t_R = 6.31 min (cation), 18.71 min (anion), Purity: 99.1%.

4.2.3. Dicyanamides

Compounds 1c–5c were obtained according to a slightly modified previous reported procedure. A precisely weighed amount of corresponding bromide (1a–5a, 1.0 eq.) was placed in the flask and dissolved in deionized water (30.0 mL). Then, freshly prepared silver dicyanamide was added (1.0 eq.) and the mixture was stirred on a magnetic stirrer, protected from a sunlight. The exchange process was monitored by ion chromatography. After completion of the reaction, the yellow precipitate of silver bromide was separated on a sintered funnel using a glass fiber filter, and washed with water (3 × 15.0 mL). The filtrate was evaporated in a vacuum evaporator, and then the obtained compound was purified by dissolution in anhydrous methylene chloride and repeated filtration. To obtain pure product (1c–5c), methylene chloride was evaporated under reduced pressure and dicyanamide ionic liquid was dried under vacuum (24 h).

Compound 1c

Product form: yellow solid; yield: 96.2%; water content: 0.73 wt%; IC: t_R = 4.59 min (cation), 6.66 min (anion), Purity: 98.1%.

Compound 2c

Product form: slightly yellow solid; yield: 87.9%; water content: 0.43 wt%; IC: t_R = 4.92 min (cation), 6.59 min (anion), Purity: 97.9%.

Compound 3c

Product form: yellow solid; yield: 97.5%; water content: 0.86 wt%; IC: t_R = 5.16 min (cation), 6.58 min (anion), Purity: 98.1%.

Compound 4c

Product form: yellow, viscous solid; yield: 93.9%; water content: 0.69 wt%; IC: t_R = 5.83 min (cation), 6.63 min (anion), Purity: 97.8%.

Compound 5c

Product form: yellow, viscous solid; yield: 97.0%; water content: 0.33 wt%; IC: t_R = 6.29 min (cation), 6.60 min (anion), Purity: 97.5%.

4.3. Ion Chromatography

All experiments were conducted with the use of ion chromatography, both regarding the control of the reaction course and the determination of the purity of the obtained compounds, which were performed using Eco-IC apparatus (Metrohm, Herisau, Switzerland) in accordance with the procedures published in previous works [12,21].

4.4. Thermal Analysis

4.4.1. Differential Scanning Calorimetry (DSC) for ILs

Melting point/glass transition/decomposition temperature analyses were determined using the DSC 3 apparatus (Mettler Toledo, Greifensee, Switzerland), according to the method reported in previous work [12,21] and described in Supplementary Materials.

4.4.2. Differential Scanning Calorimetry (DSC) for Epoxy Systems

The cross-linking process of the epoxy resin was tested using a DSC 8500 apparatus (PerkinElmer, Waltham, MA, USA), according to a slightly modified method described in the previous work [21] and described in Supplementary Materials.

4.5. Water Content

Determination of the water content in the obtained compounds was carried out using the Karl Fischer titration apparatus (Metrohm, Herisau, Switzerland) in accordance with the method described in previous works [12,21].

4.6. NMR Analysis

NMR spectra were recorded using a 400 MHz (^{1}H) and 100 MHz (^{13}C) Ascend spectrometer (Bruker, Billerica, MA, USA) in commercially available D_2O solvent. The value of standard measurement uncertainty did not exceed ±0.02 ppm.

4.7. ESI-MS Analysis

ESI-MS analysis were conducted using a HPLC/MS apparatus (Waters Corporation, Milford, MA, USA), according to the procedure described in the previous work [12].

4.8. Curing Process

All experiments regarding the curing process were performed in accordance with the procedure described in the previous work [21]. For microscopic documentation of cured samples see the Supplementary Materials.

4.9. Hardness Measurments

All experiments regarding the hardness measurements were performed in accordance with the procedure described in the previous work [21].

Author Contributions

Conceptualization, D.Z. and A.S.; Formal analysis, D.Z.; Funding acquisition, M.S.; Investigation, D.Z., A.S. and P.H.; Methodology, D.Z. and A.S.; Software, D.Z. and A.S.; Supervision, M.S.; Validation, D.Z.; Visualization, D.Z.; Writing—original draft, D.Z.; Writing—review & editing, D.Z., A.S. and M.S. All authors have read and agreed to the published version of the manuscript.

REFERENCES

1. Jin, F.-L.; Li, X.; Park, S.-J. Synthesis and Application of Epoxy Resins: A Review. J. Ind. Eng. Chem. 2015, 29, 1–11.
2. Rafique, I.; Kausar, A.; Muhammad, B. Epoxy Resin Composite Reinforced with Carbon Fiber and Inorganic Filler: Overview on Preparation and Properties. Polym.-Plast. Technol. Eng. 2016, 55, 1653–1672.
3. Rezaei Motlagh, S.; Harun, R.; Awang Biak, D.R.; Hussain, S.A.; Wan Ab Karim Ghani, W.A.; Khezri, R.; Wilfred, C.D.; Elgharbawy, A.A.M. Screening of Suitable Ionic Liquids as Green Solvents for Extraction of Eicosapentaenoic Acid (EPA) from Microalgae Biomass Using COSMO-RS Model. Molecules 2019, 24, 713.
4. Khoo, K.S.; Tan, X.; Ooi, C.W.; Chew, K.W.; Leong, W.H.; Chai, Y.H.; Ho, S.-H.; Show, P.L. How Does Ionic Liquid Play a Role in Sustainability of Biomass Processing? J. Clean. Prod. 2021, 284, 124772.
5. Rieland, J.M.; Love, B.J. Ionic Liquids: A Milestone on the Pathway to Greener Recycling of Cellulose from Biomass. Resour. Conserv. Recycl. 2020, 155, 104678.
6. Wolny, A.; Chrobok, A. Silica-Based Supported Ionic Liquid-like Phases as Heterogeneous Catalysts. Molecules 2022, 27, 5900.
7. Maciejewski, H. Ionic Liquids in Catalysis. Catalysts 2021, 11, 367.
8. Seitkalieva, M.M.; Samoylenko, D.E.; Lotsman, K.A.; Rodygin, K.S.; Ananikov, V.P. Metal Nanoparticles in Ionic Liquids: Synthesis and Catalytic Applications. Coord. Chem. Rev. 2021, 445, 213982.
9. Mota, F.A.R.; Pereira, S.A.P.; Araujo, A.R.T.S.; Saraiva, M.L.M.F.S. Evaluation of Ionic Liquids and Ionic Liquids Active Pharmaceutical Ingredients Inhibition in Elastase Enzyme Activity. Molecules 2021, 26, 200.

10. Faísca, F.; Correia, V.; Petrovski, Ž.; Branco, L.C.; Rebelo-de-Andrade, H.; Santos, M.M. Enhanced In Vitro Antiviral Activity of Hydroxychloroquine Ionic Liquids against SARS-CoV-2. Pharmaceutics 2022, 14, 877.

11. Reddy, A.V.B.; Rafiq, R.; Ahmad, A.; Maulud, A.S.; Moniruzzaman, M. Cross-Linked Ionic Liquid Polymer for the Effective Removal of Ionic Dyes from Aqueous Systems: Investigation of Kinetics and Adsorption Isotherms. Molecules 2022, 27, 7775.

12. Zajac, A.; Szpecht, A.; Zielinski, D.; Rola, K.; Hoppe, J.; Komorowska, K.; Smiglak, M. Synthesis and Characterization of Potentially Polymerizable Amine-Derived Ionic Liquids Bearing 4-Vinylbenzyl Group. J. Mol. Liq. 2019, 283, 427–439.

13. Maksym, P.; Tarnacka, M.; Bielas, R.; Hachuła, B.; Zajac, A.; Szpecht, A.; Smiglak, M.; Kaminski, K.; Paluch, M. Structure-Property Relationships of Tailored Imidazolium- and Pyrrolidinium-Based Poly(Ionic Liquid)s. Solid-like vs. Gel-like Systems. Polymer 2020, 192, 122262.

14. Ferdeghini, C.; Guazzelli, L.; Pomelli, C.S.; Ciccioli, A.; Brunetti, B.; Mezzetta, A.; Vecchio Ciprioti, S. Synthesis, Thermal Behavior and Kinetic Study of N-Morpholinium Dicationic Ionic Liquids by Thermogravimetry. J. Mol. Liq. 2021, 332, 115662.

15. Kazemiabnavi, S.; Zhang, Z.; Thornton, K.; Banerjee, S. Electrochemical Stability Window of Imidazolium-Based Ionic Liquids as Electrolytes for Lithium Batteries. J. Phys. Chem. B 2016, 120, 5691–5702.

16. Barulli, L.; Mezzetta, A.; Brunetti, B.; Guazzelli, L.; Vecchio Ciprioti, S.; Ciccioli, A. Evaporation Thermodynamics of the Tetraoctylphosphonium Bis(Trifluoromethansulfonyl)Imide([P8888]NTf2) and Tetraoctylphosphonium Nonafluorobutane-1-Sulfonate ([P8888]NFBS) Ionic Liquids. J. Mol. Liq. 2021, 333, 115892.

17. Gond, R.; van Ekeren, W.; Mogensen, R.; Naylor, A.J.; Younesi, R. Non-Flammable Liquid Electrolytes for Safe Batteries. Mater. Horiz. 2021, 8, 2913–2928.

18. Rahmathullah, M.A.M.; Jeyarajasingam, A.; Merritt, B.; VanLandingham, M.; McKnight, S.H.; Palmese, G.R. Room Temperature Ionic Liquids as Thermally Latent Initiators for Polymerization of Epoxy Resins. Macromolecules 2009, 42, 3219–3221.

19. Binks, F.C.; Cavalli, G.; Henningsen, M.; Howlin, B.J.; Hamerton, I. Investigating the Mechanism through Which Ionic Liquids Initiate the Polymerisation of Epoxy Resins. Polymer 2018, 139, 163–176.

20. Binks, F.C.; Cavalli, G.; Henningsen, M.; Howlin, B.J.; Hamerton, I. Examining the Kinetics of the Thermal Polymerisation Behaviour of Epoxy Resins Initiated with a Series of 1-Ethyl-3-Methylimidazolium Based Ionic Liquids. Thermochim. Acta 2018, 663, 19–26.

21. Zielinski, D.; Szpecht, A.; Hinc, P.; Maciejewski, H.; Smiglak, M. Mono N-Alkylated DABCO-Based Ionic Liquids and Their Application as Latent Curing Agents for Epoxy Resins. ACS Appl. Polym. Mater. 2021, 3, 5481–5493.

22. Al Hokayem, K.; El Hage, R.; Svecova, L.; Otazaghine, B.; Le Moigne, N.; Sonnier, R. Flame Retardant-Functionalized Cotton Cellulose Using Phosphonate-Based Ionic Liquids. Molecules 2020, 25, 1629.

23. Sonnier, R.; Dumazert, L.; Livi, S.; Nguyen, T.K.L.; Duchet-Rumeau, J.; Vahabi, H.; Laheurte, P. Flame Retardancy of Phosphorus-Containing Ionic Liquid Based Epoxy Networks. Polym. Degrad. Stab. 2016, 134, 186–193.
24. Xu, Y.-J.; Shi, X.-H.; Lu, J.-H.; Qi, M.; Guo, D.-M.; Chen, L.; Wang, Y.-Z. Novel Phosphorus-Containing Imidazolium as Hardener for Epoxy Resin Aiming at Controllable Latent Curing Behavior and Flame Retardancy. Compos. Part B Eng. 2020, 184, 107673.
25. Wei, R.; Yang, B.; He, C.; Jin, L.; Zhang, X.; Zhao, C. Versatile and Robust Poly(Ionic Liquid) Coatings with Intelligent Superhydrophilicity / Superhydrophobicity Switch in High-Efficient Oil-Water Separation. Sep. Purif. Technol. 2022, 282, 120100.
26. Zhang, X.; Chen, Q.; Wei, R.; Jin, L.; He, C.; Zhao, W.; Zhao, C. Design of Poly Ionic Liquids Modified Cotton Fabric with Ion Species-Triggered Bidirectional Oil-Water Separation Performance. J. Hazard. Mater. 2020, 400, 123163.
27. Kong, M.; Liu, C.; Tang, B.; Xu, W.; Huang, Y.; Li, G. Improved Mechanical and Thermal Properties of Trifunctional Epoxy Resins through Controlling Molecular Networks by Ionic Liquids. Ind. Eng. Chem. Res. 2019, 58, 8080–8089.
28. Xiao, F.; Wu, K.; Luo, F.; Guo, Y.; Zhang, S.; Du, X.; Zhu, Q.; Lu, M. An Efficient Phosphonate-Based Ionic Liquid on Flame Retardancy and Mechanical Property of Epoxy Resin. J. Mater. Sci. 2017, 52, 13992–14003.
29. Węgrzyn, M.; Rudnik, E.; Kamocka-Bronisz, R.; Kukfisz, B. Mechanical and Thermal Properties of Biocomposites Based on Polyethylene from Renewable Resources Modified with Ionic Liquids. J. Polym. Environ. 2021, 29, 1808–1816.
30. Shi, T.; Livi, S.; Duchet-Rumeau, J.; Gérard, J.-F. Enhanced Mechanical and Thermal Properties of Ionic Liquid Core / Silica Shell Microcapsules-Filled Epoxy Microcomposites. Polymer 2021, 233, 124182.
31. Sanaei-Rad, S.; Saeidiroshan, H.; Mirhosseini-Eshkevari, B.; Ghasemzadeh, M.A. Hexamethylenetetramine-Based Ionic Liquid Anchored onto the Metal–Organic Framework MIL-101(Cr) as a Superior and Reusable Heterogeneous Catalyst for the Preparation of Hexahydroquinolines. Res. Chem. Intermed. 2021, 47, 2143–2159.
32. Mirhosseini-Eshkevari, B.; Ali Ghasemzadeh, M.; Esnaashari, M.; Taghvaei Ganjali, S. Hexamethylenetetramine-Based Ionic Liquid / MIL-101(Cr) Metal–Organic Framework Composite: A Novel and Versatile Tool for the Preparation of Pyrido[2,3-D:5,6-d ']Dipyrimidines. RSC Adv. 2021, 11, 364–373.
33. Song, Y.; Tang, X.; Kong, S.; Bai, L.; He, X.; Meng, F. Synthesis and Characterization of Hexamethylenetetramine-Based Ionic Liquid Crystals. J. Mol. Struct. 2019, 1178, 135–141.
34. Jordan, A.; Huang, S.; Sneddon, H.F.; Nortcliffe, A. Assessing the Limits of Sustainability for the Delépine Reaction. ACS Sustain. Chem. Eng. 2020, 8, 12746–12754.
35. Surrey, A.R. Name Reactions in Organic Chemistry; Elsevier: Amsterdam, The Netherlands, 2013; ISBN 978-1-4832-5868-3.
36. Cheng, S.; Fu, X.; Liu, J.; Zhang, J.; Zhang, Z.; Wei, Y.; Han, B. Study of Ethylene Glycol / TX-100 / Ionic Liquid Microemulsions. Colloids Surf. Physicochem. Eng. Asp. 2007, 302, 211–215.

37. Kaur, M.; Singh, G.; Kumar, S.; Navnidhi; Kang, T.S. Thermally Stable Microemulsions Comprising Imidazolium Based Surface Active Ionic Liquids, Non-Polar Ionic Liquid and Ethylene Glycol as Polar Phase. J. Colloid Interface Sci. 2018, 511, 344–354.

38. Dewan, M.; Kumar, A.; Saxena, A.; De, A.; Mozumdar, S. Using Hydrophilic Ionic Liquid, [Bmim]BF4—Ethylene Glycol System as a Novel Media for the Rapid Synthesis of Copper Nanoparticles. PLoS ONE 2012, 7, e29131.

39. Zhao, X.; Ma, G.; Wu, D.; Cai, P.; Pan, Y. A Novel Strategy to Utilize Ethylene Glycol-Ionic Liquids for the Selective Precipitation of Polysaccharides. J. Sep. Sci. 2019, 42, 1757–1767.

40. Holbrey, J.D.; Visser, A.E.; Spear, S.K.; Reichert, W.M.; Swatloski, R.P.; Broker, G.A.; Rogers, R.D. Mercury(II) Partitioning from Aqueous Solutions with a New, Hydrophobic Ethylene-Glycol Functionalized Bis-Imidazolium Ionic Liquid. Green Chem. 2003, 5, 129–135.

41. Pal, A.; Kumar, B.; Singh Kang, T. Effect of Structural Alteration of Ionic Liquid on Their Bulk and Molecular Level Interactions with Ethylene Glycol. Fluid Phase Equilibria 2013, 358, 241–249.

42. Peng, Q.; Wei, L.; Zhang, X.; Wu, Y.; Mahmood, K.; Liu, Z.; Zhang, L. Direct Polycondensation of L-Lactic Acid in Hydrophobic Bis(Trifluoromethanesulfonyl)Imide-Anionic Ionic Liquids: A Kinetic Study. Eur. Polym. J. 2021, 158, 110692.

43. Zheng, Y.; Xuan, X.; Wang, J.; Fan, M. The Enhanced Dissolution of β-Cyclodextrin in Some Hydrophilic Ionic Liquids. J. Phys. Chem. A 2010, 114, 3926–3931.

44. Kalita, D.J.; Tarnavchyk, I.; Chisholm, B.J.; Webster, D.C. Novel Bio-Based Epoxy Resins from Eugenol as an Alternative to BPA Epoxy and High Throughput Screening of the Cured Coatings. Polymer 2021, 233, 124191.

45. Dreyfors, J.M.; Jones, S.B.; Sayed, Y. Hexamethylenetetramine: A Review. Am. Ind. Hyg. Assoc. J. 1989, 50, 579–585.

46. Hartwig, A.; MAK Commission. Hexamethylenetetramine [MAK Value Documentation, 1993]. In The MAK-Collection for Occupational Health and Safety; John Wiley & Sons, Ltd.: Hoboken, NJ, USA, 2012; pp. 356–372. ISBN 978-3-527-60041-0.

47. Paduszyński, K.; Domańska, U. Viscosity of Ionic Liquids: An Extensive Database and a New Group Contribution Model Based on a Feed-Forward Artificial Neural Network. J. Chem. Inf. Model. 2014, 54, 1311–1324.

48. Stolarska, O.; Pawlowska-Zygarowicz, A.; Soto, A.; Rodríguez, H.; Smiglak, M. Mixtures of Ionic Liquids as More Efficient Media for Cellulose Dissolution. Carbohydr. Polym. 2017, 178, 277–285.

49. Pownceby, M.I.; Jenkins, D.H.; Ruzbacky, R.; Saunders, S. Solubilities of Ammonia and Ammonium Chloride in Ammoniated and Nonammoniated Methanol and Ethylene Glycol between 298 K and 353 K. J. Chem. Eng. Data 2012, 57, 1449–1455.

50. Galvão, A.C.; Francesconi, A.Z. Solubility of Methane and Carbon Dioxide in Ethylene Glycol at Pressures up to 14MPa and Temperatures Ranging from (303 to 423)K. J. Chem. Thermodyn. 2010, 42, 684–688.

7 THERMAL CONDUCTIVITY OF HELIUM AND ARGON AT HIGH PRESSURE AND HIGH TEMPERATURE

Wen-Pin Hsieh [1,2], **Yi-Chi Tsao** [1] **and Chun-Hung Lin** [1]

[1]*Institute of Earth Sciences, Academia Sinica, Nankang, Taipei 11529, Taiwan*

[2]*Department of Geosciences, National Taiwan University, Taipei 10617, Taiwan*

ABSTRACT

Helium (He) and argon (Ar) are important rare gases and pressure media used in diamond-anvil cell (DAC) experiments. Their thermal conductivity at high pressure–temperature (P-T) conditions is a crucial parameter for modeling heat conduction and temperature distribution within a DAC. Here we report the thermal conductivity of He and Ar over a wide range of high P-T conditions using ultrafast time-domain thermoreflectance coupled with an externally heated DAC. We find that at room temperature the thermal conductivity of liquid and solid He shows a pressure dependence of $P^{0.86}$ and $P^{0.72}$, respectively; upon heating the liquid, He at 10.2 GPa follows a $T^{0.45}$ dependence. By contrast, the thermal conductivity of solid Ar at room temperature has a pressure dependence of $P^{1.25}$, while a $T^{-1.37}$ dependence is observed for solid Ar at 19 GPa. Our results not only provide crucial bases for further investigation into the physical mechanisms of heat transport in He and Ar under extremes, but also substantially improve the accuracy of modeling the temperature profile within a DAC loaded with He or Ar. The P-T dependences of the thermal conductivity of He are important to better model and constrain the structural and thermal evolution of gas giant planets containing He.

Keywords: *high pressure; thermal conductivity; helium; argon*

1. INTRODUCTION

Helium (He), behind hydrogen, is the second most-abundant element in our Universe. Physical properties of ^{4}He (the isotope studied in this work) under extreme pressure (P) and temperature (T) conditions are of great interest in order to understand not only the fundamental physics of such simple molecular systems, but also the structure and evolution of gas giant planets, such as Jupiter and Saturn, that could contain ~20–25 wt.% He [1]. As quantum matter, ^{4}He crystalizes in the hexagonal-close-packed (hcp) structure at low and moderate P-T conditions (e.g., ~11.6 GPa at room temperature), except for a narrow P-T range near its melting curve wherein the body-centered-cubic or face-centered-cubic structure is present [2,3]. In addition to its novel properties [1], ^{4}He has also been used as a pressure medium in high P-T diamond-anvil cell (DAC) experiments, as it provides nearly hydrostatic conditions within the sample chamber [4]. In past decades, a number of physical properties of ^{4}He under various P-T conditions have been experimentally and theoretically explored, including the equation of the state and phase diagram [2,3,5,6], melting curve [6,7,8], sound velocity and elastic constants [9,10,11,12,13], refractive index [14], and Raman spectrum [15], etc. Prior studies on the thermal conductivity of He, however, have been limited to ambient or low pressures at a variety of temperature conditions [16,17,18]. A precise determination of He's thermal conductivity at high P-T conditions is critically needed, because it offers an excellent platform from which to explore the physics of thermal transport and phonon interactions in such quantum systems. In addition, it also enables better modeling of He's thermal conductivity under relevant P-T conditions within gas giant planets, advancing our understanding of their thermal and structural evolution dynamics. Furthermore, in high P-T DAC experiments loaded with He as a pressure medium, the thermal conductivity of He at a certain P-T condition is a key parameter to simulate the heat transfer dynamics and temperature distribution within the DAC. Similar to other pressure media [19,20], such a thermal transport property is crucial for understanding various properties of interest in DAC experiments, such as the thermal conductivity of the sample to be studied.

On the other hand, Ar is also a rare gas at ambient conditions and frequently used as a pressure medium in high P-T DAC experiments. Due to its relatively simple interatomic potential, Ar has served as an example material system being used to test the validity of theoretical models. Therefore, various physical properties of Ar under extreme conditions have been extensively studied, see, e.g., [8 ,21,22,23,24,25,26,27,28,29,30]. In contrast to the largely unknown thermal conductivity of He at high P-T conditions, the thermal conductivity of Ar at variable P-T conditions has been investigated experimentally and computationally [24,25,29,30]. The thermal conductivity of Ar over a wide range of P-T conditions was consistently determined by Green-Kubo molecular dynamics simulations [24] and first-principles calculations [30]. Though optical pump–probe measurements within a DAC [25] showed similar P-T dependences to those obtained by the aforementioned computational results, the measured thermal conductivity values are higher. Such a discrepancy may arise from the experimental method or from the interatomic potential or formalism used in the calculations. As a result, a separate, accurate determination of Ar's thermal conductivity at relevant P-T conditions is required to decipher the inconsistency between previous experimental and computational results.

In this paper, we report the thermal conductivity of He and Ar at high P-T conditions, such as up to ~50–55 GPa at 300 K, as well as at ~10–19 GPa up to 973 K. These measurements over a wide range of P-T conditions enable us to track how the thermal conductivity of He and Ar changes, respectively, upon compression and heating. Interestingly, the thermal conductivity of liquid and solid He scales with a pressure dependence of $P^{0.86}$ and $P^{0.72}$, respectively, while that of solid Ar scales with a larger pressure slope, $P^{1.25}$. Over the temperature range we studied (300–973 K), the thermal conductivity of liquid He at 10.2 GPa increases with temperature via a $T^{0.45}$ dependence, whereas that of solid Ar at 19 GPa decreases with temperature via $T^{-1.37}$. Our data for solid Ar are in line with previous theoretical calculations. The present results for the P-T dependences of the thermal conductivity of He and Ar are significantly important and benefit a variety of studies in materials physics under extreme conditions as well as planetary sciences.

2. METHODS

2.1. Sample Preparation

For high-pressure, room-temperature thermal conductivity measurements, we first polished a thin sheet of borosilicate glass (D 263® T eco from Schott AG, Mainz, Germany), serving as a reference substrate, to ~10 μm thick. We then thermally evaporated ~90 nm-thick Al film as the thermal transducer on the glass substrate, and loaded the glass into a symmetric DAC (300 μm culet) equipped with a Re gasket. The glass substrate was compressed by loading with high-pressure gas of He or Ar (both with a purity of 99.9999%) which serves as the material of interest as well as the pressure medium. To characterize the pressure in the sample chamber of the DAC, we also placed a few ruby balls next to the glass substrate and monitored their pressure-shifted fluorescence or Raman spectrum [31]. The uncertainty of the pressure measurements at room temperature was typically < 5%, depending on the pressure range and medium. As the pressure approached~40–50 GPa, we also compared the pressures determined by the Raman spectra of the ruby and diamond anvil [32], where the difference is typically < 2–3 GPa, depending on the pressure medium. The experimental setup and sample geometry are essentially the same as those in previous studies, see, e.g., [19,20,33].

To create simultaneous high P-T conditions, we used an externally heated DAC (EHDAC) [19,34] along with a gas membrane that in situ controls the pressure within the EHDAC during heating. The combination of the EHDAC and gas membrane enabled us to minimize the effects of thermal pressure and study the temperature dependence of the thermal conductivity of Ar and He at a fixed pressure of 19 GPa and 10.2 GPa, respectively, up to ~973 K. Details of the EHDAC assemblage and temperature measurement were described in [19,34].

2.2. Thermal Conductivity Measurements

We employed time-domain thermoreflectance (TDTR) to measure the thermal conductivity of He and Ar at high pressures and variable temperatures. In past decades, TDTR has enabled high-precision thermal conductivity measurements of various materials under a wide range of pressure and temperature conditions, with critical applications to physics, chemistry, materials science, and geosciences [35,36,37,38,39]. It is an ultrafast optical pump–probe method, where the output of a Ti:sapphire laser with a central wavelength of 785 nm and a repetition rate of 80 MHz was split into pump and probe beams. In our setup, the pump beam modulated at 8.7 MHz using an electro-optical modulator first passed through a moving mechanical stage that changed its optical beam path. The pump beam was then focused on the Al film coated on the glass substrate, causing variations in Al's temperature and optical reflectivity. To monitor the heat diffusion dynamics from the Al into the pressure medium and substrate, we focused the probe beam on the Al film and then measured the variation of its reflected intensity, including the in-phase V_{in} and out-of-phase V_{out} components, using a fast Si photodiode followed by a lock-in amplifier. More details of the operation principle and setup of TDTR can be found in [40,41].

TDTR data analysis, thermal model fitting, and determination of the thermal conductivity of the sample of interest have been detailed in literature, see, e.g., [19,20,33,37,38] and references therein. For reference, we show a set of representative TDTR data for solid He at 36.1 GPa and room temperature as well as thermal model fittings in Supplementary Materials Figure S1. Note that in our data analysis, the volumetric heat capacity of each material layer (glass, Al, and He or Ar) within the DAC is an important parameter in the thermal model. Since the volumetric heat capacity of liquid He at high pressure and temperature conditions is not known, it was assumed to be a constant of 4 J cm^{-3} K^{-1}, typical of a liquid at high pressure, e.g., [42]. The volumetric heat capacity of solid He at high pressure and room temperature was taken from [9], and its temperature dependence is estimated following a method described in [33]. Compared to the He and Ar, the relatively low thermal conductivity of borosilicate glass that we used as a reference substrate significantly reduces the uncertainty when deriving the thermal conductivity of He and Ar. It is noteworthy that the major uncertainty in our data is from the analysis uncertainty, not the measurement uncertainty. We estimated that the uncertainties in all the parameters used in our thermal model would propagate ~13% error in the derived thermal conductivity of He and Ar before 30 GPa, and ~20% error at 30–50 GPa. An example of detailed TDTR sensitivity analysis and uncertainty evaluations are shown in Supplementary Materials Figures S1 and S2 and References [43,44].

3. RESULTS AND DISCUSSIONS

Figure 1a presents the pressure dependence of the thermal conductivity $\Lambda(P)$ of He at room temperature. Our measurements started from the liquid phase at 1.3 GPa where $\Lambda = 0.4$ W m^{-1} K^{-1}. Upon compression, the Λ of the liquid phase increases with pressure to 2.5 W m^{-1} K^{-1} at 11.4 GPa. A pressure-induced solidification of He at ~12 GPa leads to an ~30% increase in the thermal conductivity. The phase of the He (liquid or solid) was in situ identified by time-domain stimulated Brillouin scattering [19] (Supplementary Materials Figure S3). Similar to the liquid phase, the Λ of polycrystalline, hexagonal-close-packed (hcp) phase increases monotonically with pressure and reaches 8.8 W m^{-1} K^{-1} at 55.2 GPa. Note that compared to the liquid phase, the Λ of the hcp phase shows a slight scattering among different runs of measurement. Such behavior may be associated with the potential anisotropic thermal conductivity of the hcp phase along different orientations of the polycrystalline sample. If we assume that the thermal conductivity of He can be phenomenologically modelled as $\Lambda_{He}(P) = \alpha_{He}P^n$, where α_{He} is a normalization constant, then by fitting all the data for each phase in a lnΛ–lnP plot, the solid hcp phase has a slightly smaller pressure slope [n_S = 0.72(±0.045)] than the liquid phase [n_L = 0.86(±0.048)], potentially due to the smaller compressibility and anharmonicity in the solid phase.

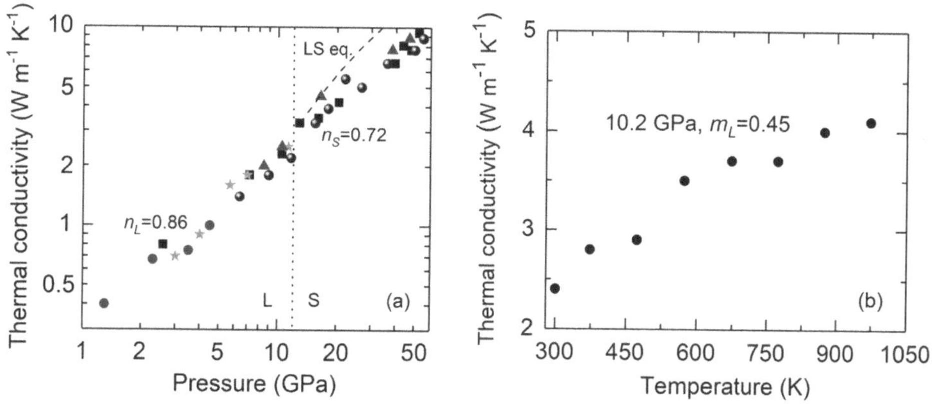

Figure 1.: *(a) Thermal conductivity of He at high pressure and room temperature. Several runs of measurements (represented by different symbols) show consistent results. In both liquid (L) and solid (S) phase, the thermal conductivity increases rapidly with pressure, with a slight increase of ~30% across the liquid–solid transition at ~12 GPa (vertical dotted line). The increasing rate of thermal conductivity with pressure in the solid phase [n_S = 0.72(±0.045)] is slightly smaller than that in the liquid phase [n_L = 0.86 (±0.048)]. The uncertainty is typically ~13% before 30 GPa and ~20% at 30–50 GPa. Compared to our experimental data, the LS equation (dashed curve) predicts a higher thermal conductivity of solid He with a larger pressure slope. (b) Temperature dependence of the thermal conductivity of liquid He at 10.2 GPa. The thermal conductivity of the liquid phase (with an analysis uncertainty of ~15%) scales with T^{m_L}, where m_L = 0.45(±0.036).*

To understand how the thermal conductivity of He changes with temperature, we have also performed simultaneous high P-T measurements (10.2 GPa and 300–973 K) on the liquid phase, see Figure 1b. (We did not perform high P-T measurements on the solid phase due to its relatively low melting line [7], where the limited heating range for the solid phase makes a precise determination of the temperature dependence difficult.) Interestingly, the Λ of the liquid phase increases with temperature, consistent with the results at ambient and lower pressures [17,18]. Such behavior, however, is in contrast to the temperature dependence of melts and glasses at higher temperatures [45] and crystalline dielectrics (see, e.g., [46] and Ar below), where their Λ typically decreases as temperature increases. Assuming the thermal conductivity of liquid He scales with T^{m_L}, the exponent value m_L is then determined to be 0.45(±0.036) by a linear slope in its lnΛ-lnT plot. Combined with previous temperature-dependence measurements at ambient and lower pressures [17,18], the temperature slope is found to decrease as the pressure increases.

The thermal conductivity of the Ar, on the other hand, shows distinct pressure and temperature dependences compared to the He. Figure 2a presents the $\Lambda(P)$ of Ar at room temperature up to 49 GPa (solid circles). Here, we focus on the solid phase where its $\Lambda = 0.6$ W m^{-1} K^{-1} at 2.1 GPa, typical of the relatively low thermal conductivity of molecular solids at similar pressures, see, e.g., CO_2 [47] and CH_4 [48]. The Λ increases rapidly with pressure to 30 W m^{-1} K^{-1} at 49 GPa, with a pressure slope of $\alpha = 1.25(\pm0.025)$ in the lnΛ–lnP plot. Our data for the values of $\Lambda(P)$ at room temperature and the pressure slope (1.25) are both in good agreement with previous theoretical calculations based on Green-Kubo molecular dynamics simulations ($\alpha = 1.29$) [24] and the Boltzmann transport equation ($\alpha = 1.39$) [30]. Nevertheless, our values for $\Lambda(P)$ are ~50–60% smaller over the pressure range we studied than those calculated by kinetic theory [24] and measured by a similar optical pump–probe technique [25]. In addition, we have further performed high P-T measurements to obtain the $\Lambda(T)$ of Ar, see Figure 2b. Two sets of data at 15.8 GPa and 19 GPa up to 973 K consistently show a temperature slope of $\beta = -1.26(\pm0.01)$ and $-1.37(\pm0.005)$ in the lnΛ–lnT plot, well in line with the temperature dependence reported in literature (β ranges from −1.31 to −1.4) [24,25,30]. The deviation of such temperature dependences from the typical T^{-1} has been experimentally observed in solid Ar at ambient [29] and high pressures [25]; such behavior was proposed to be associated with the effects of thermal expansion on the phonon vibrational frequencies [29]. Our results also suggest that the higher pressure may slightly enhance such deviation (from −1.26 to −1.37 when pressure increases from 15.8 GPa to 19 GPa), as discussed in [25].

Note that the opposite temperature dependence of thermal conductivity for liquid He and crystalline Ar originates from different physical mechanisms. The positive temperature dependence for the liquid He is due to the fact that, in liquids, the thermal energy is typically transported by the diffusion of heat from atom to atom, based on the so-called minimum thermal conductivity model, see [49] and references therein. With higher temperature, the stronger vibration of atoms enhances the heat capacity and efficiency of energy transfer between atoms, thus enhancing the thermal conductivity. Moreover, the observation that the pressure decreases the temperature slope for liquid He may be correlated with the increasing fraction of wave-like vibrations (propagons, analogue of the phonons in crystals [49]) that start to contribute to the heat transfer upon compression. The high pressure not only increases the atomic density of the liquid, but also enhances the interatomic interactions. Thus, under compression, the heat transfer mechanism could have transitioned from being primarily the local vibration of atoms, to the combination of it with wave-like vibrations that could propagate through the material. As such, the higher pressure would enhance the scattering of the propagating wave-like vibrations, and therefore suppress the increase in thermal conductivity with temperature (smaller temperature slope). On the other hand, the negative temperature dependence of crystalline Ar is a typical behavior for crystalline materials, where the heat is primarily transported through the phonon scatterings. When the temperature increases, the scattering effects are enhanced, and thus the thermal conductivity is reduced.

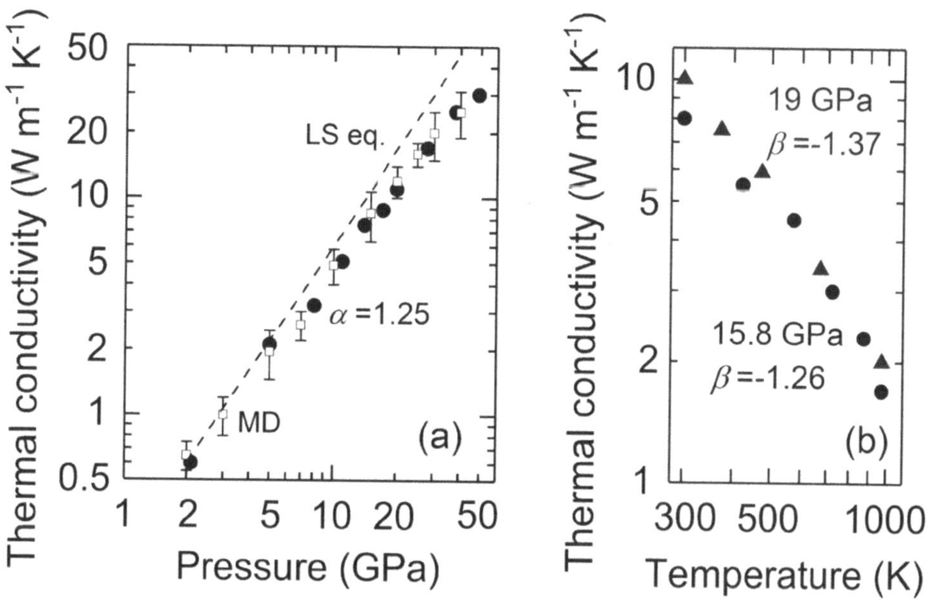

Figure 2.: Thermal conductivity of Ar at (a) high pressure, room temperature conditions and (b) high pressure–temperature conditions. At room temperature, the thermal conductivity (solid circles) increases with pressure with a slope $\alpha = 1.25$ (±0.025) in the lnΛ–lnP plot. The open squares and dashed line in (a) are results from Green-Kubo molecular dynamics (MD) simulations [24] and the LS equation. Upon heating, the thermal conductivity decreases with temperature following a dependence of T^{β}, where $\beta = -1.26$ (±0.01) and −1.37 (±0.005) at 15.8 GPa (black circles) and 19 GPa (blue triangles), respectively.

A number of phenomenological models have been proposed to describe how the thermal conductivity of a material changes upon compression at variable temperatures. For instance, under the Debye approximation [50,51], thermal conductivity can be expressed as a function of density ρ and temperature ρ:

$$\Lambda(\rho, T) = \Lambda_0 \left(\frac{\rho}{\rho_0}\right)^g \left(\frac{T_0}{T}\right)^n, \tag{1}$$

where the Λ0, ρ0, and ρ0 are the reference thermal conductivity, density, and temperature, respectively. Note that the exponent $g = 3\gamma + 2q - 1/3$ [50,51], where the Grüneisen parameter $\gamma = (\partial \ln \nu / \partial \ln \varrho)_{T}$, ν is the phonon vibrational frequency, and $q = -(\partial \ln \gamma / \partial \ln \rho)_T$. Based on our experimental results, $n = -0.45$ and about 1.3 for liquid He and solid Ar, respectively, (Figure 1b and Figure 2b). As shown in Figure 3, we derive the g values for liquid He, solid He, and solid Ar by plotting their $\ln(\Lambda/\Lambda_0) - \ln(\rho/\rho_0)$ relations, where the Λ0 and ρ0 are the thermal conductivity (Figure 1a and Figure 2a) and density (from [9,23,52]) at 1.3 GPa, 12.9 GPa, and 2.1 GPa for liquid He, solid He, and solid Ar, respectively. As expected, due to the more harmonic interatomic potentials, the $g = 2.33(\pm 0.15)$ for the solid He, slightly smaller than that of the liquid He, g = 2.57(±0.15). By contrast, the solid Ar has a g value of 5.49(±0.2), which is in reasonable agreement with that obtained by molecular dynamics calculations [24] and comparable to that of typical dielectrics, such as MgO [51] and NaCl [19].

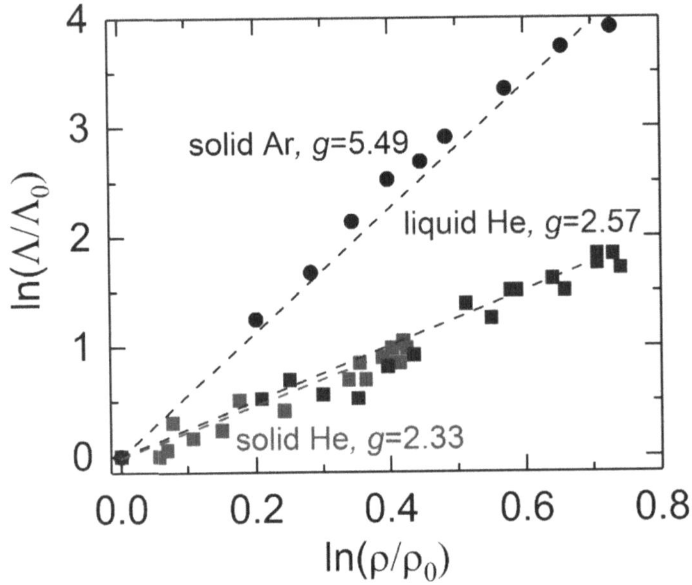

Figure 3.: *Logarithmic normalized thermal conductivity as a function of logarithmic normalized density for liquid He, solid He, and solid Ar. The blue, red, and black symbols are the present results for liquid He, solid He, and solid Ar, respectively. The dashed line for each set of data is a linear fit with a slope (g) of 2.57 (±0.15) (liquid He), 2.33 (±0.15) (solid He), and 5.49 (±0.2) (solid Ar).*

Alternatively, the effects of pressure and temperature on the thermal conductivity Λ of a dielectric is often described by the Leibfried-Schlömann (LS) equation [46,53]:

$$\Lambda = A\frac{V^{1/3}{\omega_D}^3}{\gamma^2 T}, \quad (2)$$

where A is a normalization constant, V the volume, w_D the Debye frequency, r the Grüneisen parameter, and T the temperature. The LS equation is a simple physical model originally formulated for pure, isotropic dielectric crystals. It assumes that heat is predominantly transported by acoustic phonons via the anharmonic three phonon scattering mechanism. Prior studies have shown that the LS equation adequately describes the pressure dependence of Λ for H_2O ice VII [42], MgO [51], solid Ar [25], and NaCl in both the B1 and B2 phase [19], at room temperature and over a large range of volume compression (~20–35%). We here also test the validity of the LS equation on the Λ(P) of solid He and solid Ar (from 2.1 GPa to 49 GPa with a volume compression of ~50%, larger than that in [25]). As described in

[19,42], for simplicity, we assume that the Grüneisen parameter, Poisson ratio, and elastic anisotropy parameter are nearly pressure independent [11,23]. As such, the $\Lambda(P)$ can be simplified to $\Lambda(P) = A'V^{5/6}K_T^{3/2}$ at a constant temperature condition, where K_T is the isothermal bulk modulus and A' is a normalization constant that will be determined by fitting the present thermal conductivity data at 12.9 GPa and 2.1 GPa for solid He and solid Ar, respectively. Using the equation of state and elastic constants of solid He [9] and solid Ar [23], we plot the $\Lambda(P)$ predicted by the LS equation at room temperature for solid He as the dashed line in Figure 1a, and for solid Ar as the dashed line in Figure 2a. In both cases, the LS equation predicts a larger pressure slope, and its overprediction increases with higher pressure. For instance, the Λ of solid He predicted by the LS equation is ~27%, ~42%, and ~78% larger than our data at 16.1 GPa, 20.5 GPa, and 50 GPa, respectively. Moreover, the predicted Λ for solid Ar based on the LS equation is ~4%, ~35%, and ~72% larger than our data at 5 GPa, 11 GPa, and 38 GPa, respectively. Such differences for solid Ar between the LS prediction and our data at 11–49 GPa are much larger than the uncertainties in our data. Since the Grüneisen parameter γ typically decreases with higher pressure [9,24], the overprediction is expected to be further enlarged. The overprediction of the solid Ar's thermal conductivity by the LS equation we find here is at odds with the results reported in [25], where the LS prediction roughly captured the trend of their experimental data at 10–50 GPa. The overprediction of the LS equation for the $\Lambda(P)$ of solid He and solid Ar suggests that besides the three phonon anharmonic scattering between acoustic modes, as pressure increases, additional scattering mechanisms that suppress the phonon transport may play an influential role in heat conduction in these materials. Future, more sophisticated computational studies are required to identify their detailed physics of phonon transport.

4. CONCLUSIONS

We have coupled the ultrafast TDTR and EHDAC to precisely determine the thermal conductivity of He and Ar at high *P-T* conditions. The effects of the density (pressure) and temperature on their thermal conductivity not only shed light on the physical mechanisms of heat transport in these materials, but also enable modelling of their thermal conductivity at further extreme conditions. Our results for liquid He offer critical information to better simulate the thermal conductivity of gas giant planets over a depth range where liquid He is present, such as Jupiter and Saturn which could contain ~20–25 wt.% He and ~75–80 wt.% H_2 in their interiors. Future studies on the high *P-T* thermal conductivity of H_2 will significantly improve our understanding of the fundamental physics of heat transport in such simple, quantum systems as well as the thermal evolution of the aforementioned gas giant planets. In addition to He and Ar, knowledge about the thermal conductivity of Ne will also be required. A comprehensive documentation of the thermal conductivity of these important, common pressure media used in high *P-T* DAC experiments will substantially advance research in Earth and planetary sciences, as well as materials physics and chemistry under extreme conditions.

Supplementary Materials

The following supporting information can be downloaded at: https://www.mdpi.com/article/10.3390/ma15196681/s1, Figure S1: Representative TDTR data for the solid He at 36.1 GPa and room temperature. The temporal evolution of the ratio $-V_{in}/V_{out}$ (open circles) are fitted by the thermal model calculations (color curves). A thermal conductivity of Λ_{He} = 6.6 W m^{-1} K^{-1} (red curve) gives a best-fit to the data using input parameters listed in Table S1. A testing 10% change in Λ (green and blue curves) causes the model calculation deviating from the data. This indicates a high sensitivity of our thermal model fitting and a precisely derived Λ_{He} due to our high-quality data. Figure S2: Sensitivity tests of the thermal model to each input parameter for the solid He at 36.1 GPa. As derived in Figure S1, here the thermal conductivity of He is fixed at 6.6 W m^{-1} K^{-1}. (a) and (b) The thermal model calculations remain the same even when the thicknesses of He (h_{He}) and glass substrate (h_{glass}) change by 50%, respectively. Thus the uncertainties in the h_{He} and h_{glass} have no effects on the thermal conductivity of He (Λ_{He}). (c) The high thermal conductivity of Al film, Λ_{Al}, has a very minor influence on the Λ_{He}, even if a 50% change is applied. (d) If the thermal effusivity of the glass substrate, $e_{g-lass}=(\Lambda_{glassCglass})^{1/2}$, is uncertain by 10%, it slightly shifts the model calculation, and requires Λ_{He} to decrease slightly from 6.6 to 6.2 W m^{-1} K^{-1} to fit the data again, i.e., translating ~6% error to the Λ_{He}. (e) Assuming the volumetric heat capacity of He, C_{He}, has 5% uncertainty, to fit the data the Λ_{He} needs to slightly increase to 6.8 W m^{-1} K^{-1}, ~3% uncertainty. (f) The major uncertainty source in our data analysis is from the uncertainty in Al heat capacity per unit area (product of its volumetric heat capacity and thickness, C_{Al} h_{Al}). Assuming a 5% uncertainty, it requires ~11% shift in the Λ_{He} to fit the data. (g) Even if the laser spot size is changed by 10%, it does not affect the model calculation. (h) A change in the thermal conductance of Al/He interface and Al/glass interface, *G*, by 10% slightly shifts the calculation, having very minor effect on the Λ_{He}. Figure S3: Representative time-domain stimulated Brillouin scattering spectra [19,33] for the liquid He at 7 GPa and solid He at 15 GPa and room temperature. The Brillouin frequency *f* at 7 GPa and 15 GPa was determined to be 17.1 GHz and 25.6 GHz, respectively, which were used to confirm the phase of the He [12]. Table S1: Input parameters for the bi-directional thermal model calculation at 36.1 GPa.

REFERENCES

1. McMahon, J.M.; Morales, M.A.; Pierleoni, C.; Ceperley, D.M. The properties of hydrogen and helium under extreme conditions. Rev. Mod. Phys. 2012, 84, 1607.
2. Loubeyre, P.; Letoullec, R.; Pinceaux, J.P. Equation of state and phase diagram of solid He from single-crystal. Phys. Rev. Lett. 1993, 71, 2272–2275.
3. Mao, H.K.; Hemley, R.J.; Wu, Y.; Jephcoat, A.P.; Finger, L.W.; Zha, C.S.; Bassett, W.A. High-Pressure Phase Diagram and Equation of State of Solid Helium from Single-Crystal X-Ray Diffraction to 23.3 GPa. Phys. Rev. Lett. 1988, 60, 2649.
4. Klotz, S.; Chervin, J.-C.; Munsch, P.; Le Marchand, G. Hydrostatic limits of 11 pressure transmitting media. J. Phys. D. Appl. Phys. 2009, 42, 075413.

5. Lesar, R. Equation of state of dense helium. Phys. Rev. Lett. 1988, 61, 2121–2124.
6. Young, D.A.; McMahan, A.K.; Ross, M. Equation of state and melting curve of helium to very high pressure. Phys. Rev. B 1981, 24, 5119–5127.
7. Santamaría-Pérez, D.; Mukherjee, G.D.; Schwager, B.; Boehler, R. High-pressure melting curve of helium and neon: Deviations from corresponding states theory. Phys. Rev. B—Condens. Matter Mater. Phys. 2010, 81, 214101.
8. Datchi, F.; Loubeyre, P.; LeToullec, R. Extended and accurate determination of the melting curves of argon, helium, ice and hydrogen. Phys. Rev. B 2000, 61, 6535–6546.
9. Zha, C.S.; Mao, H.K.; Hemley, R.J. Elasticity of dense helium. Phys. Rev. B—Condens. Matter Mater. Phys. 2004, 70, 174107.
10. Freiman, Y.A.; Grechnev, A.; Tretyak, S.M.; Goncharov, A.F.; Zha, C.S.; Hemley, R.J. Sound velocities of hexagonal close-packed H2 and He under pressure. Phys. Rev. B 2013, 88, 214501.
11. Grechnev, A.; Tretyak, S.M.; Freiman, Y.A.; Goncharov, A.F.; Gregoryanz, E. Elastic anisotropy and Poisson's ratio of solid helium under pressure. Phys. Rev. B 2015, 92, 024102.
12. Polian, A.; Grimsditch, M. Elastic properties and density of helium up to 20 gpa. Eur. Lett. 1986, 2, 849–855.
13. Herrero, C.P. Compressibility of solid helium. J. Phys. Condens. Matter 2008, 20, 295230.
14. Dewaele, A.; Eggert, J.H.; Loubeyre, P.; Le Toullec, R. Measurement of refractive index and equation of state in dense He, H2, H2O, and Ne under high pressure in a diamond anvil cell. Phys. Rev. B 2003, 67, 094112.
15. Watson, G.H.; Daniels, W.B. Raman scattering from solid helium at high pressure. Phys. Rev. B 1985, 31, 4705–4707.
16. Chase, C.E. Thermal conduction in liquid helium II. I. Temperature dependence. Phys. Rev. 1962, 127, 361–370.
17. Arp, V.; McCarty, R. Thermophysical properties of helium-4 from 0.8 to 1500 K with pressures to 2000 MPa. NIST Tech. Note 1989, 1334, 101–131.
18. Blais, N.C.; Mann, J.B. Thermal conductivity of helium and hydrogen at high temperatures. J. Chem. Phys. 1960, 32, 1459–1465.
19. Hsieh, W.-P. High-pressure thermal conductivity and compressional velocity of NaCl in B1 and B2 phase. Sci. Rep. 2021, 11, 21321.
20. Hsieh, W.-P. Thermal conductivity of methanol-ethanol mixture and silicone oil at high pressures. J. Appl. Phys. 2015, 117, 235901.
21. Ross, M.; Mao, H.K.; Bell, P.; Xu, J.A. The equation of state of dense argon: A comparison of shock and static studies. J. Chem. Phys. 1986, 85, 1028.
22. Grimsditch, M.; Loubeyre, P.; Polian, A. Brillouin scattering and three-body forces in argon at high pressures. Phys. Rev. B 1986, 33, 7192–7200.
23. Shimizu, H.; Tashiro, H.; Kume, T.; Sasaki, S. High-pressure elastic properties of solid argon to 70 GPa. Phys. Rev. Lett. 2001, 86, 4568–4571.

24. Tretiakov, K.V.; Scandolo, S. Thermal conductivity of solid argon at high pressure and high temperature: A molecular dynamics study. J. Chem. Phys. 2004, 121, 11177–11182.
25. Goncharov, A.F.; Wong, M.; Allen Dalton, D.; Ojwang, J.G.O.; Struzhkin, V.V.; Konôpková, Z.; Lazor, P. Thermal conductivity of argon at high pressures and high temperatures. J. Appl. Phys. 2012, 111, 112609.
26. Gewurtz, S.; Stoicheff, B. Elastic constant of Ar single crystal by Brillouin scattering. Phys. Rev. B 1974, 10, 3487.
27. Freiman, Y.A.; Goncharov, A.F.; Tretyak, S.M.; Grechnev, A.; Tse, J.S.; Errandonea, D.; Mao, H.K.; Hemley, R.J. Raman scattering in hcp rare gas solids under pressure. Phys. Rev. B 2008, 78, 014301.
28. Finger, L.; Hazen, R.; Zou, G.; Mao, H.K.; Bell, P. Structure and compression of crystalline argon and neon at high pressure and room temperature. Appl. Phys. Lett. 1981, 39, 892.
29. Christen, D.; Pollack, G. Thermal conductivity of solid argon. Phys. Rev. B 1975, 12, 3380.
30. Chernatynskiy, A.; Phillpot, S.R. Thermal conductivity of argon at high pressure from first principles calculations. J. Appl. Phys. 2013, 114, 064902.
31. Dewaele, A.; Loubeyre, P.; Mezouar, M. Equations of state of six metals above 94 GPa. Phys. Rev. B 2004, 70, 094112.
32. Akahama, Y.; Kawamura, H. High-pressure Raman spectroscopy of diamond anvils to 250 GPa: Method for pressure determination in the multimegabar pressure range. J. Appl. Phys. 2004, 96, 3748–3751.
33. Hsieh, W.-P.; Chen, B.; Li, J.; Keblinski, P.; Cahill, D.G. Pressure tuning of the thermal conductivity of the layered muscovite crystal. Phys. Rev. B 2009, 80, 180302.
34. Lai, X.; Zhu, F.; Zhang, J.S.; Zhang, D.; Tkachev, S.; Prakapenka, V.B.; Chen, B. An externally-heated diamond anvil cell for synthesis and single-crystal elasticity determination of Ice-VII at high pressure-temperature conditions. J. Vis. Exp. 2020, 160, e61389.
35. Cahill, D.G.; Ford, W.K.; Goodson, K.E.; Ford, W.K.; Maris, H.J. Nanoscale thermal transport. J. Appl. Phys. 2003, 93, 793.
36. Cahill, D.G.; Braun, P.V.; Chen, G.; Clarke, D.R.; Fan, S.; Goodson, K.E.; Keblinski, P.; King, W.P.; Mahan, G.D.; Majumdar, A.; et al. Nanoscale thermal transport. II. 2003–2012. Appl. Phys. Rev. 2014, 1, 011305.
37. Hsieh, W.P.; Goncharov, A.F.; Labrosse, S.; Holtgrewe, N.; Lobanov, S.S.; Chuvashova, I.; Deschamps, F.; Lin, J.F. Low thermal conductivity of iron-silicon alloys at Earth's core conditions with implications for the geodynamo. Nat. Commun. 2020, 11, 3332.
38. Hsieh, W.-P.; Deschamps, F.; Okuchi, T.; Lin, J.-F. Effects of iron on the lattice thermal conductivity of Earth's deep mantle and implications for mantle dynamics. Proc. Natl. Acad. Sci. USA 2018, 115, 4099.
39. Zhou, Y.; Dong, Z.Y.; Hsieh, W.P.; Goncharov, A.F.; Chen, X.J. Thermal conductivity of materials under pressure. Nat. Rev. Phys. 2022, 4, 319–335.

40. Kang, K.; Koh, Y.K.; Chiritescu, C.; Zheng, X.; Cahill, D.G. Two-tint pump-probe measurements using a femtosecond laser oscillator and sharp-edged optical filters. Rev. Sci. Instrum. 2008, 79, 114901.
41. Cahill, D.G. Analysis of heat flow in layered structures for time-domain thermoreflectance. Rev. Sci. Instrum. 2004, 75, 5119–5122.
42. Chen, B.; Hsieh, W.-P.; Cahill, D.G.; Trinkle, D.R.; Li, J. Thermal conductivity of compressed H2O to 22 GPa: A test of the Leibfried-Schlömann equation. Phys. Rev. B 2011, 83, 132301.
43. Cahill, D.G.; Watanabe, F. Thermal conductivity of isotopically pure and Ge-doped Si epitaxial layers from 300 to 550 K. Phys. Rev. B 2004, 70, 235322.
44. Zheng, X.; Cahill, D.G.; Krasnochtchekov, P.; Averback, R.S.; Zhao, J.C. High-throughput thermal conductivity measurements of nickel solid solutions and the applicability of the Wiedemann-Franz law. Acta Mater. 2007, 55, 5177–5185.
45. Tikunoff, D.; Spera, F. Thermal conductivity of molten and glassy NaAlSi3O8, CaMgSi2O6, and Mg2SiO4 by non-equilibrium molecular dynamics at elevated temperature and pressure. Am. Mineral. 2014, 99, 2328–2336.
46. Roufosse, M.; Klemens, P.G. Thermal Conductivity of Complex Dielectric Crystals. Phys. Rev. B 1973, 7, 5379.
47. Shieh, S.R.; Hsieh, W.P.; Tsao, Y.C.; Crisostomo, C.; Hsu, H. Low Thermal Conductivity of Carbon Dioxide at High Pressure: Implications for Icy Planetary Interiors. J. Geophys. Res. Planets 2022, 127, e2022JE007180.
48. Meyer, D.W.; Hsieh, W.P.; Hsu, H.; Kuo, C.Y.; Lin, J.F. Thermal Conductivity and Compressional Velocity of Methane at High Pressure: Insights Into Thermal Transport Properties of Icy Planet Interiors. J. Geophys. Res. Planets 2022, 127, e2021JE007059.
49. Hsieh, W.-P.; Losego, M.D.; Braun, P.V.; Shenogin, S.; Keblinski, P.; Cahill, D.G. Testing the minimum thermal conductivity model for amorphous polymers using high pressure. Phys. Rev. B 2011, 83, 174205.
50. de Koker, N. Thermal conductivity of MgO periclase at high pressure: Implications for the D″ region. Earth Planet. Sci. Lett. 2010, 292, 392–398.
51. Dalton, D.A.; Hsieh, W.-P.; Hohensee, G.T.; Cahill, D.G.; Goncharov, A.F. Effect of mass disorder on the lattice thermal conductivity of MgO periclase under pressure. Sci. Rep. 2013, 3, 2400.
52. Le Toullec, R.; Loubeyre, P.; Pinceaux, J.P. Refractive-index measurements of dense helium up to 16 GPa at T=298 K: Analysis of its thermodynamic and electronic properties. Phys. Rev. B 1989, 40, 2368–2378.
53. Slack, G.A. Solid State Physics; Academic: New York, NY, USA, 1979; Volume 34, p. 35.

8

SYNTHESIS AND CHARACTERIZATION OF NOVEL MOLYBDOVANADOPHOSPHORIC ACID SUPPORTED KAOLIN HYBRID CATALYST FOR CHROMOTROPE 2R DYE DEGRADATION IN WATER

Taral Patel[1], Vishal J. Mayani[2] and Suranjana V. Mayani[3]

[1]*Department of Physical Chemistry and Technology of Polymers, Faculty of Chemistry, Silesian University of Technology, Ul. M. Strzody 9, 44-100, Gliwice, Poland*

[2]*Hansgold ChemDiscovery Center (HCC), Hansgold ChemDiscoveries Pvt. Ltd., Rajkot, Gujarat, India*

[3]*Department of Chemistry, Marwadi University, Rajkot-Morbi Road, P.O. Gauridad, Rajkot, Gujarat, 360003, India*

ABSTRACT

The present study provides a complete perspective of recyclable molybdovanadophosphoric acid immobilized on kaolin (MVPA/Kaolin), a hybrid material for catalytic degradation of the dye, Chromotrope 2R in water under mild reaction conditions. Novel MVPA/Kaolin was synthesized with an advanced impregnation method. The material was characterized with PXRD, FTIR, ICP-OES, SEM–EDX, TEM, TGA, surface area, porosity, and ^{31}P NMR measurements. The material that can be classified as a Keggin type catalyst, could be used for hydrogen peroxide oxidation of aromatic azo (–N=N–) dyes with–OH and $-SO_3H$ functional groups. The degradation showed excellent conversion of 99.1% under optimum conditions of normal pressure and 60 °C temperature. Products like malonic acid and oxalic acid have been identified as the final products with GC–MS analysis. The present work has demonstrated the application of a simple and effective catalyst for degradation of azo dyes and the catalyst had the reusability capacity of up to six cycles without loss in performance.

Keywords: *Chromotrope 2R, MVPA/Kaolin, Heterogeneous catalyst, Catalytic degradation, Azo dye*

1 INTRODUCTION

Heteropoly acids (HPA) with their superior redox characteristics and unique Keggin structure, have played an important role in synthesis and applications.

These materials contain mesoporous and microporous heteropoly materials and heteropoly anions with oxo-bridged multiple centers. The heteropoly acids have the advantage that they could be used to design new catalytic materials. The only limitation in their utilization is the uncertain surface area and the lack of thermal stability. Many studies have revealed that supporting HPAs on acceptable carriers and impregnating them with appropriate chemical species boost their functionality significantly [1,2,3,4]. The acid and redox characteristics of metals inserted into the HPA framework could be altered to ensure their utilization for varied purposes [2]. For example, their selective oxidation properties could be exploited to create fine chemicals and pharmaceuticals. The vanadium-substituted molybdophosphoric HPAs have shown a good potential for use as catalysts. It has been shown that incorporating vanadium (V) into the Keggin framework results in getting significantly advantageous redox catalysis [5,6,7,8,9,10]. Molybdovanadophosphoric acid, supported on mesoporous material, MCM-41, nano silica ball (NSB), mesoporous carbon (MC), and SBA-15, could be utilized as catalysts in oxidation reactions using the harmless H_2O_2 as the oxidizing agent [4, 11]. Traditional oxidation procedures employ expensive chemicals and operate at high temperatures with yields being very often limited [12]. This has led to a continuous search for better catalytic materials that could be used under mild reaction conditions for degrading hazardous chemicals in water.

Dyes and pigments are routinely utilized in a large number of chemical industries including textiles, leather, food, paper, plastic and other industries. The effluents from these industries carry a high concentration of these chemicals, many of which are harmful to the environment [13]. The removal of dyes and associated chemicals from these wastewaters has long been a key industrial concern. The current study had the limited goal of degrading the dye, Chromotrope 2R (Acid Red 29) via oxidation in the presence of a catalyst. Chromotrope 2R is an azo dye with the azo group (N=N) as the chromophore and an aromatic structure with functional groups OH and $-SO_3H$. In nature, azo dyes have been recognized as harmful as they are known to be carcinogenic with major environmental effects [14,15,16]. For treatment of such dye-containing wastewater, advanced oxidative degradation has always been considered as an effective, green and sustainable way to combat pollution. Simple Fenton type oxidation processes have been tried very often but have limited application potential. Photochemical oxidation processes such as those utilizing Gd2O3 (15%)–CdO nanocomposite as the catalyst have been tried for the catalytic degradation of Methylene blue [17]. Pure ZnO nanoparticles (NP) and mixed Eu_2O_3 and ZnO have also been tried for the photocatalytic decomposition of methylene blue and the results showed that 5% Eu_2O_3 and ZnO NPs showed the highest photocatalytic activity of 96.5% under 180 min of UV irradiation [18]. The degradation pathway of the dye under UV irradiation with synthesized α-Bi_2O_3 nanorods has also been investigated [19]. The present work on advanced oxidation was undertaken to investigate photocatalytic degradation/mineralization of an azo dye in water. It is to be noted that the photocatalytic process transforms the azo dye into numerous biodegradable intermediates or may completely destroy it, resulting in complete mineralization, yielding the ultimate products, CO_2, H_2O, and some inorganic salts, making it possible to discharge the final products safely into the environment. The final oxidation produces may include low molecular weight oxygenated molecules, like oxalic acid, acetic acid, propionic acid, ethanol, and others [12].

Adsorption, coagulation, oxidation, reduction, microbiological oxidation, photo catalysis, and biodegradation are some of the well-known strategies for removing colors and pigments from wastewater [9, 18,19,20]. Banerjee and Chattopadhyaya (2017) have described the adsorption process as the effective process for removal of toxic dye like tartrazine (97%) using Agriculture by-product like saw dust [20]. Although these approaches are successful in entirely eliminating or destroying pollutants, there may be issues with incomplete removal, excessive chemical use, being highly expensive and kinetically sluggish to employ, and occasionally resulting in the formation of harmful secondary pollutants [21]. Modified titanium dioxide (TiO_2) has established itself as a key photocatalyst in these oxidation processes. However, there are significant drawbacks to TiO_2-based photocatalysis that keep it from being widely used. There have been reports of (i) insufficient use of visible light, (ii) hydrophobic pollutants with limited adsorption potential, (iii) a high tendency for products to agglomerate, and (iv) problems with catalyst recovery and reuse in subsequent catalytic cycles [21]. Rafaqat et al. (2022) have well described about the recent progress in treatment of dyes wastewater using microbial-electro-Fenton technology [22].

Biological oxidation techniques for decolorization of industrial effluents are efficient only at low pollutant concentration and cannot be employed for non-biodegradable, poisonous, or inhibiting contaminants. Alkalinity and temperature of the effluents also influence these processes. Although researchers are working to solve the insufficiencies. Thanavel et al. (2018) have worked with Aeromonas hydrophila SK16 and AOPs-H_2O_2 (4%) were used to assess the effectiveness of a combined biological and Advanced Oxidation Process (Bio-AOP) for the remediation of textile dyes. Reactive Red 180 (RR 180), Reactive Black 5 (RB 5) and Remazol Red (RR) all showed 100% decolorization after Bio-AOP treatment, compared to 72% decolorization after separate treatments [23]. A wide spectrum of organic reactions could be catalyzed using molybdovanadophosphoric acid-based catalysts in the presence of hydrogen peroxide [4]. Hydrogen peroxide is not only an effective oxidant, but is also environmentally beneficial, as its degradation products are solely H_2O and O_2. Advanced oxidation techniques based on the use of hydrogen peroxide as the low-cost, environmentally benign, highly efficient oxidation agent with an appropriate catalyst (chemical or photochemical oxidation) have become popular in recent years for direct wastewater treatment.

A novel, eco-friendly method is presented here for synthesizing molybdovanadophosphoric acid (MVPA) with kaolin as the support (MVPA/Kaolin) to be used as a catalyst in Chromotrope 2R dye degradation in aqueous solution. The study is reported in several sections. Section 2 describes the experimental sequences of materials, preparation and characterization of MVPA/Kaolin and the actual catalytic degradation process. Section 3 represents the detailed characterization of the catalyst and discusses the catalytic degradation reactions of the dye with process variables like the reaction time, the catalyst loading, etc., along with the suggestion of a dye-degradation mechanism and the reusability of the catalyst. The last Sect. 4 summarizes the findings of the study.

2 MATERIALS AND METHODS

2.1 Materials

Reagents such as, MoO_3 (Loba Chemie Pvt. Ltd., Mumbai, India; 99%), V_2O_5 (Loba Chemie Pvt. Ltd., Mumbai, India; 99%), phosphoric acid (Loba Chemie Pvt. Ltd., Mumbai, India; 90%) and Kaolin (Molychem, Mumbai, India; 95%), hydrogen peroxide (Loba Chemie Pvt. Ltd., Mumbai, India; 30%) were used without further purification.

2.2 Characterization of the hybrid material

The hybrid composites were characterized for two dimensional crystalline structure by means of Powder X-ray diffraction (PXRD; Philips X'pert MPD diffractometer in Almelo, the Netherlands; 2θ 1–10 and 10–80), Fourier Transform Infrared spectroscopy (FT-IR, Perkin–Elmer spectrometer) with pellets made in potassium bromide (KBr), Inductively coupled plasma optical emission spectroscopy (ICP–OES, JY Ultima 2CHR) to determine the metals entering kaolin, N_2 adsorption–desorption study at 77 K for BET surface area and pore volume measurement (Micromeritics ASAP-2010, USA), thermogravimetric analysis (TGA, SDT600, TA instrument, USA) using platinum crucibles to quantify the temperature and microstructure of the samples over the range 30–800 °C at a heating rate of 10 °C min^{-1} in an air atmosphere with a flow rate of 60 cm^3 min^{-1}, Scanning electron microscopy-Energy dispersive X-ray (SEM-EDX, LEO–1430, VP, UK) and Transmission electron microscopy (TEM; JEM 2011, Jeol Corporation, Japan) to determine the surface morphology of the material, and ^{31}P Nuclear Magnetic Resonance (^{31}P NMR) for characteristic confirmation.

2.3 Synthesis of molybdovanadophosphoric acid supported on kaolin catalyst (MVPA/Kaolin)

The manufacture of molybdovanadophosphoric acid supported kaolin (MVPA/Kaolin) catalyst is schematically presented in Fig. 1. During the synthesis of MVPA, in a round bottom flask, 22.3 g MoO_3 (Loba Chemie, Mumbai, India) and 1.3 g V_2O_5 (Loba Chemie, Mumbai, India) were combined with 350 ml distilled water. The mixture was then refluxed for 1 h before being added to the flask in three equal aliquots of 1.7 g phosphoric acid (Loba Chemie, Mumbai, India) and 150 ml distilled water. A brilliant orange tinted solution was generated after the mixture was refluxed continuously for 15 h. Then, a homogeneous mixture was made by constantly stirring an aqueous solution of MVPA (0.50 g in 50 ml distilled water) and 1.0 g of kaolin, followed by complete evaporation of the water at 100 °C. The material was dried in an oven at 110 °C for 24 h before being calcined at 500 °C for 2 h. The yield was 3.3 g of MVPA/Kaolin. The elemental composition of the catalyst was determined using inductively coupled plasma-optical emission spectroscopy (ICP-OES). MVPA has a Mo/P ratio of 11.0, indicating a relatively pure heteropoly acid phase [3]. The well-characterized FTIR (KBr) bands were MVPA/Kaolin: 793, 869, 962, 1115, 1404, 1613, 3437 cm^{-1} and MVPA: 777, 866, 960, 1062, 1615, 1932, 3390 cm^{-1} MVPA/Kaolin: 793, 869, 962, 1115, 1404, 1613, 3437 cm^{-1}.

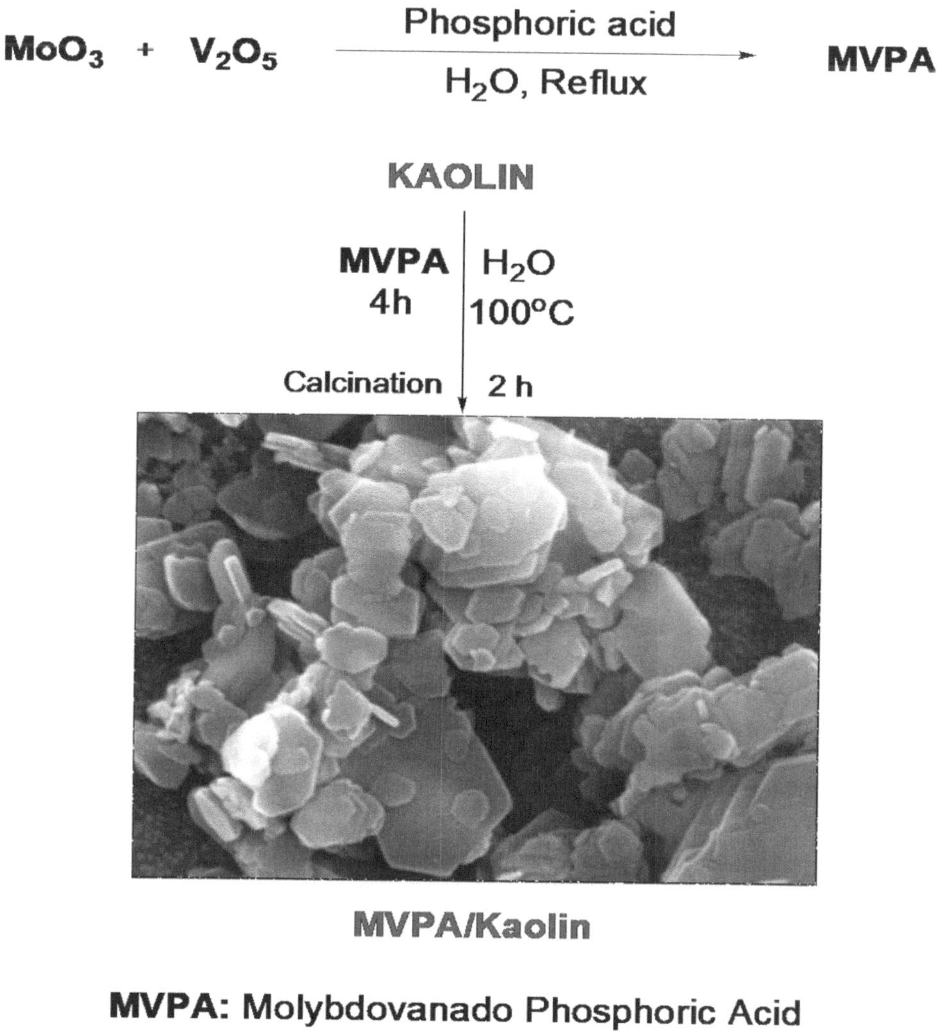

Fig. 1.: *Schematic diagram for synthesis of molybdovanado phosphoric acid/kaolin catalyst*

2.4 Catalytic oxidation of Chromotrope 2R

The oxidation and subsequent degradation of Chromotrope 2R were carried out in a glass reactor fitted with a con``denser. The reaction mixture consisted of (a) 50 ml Chromotrope 2R aqueous solution (2.0×10^{-4} mol L^{-1}), (b) 50 mL H_2O_2 (10^{-4} mol L^{-1}), and (c) 5 g L^{-1} MVPA/Kaolin catalyst. Under normal atmospheric pressure, the mixture was stirred for 60 min at 180 rpm and 60 °C. The mixture was filtered after the reaction was completed to separate the catalyst MVPA/Kaolin. The degradation of the dye was monitored at λ_{max} = 508 nm in the filtrate with UV–Visible spectroscopy (UV-1800, Shimadzu corporation, Japan). The degradation of Chromotrope 2R in percentage was calculated according to the relation:

$$Deagradation(\%)=[(C0-Ct)/C0]\times 100Deagradation(\%)=[(C0-Ct)/C0]\times 100$$

where C_0 and C_t were the initial concentration and the concentration at any time (mol L^{-1}) after the reaction was run for t (min).

Gas chromatography-mass spectroscopy (GC–MS, Agilent, US) measurements were used to identify the products in the reaction mixture. After each reaction, the products were extracted using equal quantities of chloroform and were sonicated for a period of time to separate the organic and aqueous layers. 0.6 μL of the chloroform layer was then injected into the GC–MS for analysis (initial temperature 50 °C, kept for 5 min, increased to 180 °C at 10 °C min^{-1}, kept for 5 min, increased to 250 °C at 10 °C min^{-1}, and kept for 5 min at 250 °C).

3 RESULTS AND DISCUSSION

3.1 Characterization

3.1.1 Powder X-ray diffraction (XRD)

The results of low angle and high angle mode of powder X-ray diffraction (PXRD) are presented in Fig. 2 showing the characteristic patterns of MVPA [5] and MVPA/Kaolin. The low angle XRD of MVPA revealed a strong XRD peak at 2θ of 0.2° (Fig. 2), while MVPA/kaolin revealed additional peaks attributable to kaolin (0.1°) [3, 8]. Functionalization with kaolin had no effect on the principal reflections due to MVPA, as seen by the wide angle XRD (Fig. 2). XRD peaks at 2θ of 18.1°, 20.3°, 22.6°, 26.9°, 29.1°, 31.6°, 38.0°, 54.3°, 59.9° indicate a significant degree of crystallinity in the MVPA. The MVPA/Kaolin had diffraction peaks at 2θ of 10.7°, 15.2°, 18.6°, 24.1°, 30.7°, 32.6°, 36.1°, 39.4°, 43.9°, 48.1°, 55.7°, 62.7°, 64.9° showing that the material remained crystalline at the atomic level giving rise to reflections at wide angle. Because all of the principal reflections caused by kaolin are also visible in MVPA/Kaolin, it may be resolved that the structural properties of the support material (kaolin) were preserved. This indicates that the crystalline structure of the heterogeneous hybrid material was retained, which would be beneficial in explaining its activity in the subsequent applications.

3.1.2 Scanning Electron Microscopy (SEM)

The SEM micrographs of MVPA and MVPA/Kaolin are shown in Fig. 3. The SEM of MVPA revealed the usual MVPA structural features. MVPA covered the well-ordered layered crystal structure due to kaolin (Fig. 3) with particle size 170 nm in the hybrid composite MVPA/Kaolin. The complex, multi-layered structure is unaltered. Even after incorporating MVPA, the fundamental morphology of the support material might be preserved in the hybrid composite materials.

Strong signals for vanadium in the kaolin base could be found in the EDX spectrum of MVPA/Kaolin (Fig. 3d) The layered structure of the catalyst material was well-preserved and the well-organized morphology of the support material could be retained during the fabrication of the catalyst material as well as the subsequent catalytic oxidation of the dye.

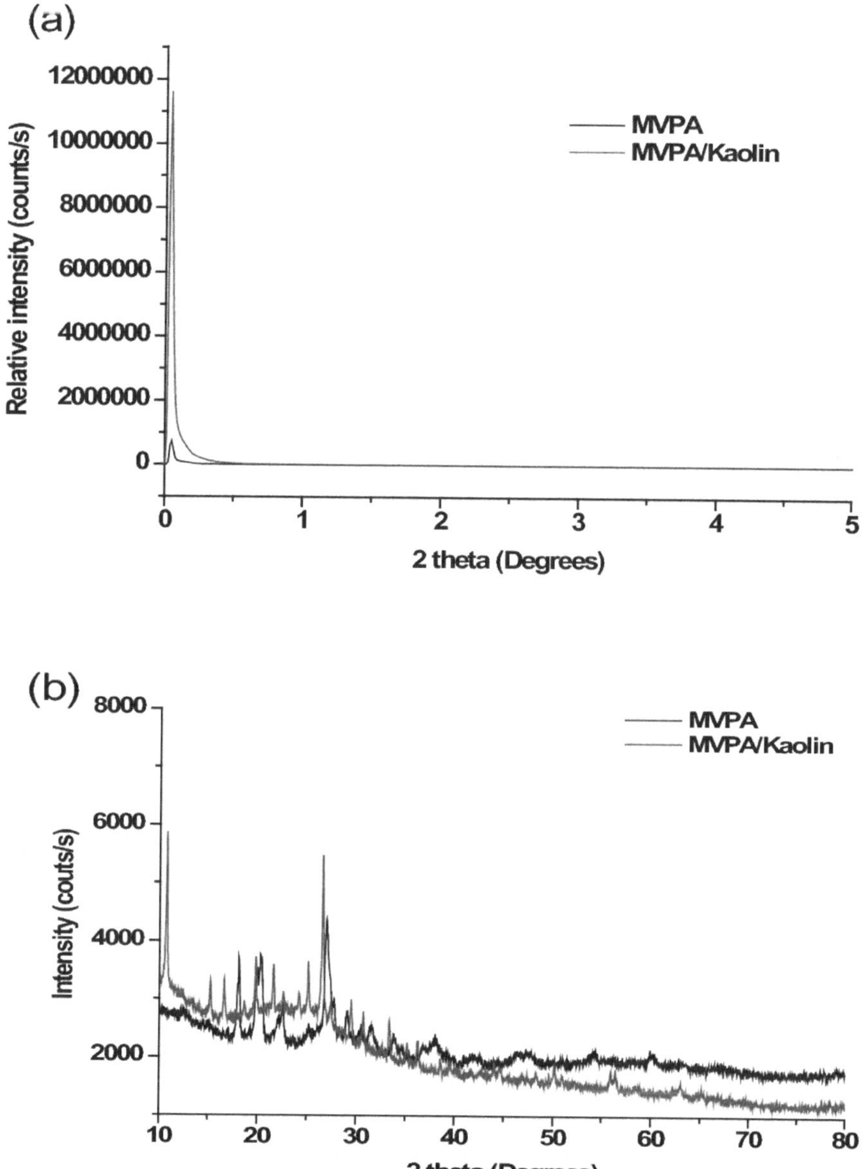

Fig. 2.: *X-ray diffraction pattern of MVPA and MVPA/Kaolin [a low-angle and b wide-angle diffractions*

Fig. 3.: SEM images of a MVPA, b, c: MVPA/Kaolin d SEM–EDX spectra of MVPA/Kaolin

3.1.3 Transmission Electron Microscopy (TEM)

The TEM images of MVPA showed well-arranged structure of the material (Fig. 4). The composite material MVPA/Kaolin provides characteristic well-ordered morphology which provide strong evidence that the shape of the support, Kaolin, is retained upon MVPA functionalization [5].

3.1.4 Thermogravimetric analysis (TGA)

The thermo gravimetric curves of the material showed modest weight losses of about 61.5 percent for MVPA in the temperature range 25 to 900 °C (Fig. 5.) The weight loss could be observed as 24.2 percent on loading of MVPA as a hybrid composite in MVPA/Kaolin. In accordance with the literature, ICP data for MVPA and MVPA/Kaolin revealed Mo/P ratios of 11.0 and 12.0 wt%, which are required to manufacture heteropoly acid composite materials [4].

3.1.5 Surface area and pore volume

MVPA and MVPA/Kaolin have BET surface areas of 2.7 and 15.2 $m^2\ g^{-1}$, respectively (Table 1). MVPA and MVPA/Kaolin have pore sizes of 85.1 and 162.2 (Å) respectively, and pore volumes of 0.006 and 0.089 $cm^3\ g^{-1}$. The BET surface area of MVPA/Kaolin is determined by the support material's basic structure, kaolin. The experimental results might be understood to mean that in the MVPA-modified material, the internal pores and surface of kaolin are filled with MVPA, and the structure of the support is unchanged even after MVPA functionalization

[4, 24, 25]. Figure 6 shows the N_2 adsorption–desorption isotherms of MVPA and MVPA/Kaolin.

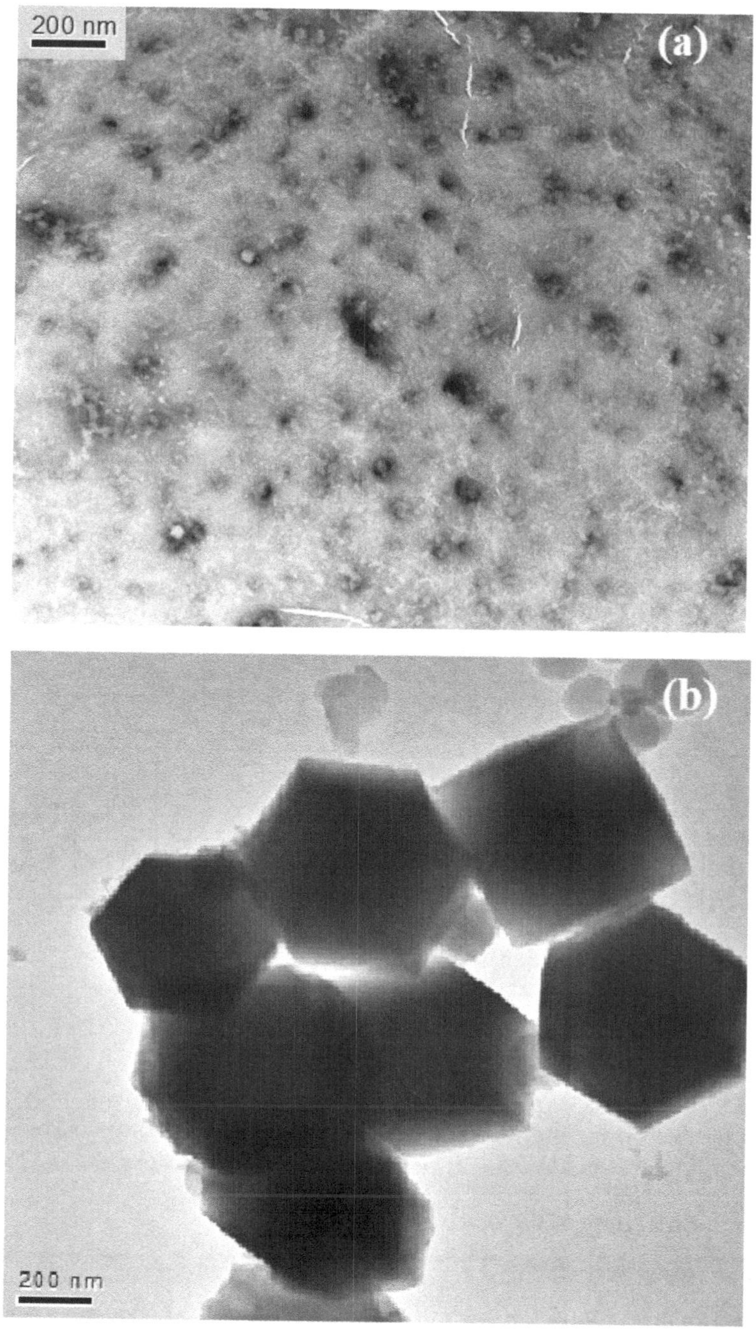

Fig. 4.: *TEM images of a MVPA, b: MVPA/Kaolin*

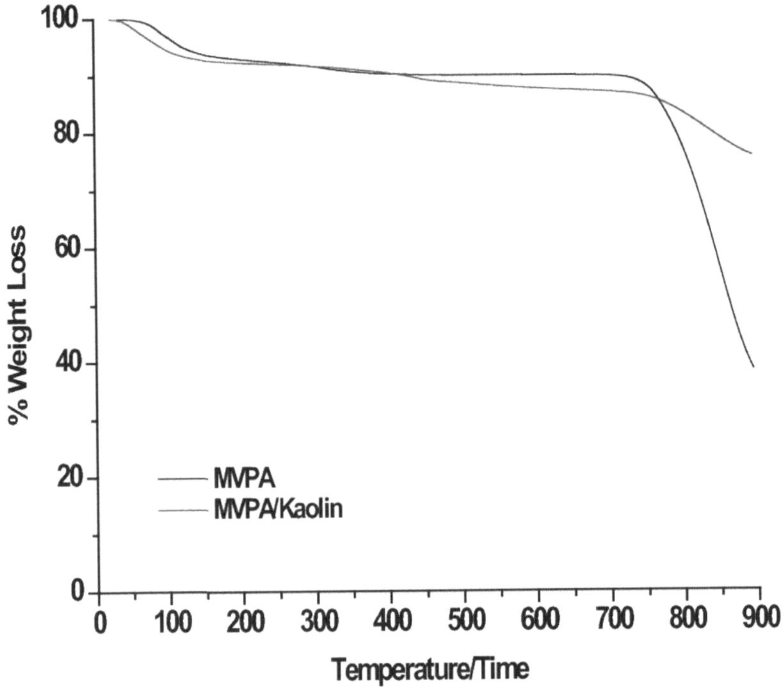

Fig. 5.: *Thermogravimetric analysis (TGA) curves of MVPA and MVPA/Kaolin*

Table 1.: *Important adsorption–desorption data for MVPA and MVPA/Kaolin*

Sample	BET Surface Area (m^2/g)	Total Pore Volume (cm^3/g)	BJH Pore Diameter (Å)
MVPA	2.7	0.006	85.1
MVPA/ Kaolin	15.2	0.089	162.2

3.1.6 ^{31}P Nuclear Magnetic Resonance (^{31}P NMR) spectra

^{31}P MAS NMR spectra (Fig. 7) at 10 kHz for pure MVPA showed a sharp signal, recorded at −3.28 ppm that could be assigned to P–OH groups associated with phosphomolybdates in MVPA, supported by the existing literature [4, 26].

3.2 Degradation study of Chromotrope 2R dye

3.2.1 Catalytic activity study

The catalytic oxidation of chromotrope 2R has been taken as the model application using synthesized hybrid MVPA / Kaolin catalyst. UV–Vis spectra, showing degradation of Chromotrope 2R dye in the presence of MVPA/Kaolin are presented in Fig. 8.

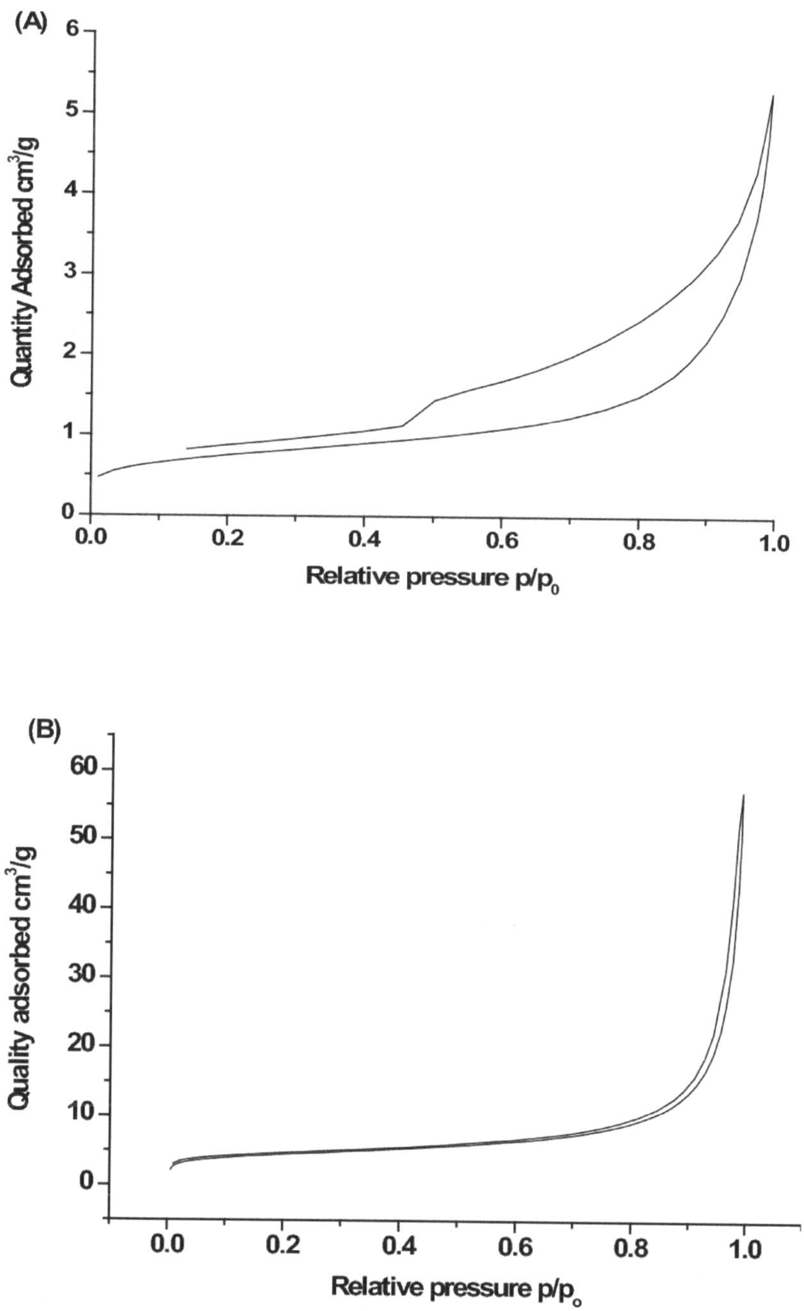

Fig. 6.: Nitrogen adsorption–desorption isotherms as per BET equation for A MVPA and B MVPA/Kaolin

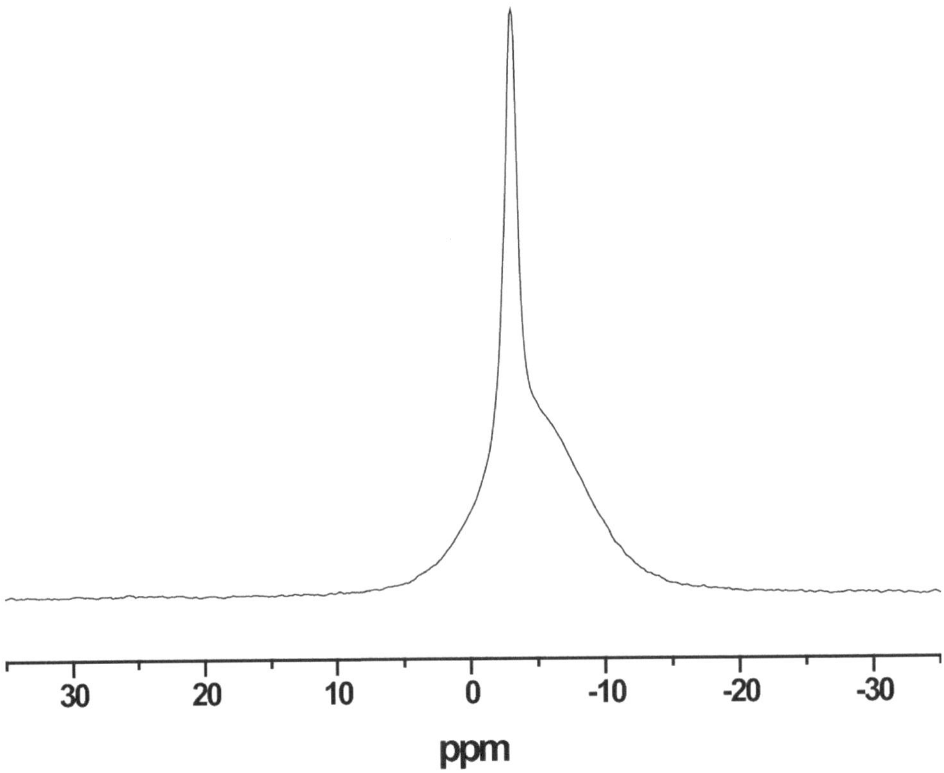

Fig. 7.: ^{31}P MAS NMR profile of MVPA

The primary absorbance band at 508 nm is visible in Chromotrope 2R. The oxidation of Chromotrope 2R with H_2O_2 occurs once the catalyst is introduced, and the absorbance band decreases, practically leveling out after 60 min (Fig. 8). As oxidation destroys the aromatic structure interacting with the azo group (N=N), OH, and $-SO_3H_9$ (GC–MS), the degradation of the dye leads to a few simpler active acid species. In approximately 60 min, 99.1% of Chromotrope 2R was degraded in the presence of the catalyst, MVPA/Kaolin. These results are far superior to those obtained from Fenton degradation of the dye. The current dye degradation method is more straightforward and efficient, and the catalyst may be recovered and reused. When compared to traditional Fenton oxidation techniques, the degradation might be accomplished in a fraction of the time [27]. In this study, the oxidation proceeds with H_2O_2 in the reactant feed, since the conversion is based on.OH radicals. The OH-radicals originate from O on the catalytic system and H atom cleaved from the substrate (either water or C2R since the reactions were carried out in aqueous solution) as dissolved oxygen is not likely to take part in the oxidation process directly. Participation of dissolved oxygen can also be ruled out since the reaction does not slow down appreciably. The hydroxyl radicals play an important role in the degradation process. The electron acceptor, hydrogen peroxide, in the solution, enhances the formation of hydroxyl radicals and inhibits the electron/hole (e^-/H^+) pair recombination. The mechanistic pathway for the degradation of the dye, chromotrope 2R, along with the formation of probable intermediates is depicted in Fig. 9. The end products of dye degradation were identified by GC–MS as (i) malonic acid, and (ii) oxalic acid, based on measurements of the reaction products, and these compounds are likely to mineralize to CO_2 and H_2O.

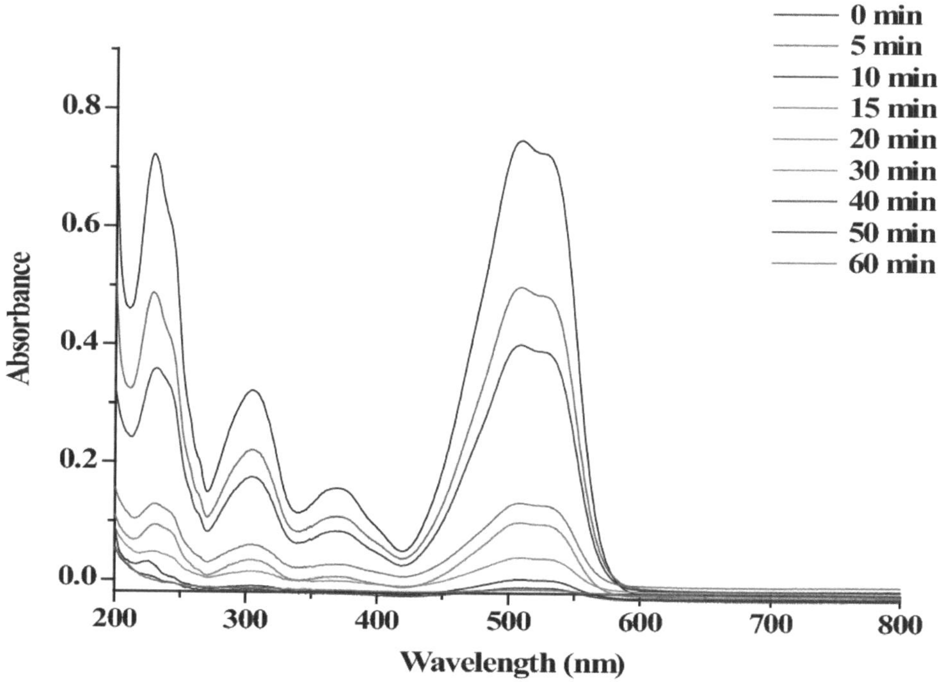

Fig. 8.: UV–Vis absorption spectra showing Chromotrope 2R oxidation (C2R: 2.0×10^{-4} mol L^{-1}; H_2O_2: 10^{-4} mol L^{-1}; MVPA/Kaolin: 5 g L^{-1}; 60 °C, normal pressure, and 180 rpm rate)

Degradation of dye pollutants by catalytic process has been regarded as the most efficient green method for removal of organic dyes from contaminated water [15, 17, 22, 23]. The improved toxicity and carcinogenicity of the components, released from industrial effluents, have always been considered as a significant environmental issue and the present work demonstrates one possible approach to control the same. As dyes are widely used in textile and paper industries, and exposure to them and their natural degradation products may lead to skin diseases, eye irritations, etc., including serious carcinogenicity, any process development to convert them to harmless products could be considered as a welcome development in environmental sustainability. The present work has developed a low-cost and easy-handling type catalyst, which has potential to be used for treatment of dyes in water.

A catalyst optimization study was carried out to determine the amount of catalyst, necessary for the degradation of Chromotrope 2R (Table 2). The degradation reaction of Chromotrope 2R was catalyzed by using 1, 3, 5, 8, 10 g L^{-1} MVPA/

kaolin that took 80, 70, 60, 60, and 60 min for completion respectively. The optimum catalyst load was 5 g L^{-1} resulting in maximum conversion of Chromotrope 2R in 60 min (Fig. 10).

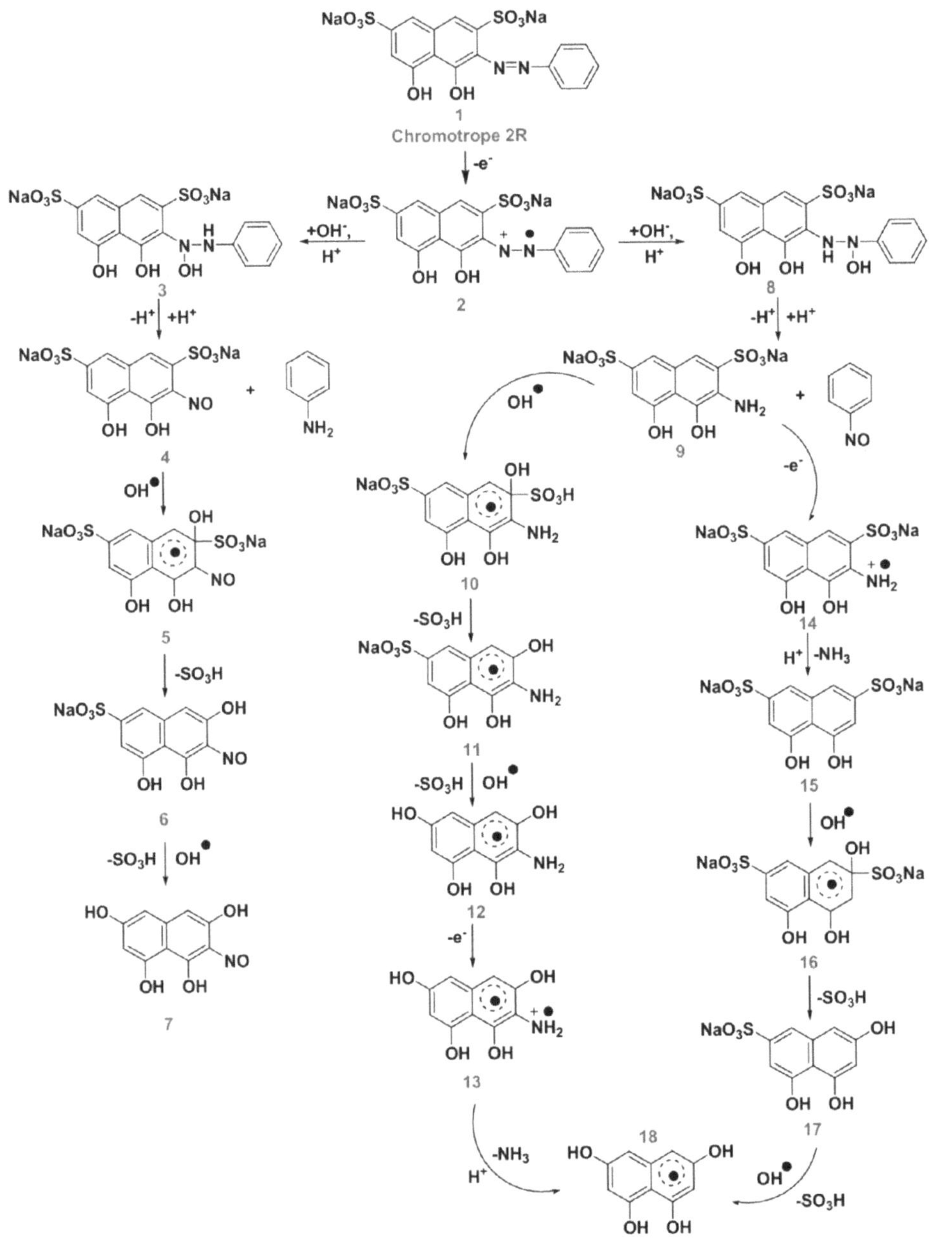

Fig. 9.: The possible degradation mechanism of Chromotrope 2R

Table 2.: *Reaction time variation with different catalyst (MVPA/Kaolin) loads for oxidation/degradation of Chromotrope 2R in the presence of hydrogen peroxide*

Catalyst load (g L^{-1})	**Chromotrope 2R**
	Reaction time (min)
1	80
3	70
5	60
8	60
10	60

3.2.2 Reusability of the catalyst

The possibility of recycling the catalyst, MVPA/Kaolin, after the initial degradation reaction of Chromotrope 2R was investigated using a set of reaction parameters (50 mL Chromotrope 2R (2.0×10^{-4} mol L^{-1}), 50 mL H_2O_2 (10^{-4} mol L^{-1}), and 5 g L^{-1} MVPA/Kaolin). This series of tests was also used to assess the stability of the catalyst in oxidative degradation reactions, a prerequisite for utilizing the catalyst in real-world applications. The reaction mixture was filtered after the first cycle of the dye catalytic degradation was completed to get the MVPA/Kaolin catalyst. The catalyst was properly rinsed with distilled water several times and the MVPA/Kaolin residue was allowed to dry overnight, calcined at 500 °C for 2 h in a muffle furnace. The MVPA/Kaolin catalyst was recovered and employed as a catalyst in a subsequent run of the catalytic reaction. It was discovered that the reaction can be repeated in six cycles with the same catalyst without much performance loss (Fig. 11).

4 CONCLUSIONS

Novel and efficient Keggin type MVPA/Kaolin was successfully synthesized with vanadium-substituted molybdophosphoric heteropoly acids. Characterization studies confirmed the regular layered crystal structure of the hybrid catalyst with thermal stability, increased surface area, pore diameter and pore volume. The catalyst exhibited a high catalytic activity for destroying up to 99.1% of Chromotrope 2R dye in aqueous media by oxidation with H_2O_2 under optimum reaction conditions. The work has revealed that MVPA/Kaolin catalyst could bring about near complete degradation of the azo dye by converting it to smaller acidic components like malonic acid and oxalic acid, that could undergo further mineralization to CO_2 and water. The work establishes MVPA/Kaolin as a potential high-performance catalyst in environmental treatment processes. The recyclability of the eco-friendly catalyst is an additional advantage for its environmental use.

The experiments carried out in the laboratory have yielded promising results. For extending the work to actual application in the field, the processes have to be upscaled for use with large volumes of dye-containing wastewater. One area needing immediate attention would be to increase the yield of the catalyst as it will have to be frequently replaced and reused.

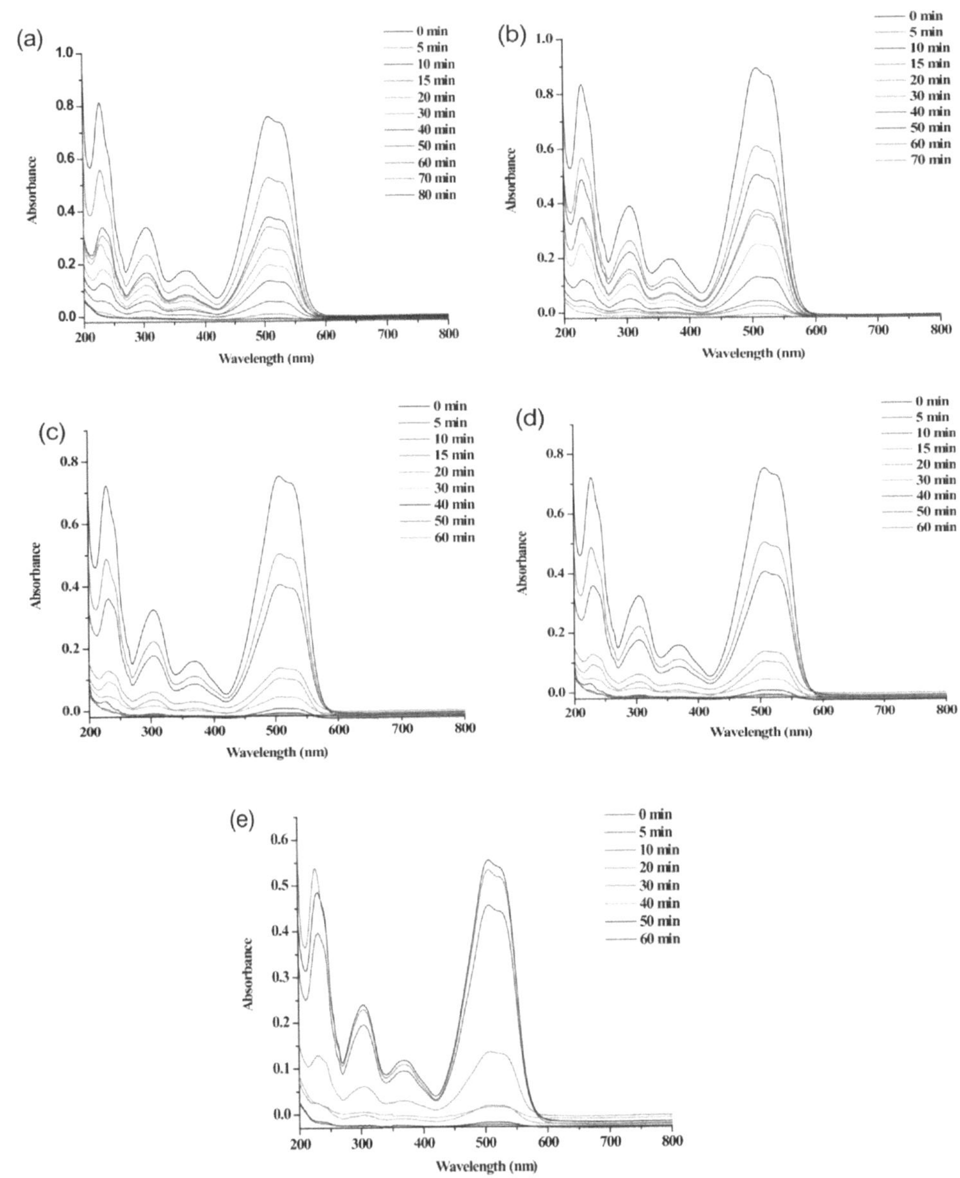

Fig. 10.: *UV–Vis absorption spectra for oxidative degradation of Chromotrope 2R with MVPA/Kaolin loads of 1 g L^{-1} (a), 3 g L^{-1}(b), 5 g L^{-1} (c), 8 g L^{-1} (d) and 10 g L^{-1} (e)*

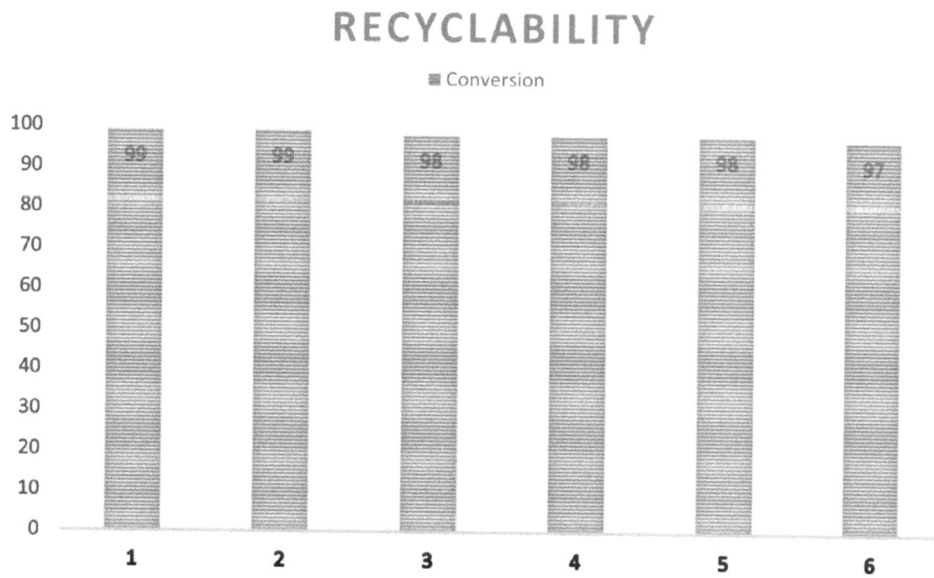

Fig. 11.: *Recyclability of the catalyst (C2R: 2.0 × 10^{-4} mol L^{-1}; H_2O_2: 10^{-4} mol L^{-1}; MVPA/Kaolin: 5 g L^{-1}; 60 °C, normal pressure, and 180 rpm rate)*

Acknowledgements

This research was supported by Department of Chemistry, Marwadi University, Rajkot, Gujarat, India.

REFERENCES

1. Fournier M, Thouvenot R, Rocchiccioli-Deltcheff C (1991) Catalysis by polyoxometalates. Part 1. Supported polyoxoanions of the Keggin structure: spectroscopic study (IR, Raman, UV) of solutions used for impregnation. J Chem Soc Faraday Trans 87(2):349–356.
2. Parida KM, Mallick S (2008) Hydroxylation of phenol over molybdovanadophosphoric acid modified zirconia. J Mol Catal A Chem 279(1):104–111.
3. Zhang Q, Lei D, Luo Q, Yang X, Wu Y, Wang J, Zhang Y (2021) MOF-derived zirconia-supported Keggin heteropoly acid nanoporous hybrids as a reusable catalyst for methyl oleate production. RSC Adv 11(2021):8117–8123.
4. Mayani SV, Mayani VJ, Kim SW (2013) Synthesis of molybdovanadophosphoric acid supported hybrid materials and their heterogeneous catalytic activity. Mater Lett 111(2013):112–115.
5. Mizuno N, Misono M (1994) Heteropolyanions in catalysis. J Mol Catal 86(1–3):319–342.

6. Hill CL (2003) Comprehensive coordination chemistry II, 2nd edn. Elsevier, Amsterdam, p 679
7. Neumann R, Khenkin AM (2006) Molecular oxygen and oxidation catalysis by phosphovanadomolybdates. Chem Commun 24:2529–2538.
8. Zhang F, Maiping G, Hanqingd GE, Wang J (2007) Hydroxylation of benzene with hydrogen peroxide over highly efficient molybdovanadophosphoric heteropoly acid catalysts. Chin J Chem Eng 15(6):895–898.
9. Villabrille P, Romanelli G, Gassa L, Vazquez P, Caceres C (2007) Synthesis and characterization of Fe- and Cu-doped molybdovanadophosphoric acids and their application in catalytic oxidation. App Catal A 324:69–76.
10. Nagaraju P, Balaraju M, Reddy KM, Prasad PS, Lingaiah N (2008) Selective oxidation of allylic alcohols catalyzed by silver exchanged molybdovanado phosphoric acid catalyst in the presence of molecular oxygen. Catal Commun 9(6):1389–1393.
11. Rana S, Mallick S, Parida KM (2012) Selective oxidation of benzaldehyde by molecular oxygen over molybdovanadophosphoric acid supported MCM-41. J Porous Mater 19(4):397–404.
12. Popova GY, Andrushkevich TV, Semionova EV, Chesalov YA, Dovliova LS, Rogov VA, Parmon VN (2008) Heterogeneous selective oxidation of formaldehyde to formic acid on V/Ti oxide catalysts: the role of vanadia species. J Mol Catal A Chem 283(1–2):146–152.
13. Mayani VJ, Mayani SV, Kim SW (2017) A sustainable nanocomposite Au(Salen)@CC for catalytic degradation of eosin Y and Chromotrope 2R dyes. Nat Sci Rep 7(2017):7239–7243.
14. Kunz A, Peralta-Zamora P, Duran SG (2002) New trends in the treatment of textile effluent. Chem New 25(1):78–82.
15. Umbuzeiro GA, Freeman HS, Warren SH, Oliveira DP, Watanade YT, Claxton LD (2004) The contribution of azo dyes to the mutagenic activity of the Cristais River. Chemosphere 60(1):55–64.
16. Sharma KP, Sharma S, Singh PK, Kumar S, Grover R, Sharma PK (2007) A comparative study on characterization of textile wastewaters (untreated and treated) toxicity by chemical and biological tests. Chemosphere 69(1):48–54.
17. Ahmed MA-D, Essawy AA, Diab AK, Mohamed WS (2020) Facile synthesis and characterization of novel Gd2O3–CdO binary mixed oxide nanocomposites of highly photocatalytic activity for wastewater remediation under solar illumination. J Phy Chem Solids.
18. Mohamed WS, Abu-Dief AM (2018) Synthesis, characterization and photocatalysis enhancement of Eu2O3-ZnO mixed oxide nanoparticles. J Phys Chem Solids.
19. Ahmed MA-D, Mohamed WS (2017) α-Bi2O3 nanorods: synthesis, characterization and UV-photocatalytic activity. Mater Res Express 4:035039.
20. Banerjee S, Chattopadhyaya MC (2017) Adsorption characteristics for the removal of a toxic dye, tartrazine from aqueous solutions by a low cost agricultural by-product. Arab J Chem 10(2):S1629–S1638.

21. Hadjltaief HB, Ameur SB, Costa PD, Zina MB, Galvez ME (2018) Photocatalytic decolorization of cationic and anionic dyes over ZnO nanoparticle immobilized on natural Tunisian clay. Appl Clay Sci 152:148–157.
22. Rafaqat S, Ali N, Torres C, Rittmann B (2022) Recent progress in treatment of dyes wastewater using microbial-electro-Fenton technology. RSC Adv 12(27):17104–17137.
23. Thanavel M, Kadam SK, Biradar SP, Govindwar SP, Jeon B-H, Sadasivam SK (2019) Combined biological and advanced oxidation process for decolorization of textile dyes. SN Appl Sci 97:1–16.
24. Bordoloi A, Lefebvre F, Halligudi S (2007) Selective oxidation of anthracene using inorganic–organic hybrid materials based on molybdovanadophosphoric acids. J Catal 47(2):166–175.
25. Mayani VJ, Mayani SV, Kim SW (2012) Development of nanocarbon gold composite for heterogeneous catalytic oxidation. Mater Lett 87:90–93.
26. Salem IA, El-Ghamry HA, El-Ghobashy MA (2015) Application of montmorillonite–Cu(II)ethylenediamine catalyst for the decolorization of Chromotrope 2R with H2O2 in aqueous solution. Spectrochim Acta Part A 139:130–137.
27. Mayani SV, Mayani VJ, Bhattacharyya KG (2011) Chemistry of phenolic compounds. In: Baruah JB (ed) State of the art, chapter 13. Nova Science Publishers Inc, New York, p 181

9 BIOPOLYMER HONEYCOMB MICROSTRUCTURES: A REVIEW

Dominik Fajstavr, Klára Fajstavrová, Bára Frýdlová, Nikola Slepičková Kasálková, Václav Švorčík and Petr Slepička

Department of Solid State Engineering, University of Chemistry and Technology Prague, Technická 3, 166 28 Prague, Czech Republic

ABSTRACT

In this review, we present a comprehensive summary of the formation of honeycomb microstructures and their applications, which include tissue engineering, antibacterial materials, replication processes or sensors. The history of the honeycomb pattern, the first experiments, which mostly involved the breath figure procedure and the improved phase separation, the most recent approach to honeycomb pattern formation, are described in detail. Subsequent surface modifications of the pattern, which involve physical and chemical modifications and further enhancement of the surface properties, are also introduced. Different aspects influencing the polymer formation, such as the substrate influence, a particular polymer or solvent, which may significantly contribute to pattern formation, and thus influence the target structural properties, are also discussed.

Keywords: *honeycomb; polymer; biopolymer; replication; tissue engineering; antibacterial properties; surface modification; morphology; breath figure; improved phase separation*

1. NATURAL HONEYCOMB

Living organisms respond to environmental conditions in such a way as to best adapt their structure and functionality. Therefore, scientists have always considered nature to be the best source of inspiration for the development of new materials [1] and applications. One idea was to artificially imitate the perfect structure of honeycombs, which is, without doubt, one of the most fascinating formations in nature [2,3]. Natural honeycomb has an efficient internal structure, with thousands of hexagonal cells connected in highly ordered arrays. In the book of the Roman scholar M. T. Varro from 36 BC., we can find one of the first discoveries regarding the shape of beehive cells [4]. Varro wrote: "Does not the chamber inside the honeycomb have six angles... The geometers Euclid and Zenodorus proved that this hexagon inscribed in a circle encloses the largest part of space." Subsequently, in the 4th century, the Greek mathematician Pappus wrote in the introduction to his book The Collection [5]: "By a certain geometrical foresight, the bees know that a hexagon is larger than a square or a triangle and will take more honey with the same consumption of material and labor."

The spatial geometry of honeycomb, which was carefully observed by J. Kepler [6] and discussed by J. Hales [7], is shown in Figure 1. It is a hexagonal prism enclosed at one end by three rhombuses. Hexagonal cells are more suitable than the round ones used, for example, by bumblebees, as gaps are created between round cells and space is wasted. Square or triangular cells are another option, which are adjacent to each other and do not create unwanted gaps. However, the bee larvae that grow in the cells do not have any of the above-mentioned shapes in the cross-section, and this would again be a waste of space, in this case inside the cells.

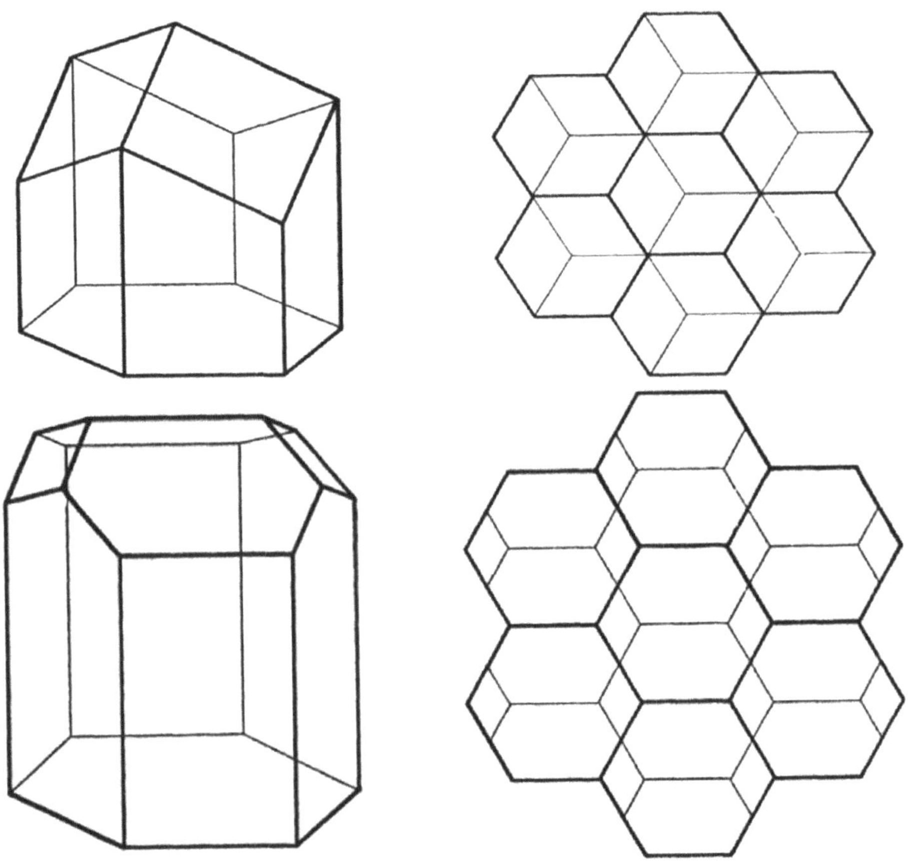

Figure 1.: *Honeycomb structures [1].*

Therefore, hexagonal cells are intuitively appropriate. Another remarkable strategy of bees, to save building material and space, is the two-layer hexagonal honeycomb. The bottom of each cell has a pyramid shape (again, a more efficient solution than a square bottom) and thanks to this, the opposite honeycombs fit perfectly [8]. Calculations and studies by many scientists have concluded that this structure is an almost optimal mathematical way to cover the maximum amount of space using the minimum amount of building materials [9,10]. In 1964, L.F. Tóth discovered the most economical ideal cell, which is hexagonal, but the bottom consists of two squares and two hexagons (see Figure 1) [1]. However, the saving was not that significant (less than 0.35%) compared to the complexity of the design. In an experiment with self-leveling soap bubbles [11], it was shown that, at a certain wall thickness, the ideal solution would shift from the optimal arrangement proposed by Toth to the arrangement preferred by the bees. Therefore, we can conclude that the structure is indeed very close to the theoretical one. Honeycomb structure with these kinds of cell interior have been presented in [9], see Figure 2.

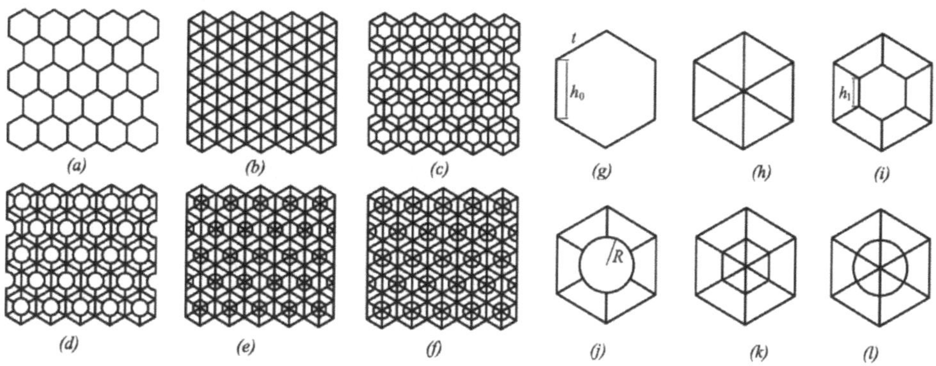

Figure 2.: *Honeycomb structures of cell interiors. (a) general structure; (b) triangular; (c) double hexagonal; (d) inside circular; (e) full double hexagonal; (f) full inside circular; (g) cell of hexagonal; (h) cell of full-triangular; (i) cell of double hexagonal; (j) cell of inside circular; (k) cell of full inside hexagonal; (l) cell of full inside circular [9].*

2. DEVELOPMENT OF ARTIFICIAL HONEYCOMB

Scientists, inspired by natural honeycombs, are trying to produce similar physical conformations in laboratories, which have been given the technical name of "honeycomb-like structures" (from the English abbreviation HCP). If we look at history and focus on the technological development of these porous structures, we can categorize this development into four periods [12]: (i) Period of first interest (60 BC-126): People were already fascinated by hexagonal honeycombs in antiquity. Several Greek mathematicians devoted themselves to the structure of the beehive, which is why the first results of their observations appear at this time [4,5]. This motif even appeared in the construction of the Pantheon in Rome, where the hexagonal structure formed the internal ribbing [13]. (ii) The period of observation and understanding (1638–1901): At first, the focus was on obtaining as much qualitative information as possible about the properties of beehives. This was also helped by the invention of the microscope in the 17th century, when, for example, the similarity between the cork structure and the honeycomb structure was observed [14]. With the advent of the industrial revolution, which connected science and technology, the demand for new lightweight materials grew, and the honeycomb structure became a hot topic. (iii) Period of application (1914–1990): It was not until the 20th century that the first attempts to artificially create a honeycomb-type structure with potential uses in many industries (aeronautics, cosmonautics, engineering, architecture, medicine [15,16]) appeared. In 1914, a structural application of honeycombs was patented [17], which was subsequently designed for construction applications. In the 1980s, the first HCPs were created from thermoplastics using the extrusion process [18]. (iv) The period of the interconnection of multiple fields and rapid development (1990-present): Due to the development and improvement of individual materials, technologies, applications, and the advent of nanotechnology, there was an exponential increase in publications on the topic of porous structures and honeycomb-like patterns [12]. At this time, the breath figure method was also discovered, thanks to which there was a rapid development of ordered HCPs from polymer materials.

3. POLYMERIC HONEYCOMBS

Polymeric materials are suitable candidates for the preparation of microporous surface structures, as their properties can mimic those of natural honeycombs. These unique properties include a large specific surface area, high mechanical strength, good structural stability, thermal and sound insulation, and low density [19]. The relationships between the topology and mechanical, acoustic, and thermal properties of honeycomb structures are discussed by Zhang [12]. The unique properties of these structures mainly depend on the structure, the material used, the preparation process, and the final application. Depending on the polymers used, various aspects can be improved, by the introduction of additional functional groups or improvements in mechanical properties. While the physical aspects (thermal resistance, mechanical properties, etc.) are linked to the type of polymer that is used, the functionalization of HCP surfaces is undoubtedly a critical aspect of the usability of the structures [20].

Standard commercial polymers were most often used for the preparation of porous structures due to their low price and availability; biopolymers may also be used [21]. These include, for example, polystyrene (PS) [22], cellulose acetate [23], polymethyl methacrylate (PMMA) [24], polylactic acid (PLA) [25], poly(e-caprolactone) (PCL). Furthermore, polymers with unique properties have been used, such as polymers that are responsive to external stimuli, biodegradable polymers, thermally and chemically stable polymers, and conductive and semi-conductive polymers. Furthermore, in addition to the type of polymer that is selected, macromolecular properties (molecular weight, polymer topology) play a key role in pore formation. Peng [26] observed the effect of molecular weight on the viscosity of the polymer solution and the subsequent surface morphology of the prepared porous membranes. When the molecular weight was low, the viscosity of the solution was too low to encapsulate the droplets needed to form ordered membranes or prevent them from coalescing. Conversely, a high molecular weight led to a highly viscous polymer solution and the droplets did not have time to wrap, resulting in gaps in the prepared membranes. The very first polymers used by the breath figure technique were star homopolymers. Thanks to the spherical shape that they form in solution, they can create highly ordered structures [27] (Figure 3). However, it was later shown that even linear homopolymers can form ordered porous films [28] and their use is undoubtedly the most widespread strategy. These include homopolymers, homopolymers with functional groups, and random and block copolymers. PS-based ones dominate the linear block copolymers because styrene is easily soluble in various solvents and is a standard precursor in the synthesis of block copolymers [3]. Other possible groups, according to topology, are grafted, branched and cyclic polymers. All these groups of polymers are suitable for the preparation of HCPs, and concrete examples were comprehensively described in the book by Rodríguez-Hernández [21]. Parameters such as average pore size or pore distribution depend to a large extent on the functional groups contained in the polymer chain. Polar groups are key in the breath figure technique because they stabilize the water droplets.

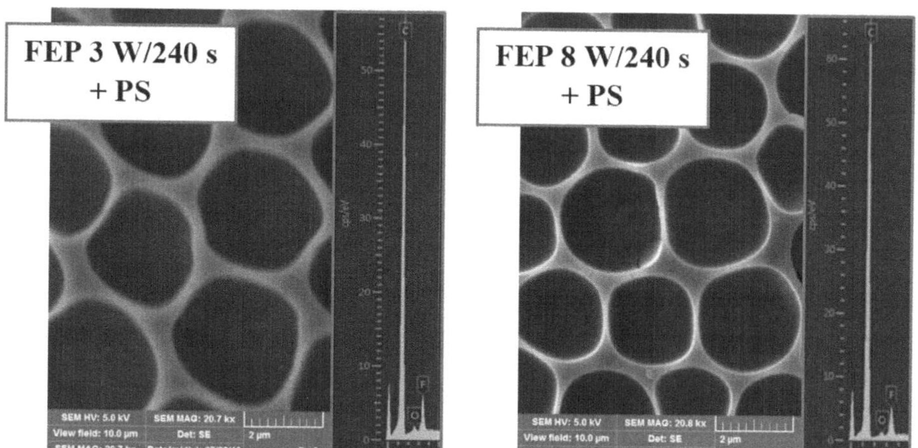

At. %		At. %
94.8	C	96.0
0.9	O	0.6
4.3	F	3.4

Figure 3.: *Scanning electron microscopy images (10 × 10 μm²) of plasma-modified fluorinated ethylene propylene (FEP) treated at different plasma discharges (3 and 8 W), with a polystyrene (PS) layer forming a honeycomb-like pattern. On the right side, corresponding energy-dispersive X-ray spectroscopy graphs and a table of element concentrations on the surface are depicted [27].*

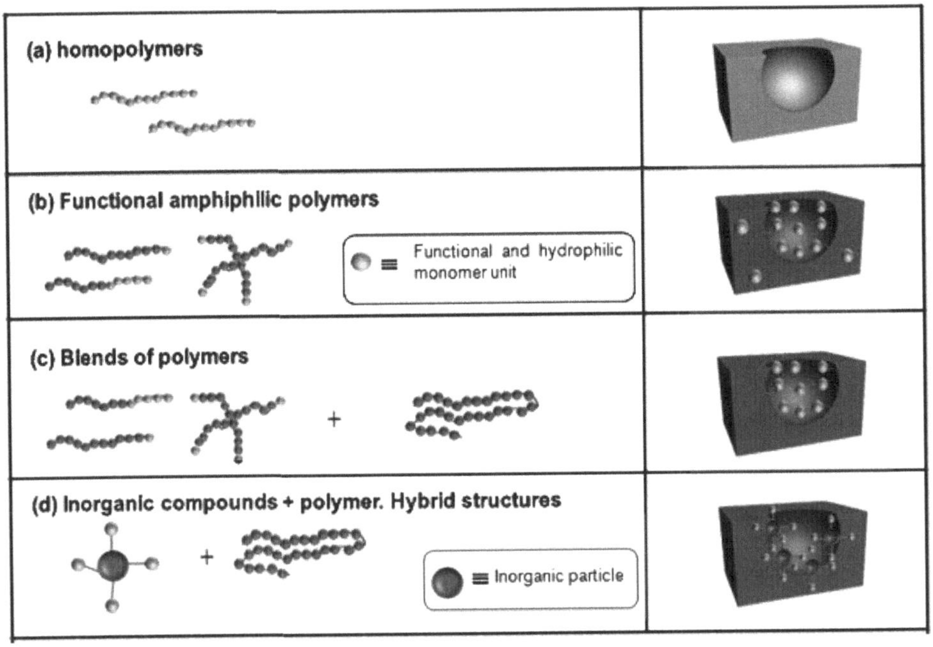

Figure 4.: Various alternatives leading to functional surfaces being obtained in situ [2].

For this reason, amphiphilic copolymers (polymers carrying hydrophilic groups) have been thoroughly studied for the preparation of HCP films, and proven to improve their construction [29]. The preferential orientation of hydrophilic groups to the pore wall occurs in the case of amphiphilic structures.

4. PREPARATION OF HCP STRUCTURES

There are several methods to prepare honeycomb-like patterned surfaces. Both ordered and disordered structures can be prepared, depending on the method used and the applications. We can encounter many successful top-down approaches for the construction of ordered HCPs with controlled pore size and shape. These include, for example, various lithographic methods (photolithography, electron lithography, soft lithography), microprinting, pressing, and lasers [32,33]. Their disadvantages often include a high price, demanding and complex procedures, and can destroy and damage the prepared pattern. From an economic point of view, their surface instability is used for the construction of porous polymer films, which is based on bottom-up wetting techniques [34], phase separation [35], and "template methods" [36]. If the solvent starts to evaporate from the polymer solution, or if the undiluted mixture is heated above the glass transition temperature, phase separation will occur. The phase separation of polymer mixtures can lead to changes in the properties of the system (molecular weight, composition, structure, solvent, film thickness) or changes in the external environment (pressure, substrate, humidity, temperature) to different surface morphologies, and the formation of pores. An important parameter is the rate of solvent evaporation, which affects the size and symmetry of the formed pores. We can also include the breath figure method, which will be described in more detail in the next section. It is known that phase separation can be achieved by introducing a so-called "bad" solvent (e.g., methanol) into the polymer solution [37,38]. There is a local concentration of methanol and, thus, a phase rich in methanol is formed by the evaporation of the good solvent (e.g., chloroform). This phased process is responsible for the separation of the two phases. Water that condenses on the surface of the polymer film from the ambient air helps the growth of methanol droplets responsible for the formation of hexagonal pores [39]. When a poor solvent is used, in combination with a good solvent, it serves a dual role: it increases the number of nucleation sites and shortens the crystallization time of the polymer by increasing the extent of phase separation. The amount of methanol is expected to be responsible for the pore size. Two types of methods use templates: hard and soft [40]. The hard method uses monodisperse particles as a template, mostly colloidal silica particles or PS particles. The advantage of this method is simplicity and price, but it has significant drawbacks. The particles must be processed until they have a uniform size, and their removal is not easy often leading to structural defects in the porous materials. The soft template consists of block copolymers, the selected blocks of which are subsequently etched or dissolved. UV radiation, ozonolysis, chemical etching, and reactive ion etching are used for removal. This method can not only control the pore size but also realize the functionalization of ordered nano-porous polymer materials. However, there are still two disadvantages of a soft template: poor thermal and solvent resistance [41].

4.1. Breath Figure

A successful method based on phase separation is the "breath figure" (BF). The name of this method refers to the fact that dew occurs when breathing on a cold surface [42]. Since 1895, when J. Aitken investigated the formation of BF, this method was considered a natural phenomenon [43]. In his observations, he outlined that the resulting patterns were caused by a complex interplay between the physical and interphase processes acting on water vapor. Half a century later, dew formation, whose thermodynamic and kinetic aspects are key to understanding BF formations in polymeric materials, was studied theoretically and experimentally [44]. A breakthrough in the application of BF was achieved in 1994, when Francois et al., came up with the fabrication of polymer films with a highly ordered HCP morphology using condensed water droplets, which served as a dynamic template [27]. Star PS or polystyrene-b-polyparaphenylene dissolved in carbon sulfide was used in the experiment. The solution was applied to a solid substrate with a stream of moist air, and a regular pattern was formed after evaporation of the water and solvent. Since then, much attention has been paid to the preparation of HCP films with the help of BF. Many researchers, such as Stenzel, have carried out detailed research and extended the BF method to a wide range of polymers [28].

The principle of the formation of porous membranes using the BF method is shown in Figure 5 [44]. As proposed by Munoz-Bonilla, the polymer is dissolved in a solvent, or mixture of solvents, and the solution is then applied to a substrate exposed to a stream of high-humidity air [2]. Evaporation of the volatile solvent lowers the surface temperature below the dew point, which leads to the rapid condensation and nucleation of water droplets on the surface of the polymer solution. Nucleation involves the formation of small droplets that do not evaporate because they are thermodynamically stable. During evaporation, the water droplets slowly grow at the expense of the vapor in the surrounding environment and, due to convection flow or capillary force, penetrating the surface of the solution, they pack into a hexagonal shape and serve as a template for the final HCP structure [45]. Once the surface temperature has equalized with the condensed water, the ambient temperature and remaining solvent evaporate to form the HCP. If the evaporation time is too short, the coalescence of water droplets occurs on the surface, and pores with random size and arrangement are formed. If the evaporation of the solvent ends before coalescence occurs, a homogeneous pore layer is formed. If the polymer stabilizes the condensed water droplets, and thus prevents precipitation, a situation may arise where the surface layer of condensed droplets sink into the solution. Whether a monolayer or a 3D porous network is formed depends on the density of the solvent used. When the solvent has a lower density than water, new droplets can condense and form another porous layer [2]. The key factors that determine the ordered structure in classical BFs are the stability of condensed water droplets and capillary force [46].

4.2. HCP Preparation Techniques Using the BF Method

Successful techniques for the preparation of porous structures using the BF method include spin-coating, drop-casting, and dip-coating [47]. The simplest and cheapest experimental technique using BF is drop-casting. The mixture of polymer and solvent is applied dropwise to the substrate. A relative humidity greater than 50% is usually required. Highly ordered structures are prepared by this technique. A modification of this basic technique has recently been developed to speed up the procedure by applying a series of drops in one step [48]. By applying an airflow with controlled humidity and flow, the rate of solvent evaporation can be controlled, and the flow indicates solution convection [49]. Compared to the direct application of the solution without airflow, there is a larger temperature gradient between the volume and the surface due to the faster evaporation. Control of the relative humidity, air flow rate, the concentration of dissolved polymer in solution or distance, and airflow angle are needed to optimize the resulting pattern. Films with lower roughness and homogeneous thickness can be prepared by the spin-coating method [50]. In this technique, the speed of rotation is an important factor. To ensure high ambient humidity during coating, containers filled with hot water are placed in the coating cell [51].

The dip-coating technique is also often used in combination with BF [52]. In this technique, the substrate is placed in a polymer solution in a vertical position. The speed of immersion and withdrawal from the solution plays a key role. Porous structures are mainly prepared on solid surfaces, including glass, silicon wafers, and polymers. An important feature of the carriers is their inertness towards solvents and suitable wettability.

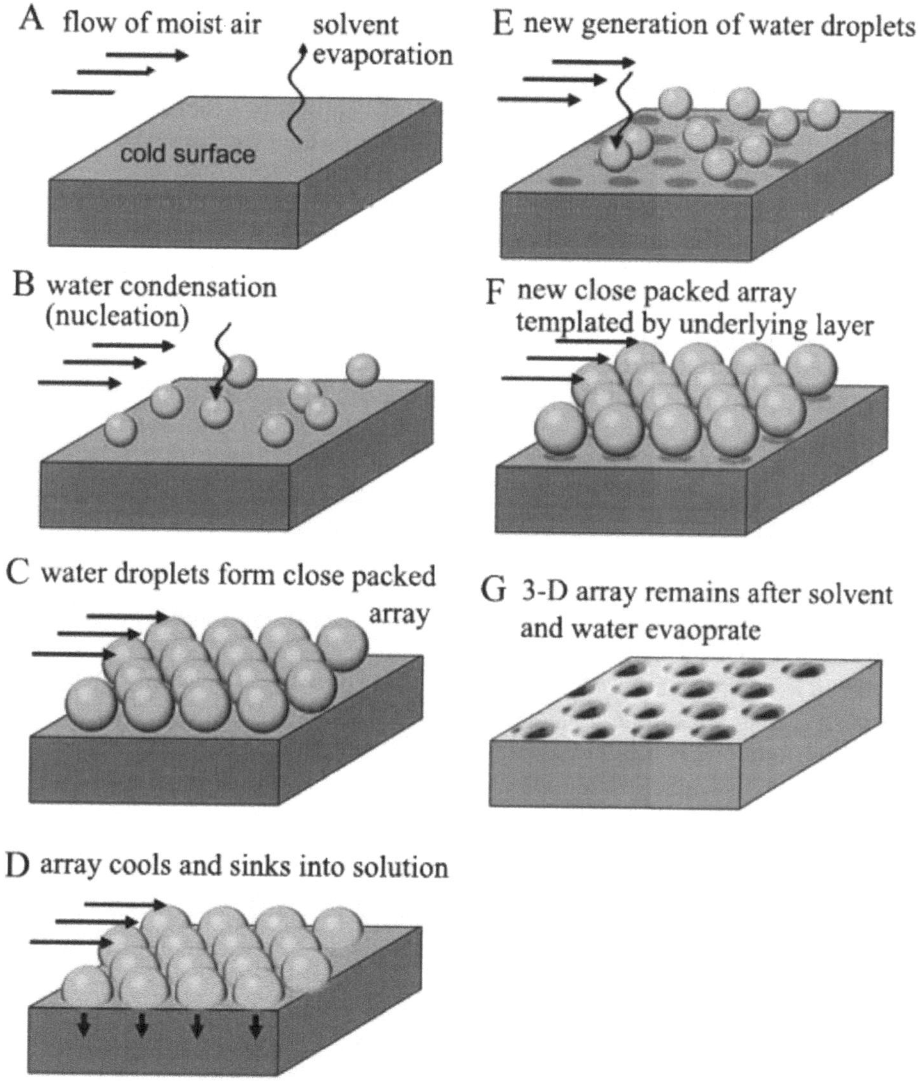

Figure 5.: *The BF method pore formation process [44].*

Plasma modification is an effective method, leading to an increase in the wettability of polymeric materials [53,54]. The action of the plasma discharge on the surface of the material results in the formation of reactive sites, cross-linking and changes in the chemical structure and molecular weight. This is a suitable strategy for incorporating functional groups, immobilizing proteins, and other biomolecules, or applying polymer coatings [55,56]. In addition, this treatment can be used to uniformly treat the surface of carriers of complex shape [57]. The air–water interface, the so-called "on-water spreading" approach, has been proven to be a suitable environment for the preparation of self-supporting structured films [58]. The morphology of the microstructure is controlled by the volume of the cast solution and the temperature of the water bath. Wan swapped water for ice and compared both interfaces, air–water, and air–ice. It was found that the polymer solution spread better at the air–ice interface [59]. To obtain highly ordered microporous films, the BF method requires the presence of specially structured polymers and appropriately designed devices that ensure precisely controlled high humidity [19]. These requirements limit the large-scale use of this method and its associated industrial applications. Farbod et al., used modified BF, known as "direct breath figure" (from the English direct breath figure, abbreviated DBF), to create a porous film. It is called direct because porous structures are created directly on the polymer substrate, and there is no need to apply an additional polymer film. It is, therefore, a top-down technique, as is the case with classic BF. The advantage of this method over classical BF is that DBF takes place without the presence of a moist surrounding atmosphere and the structure is a direct part of the substrate, retaining its mechanical stability [60]. However, the orderliness of the final pattern was not achieved, as is the case with classic BF. By combining the advantages of the BF and DBF techniques, a new breath figure method, called "semidirect breath figure" (sDBF), was created [61]. A pure solvent (in this case, chloroform) is applied to the substrate, replacing the polymer solution from classic BF. The method is as follows: (1) evaporation of chloroform in a humid environment and cooling of the substrate surface; (2) condensation of water; (3) formation of water droplets that serve as a template for the final hexagonal pattern; (4) swelling, dissolution and drying of the polymer surface; (5) evaporation of the remaining solvent and water and obtaining of a honeycomb structure, directly on the substrate.

4.3. Improved Phase Separation

Another suitable method for the preparation of HCPs is the "non-solvent-induced phase separation" (NIPS) method, which, unlike BF, enables the simple, large-scale production of porous patterns under the conditions of the surrounding atmosphere [38]. High humidity can be imitated by directly adding water or low-volatile solvents to polymer solutions [62], but the orderliness and homogeneity of the prepared films are not achieved. Bui prepared highly ordered honeycomb films from PLA without using high humidity [39]. This was achieved using a combination of the NIPS and BF methods, for which the name "improved phase separation" (from the English improved phase separation, abbreviated IPS) was adopted. This method has several advantages compared to the two mentioned above: it can be easily applied to industrial coating equipment, a large area can be covered, there is no need to control the air humidity or use surfactants, several cheap commercial polymers (PLA, PS) can be used (see Figure 6), and, in addition, the resulting morphology of the pattern can be controlled using the solvents and their ratio.

This is a simple two-step method, where, in the first step, a polymer layer is applied to a solid substrate and, in the second step, the substrate is immersed in a mixture of two solvents, "good" (chloroform) and "bad" (methanol). After extracting the substrate from the binary mixture of solvents, they evaporate into the air. As chloroform is more volatile, it evaporates faster and methanol droplets accumulate on the surface of the polymer. At the same time, based on the BF principle [19,63], the rapid evaporation of chloroform cools the polymer surface, and high relative humidity is used to condense air water vapor into the polymer phase, which is rich in methanol. Water causes an acceleration in droplet growth and an increase in their surface tension. This is essential to the conformational stability of methanol droplets, when the droplets are tightly packed due to capillary forces and thus form a porous structure. The effect of humidity on the formation of HCPs was confirmed by Bui in an experiment in a dry environment [39], where no pattern was formed. Humidity does not have to be precisely regulated, but must be kept within a certain range (25–75%). This two-step preparation can limit the uniformity of the resulting HCP on complicated, uneven substrates. For this reason, Bui came up with a simplification of the existing IPS [52] and modified the two-step method to a one-step one without losing its advantages.

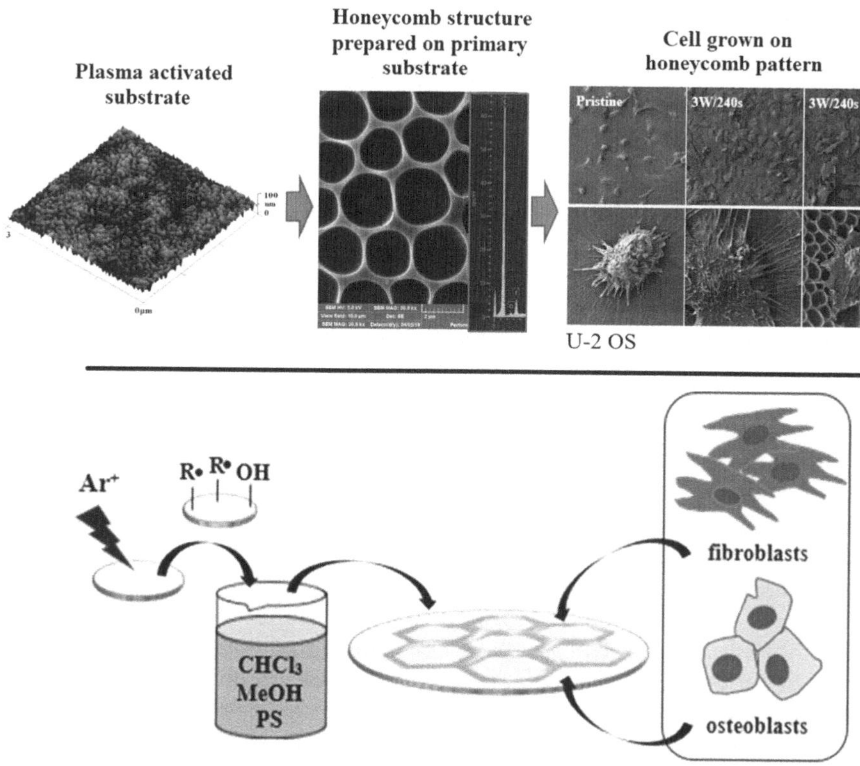

Figure 6.: Scheme of the preparation of a polymer scaffold for cell culture [27].

The principle is the direct addition of methanol to the polymer dissolved in chloroform, and the formation of a ternary system: polymer–good-solvent–bad-solvent. Subsequently, the substrate is immersed in the prepared solution and an ordered porous structure is spontaneously formed by the evaporation of the solvents in the air. This approach can be applied to various complicated surfaces.

The benefits of the suggested method depend on the relevant choice of methanol as a poor solvent, which induces phase separation under regular air and favors the creation of an ordered structure. Methanol, as a solvent, has three major impacts: (i) Methanol has a high affinity to water. Enriching the solution with methanol has a beneficial effect on the evaporation of solvents when the methanol absorbs water vapor from the surroundings. Accordingly, the phase separation that forms the resulting HCP can be induced in regular air. This preparation process has the potential to mass produce HCP films with high reproducibility and uniformity; (ii) Methanol's high solubility in water-immiscible solvents, such as carbon disulfide, chloroform and dichloromethane, enables the creation of ordered porous structures [46]. Conversely, water-miscible solvents, such as tetrahydrofuran, are poor candidates for preparing an ordered structure. Thus, if a mixture of water and water-miscible solvents is used, the formation of an inhomogeneous structure is likely to occur; (iii) The presence of methanol causes the polymer's transition to the surface of the poor solvent droplets. The formation of a gel-like layer occurs following the accumulation of polymeric material, which results in the droplet stabilizing against clotting. Therefore, a homogeneous structure can be achieved using commercial polymers. At some point, the local concentration of methanol induces a demixing process and forms two phases: (1) a polymer-rich chloroform phase containing a small amount of methanol, and (2) a methanol-rich phase containing a small amount of polymer and chloroform [64]. This phenomenon can be explained using the phase diagram of the thermodynamic behavior of the ternary-polymer–good-solvent–bad-solvent mixture [38].

5. PREPARATION TECHNIQUES AND THEIR VARIATIONS

The most distinctive feature of HCP films is their unique morphology. The parameters of the morphology of porous polymer structures play an important role in the field of bioapplications. Specifically, in tissue engineering, the size and distribution of pores determines cells' ability to attach, proliferate, and migrate across a substrate. Pore connectivity has been shown to affect cells' ability to interact with each other, their migration rate, and the nutrient/waste diffusion if cell migration into the scaffold is required [65]. The surface morphology of the honeycomb films, including the size, shape, distribution, and density of the pores, is controlled by changing the conditions of their preparation. Controlling these parameters is necessary to achieve a highly ordered hexagonal structure. Environmental conditions (the humidity, airflow, temperature, and speed of preparation technique used), polymer solution concentration, solvent, and substrate are among the most important parameters to consider when preparing polymeric HCP structures [66]. Different pore shapes can also be created using mechanical deformation. Geometric patterns such as rectangles, elongated hexagons, triangles, and squares can be obtained by compressing or stretching the hexagonal micro-polymer films [67].

5.1. Modification of BF Method

A high ambient humidity is considered essential to the formation of a regular hexagonal pattern in the BF method. The pore size becomes larger with increased humidity in the surrounding environment. This is due to the higher number of water droplets that accumulate on the surface layer of the polymer solution. However, no pores will form on the film if the relative humidity of the atmosphere is too low [68]. Smaller pores are formed if a higher flow-rate of the surrounding air is reached, which evaporates the solvent faster. However, excessive air flow cannot provide the droplets with enough time to expand and arrange into a regular array [69].

Li et al., prepared HCP films with gold nanoparticles capped with dodecanethiol and controlled the velocity and direction of the moist air flow. The result was a change in the shape of the pores, from circles to ellipses [70]. A decrease in temperature caused an increase in the viscosity of the solution and its condensation and, at the same time, decreased the rate of solvent evaporation. As a result, larger pores and highly ordered HCP structures were formed [71]. Various BF preparation techniques were mentioned in the previous section. It is important to consider their speed when applying the solution. It has been shown that a higher spinning rate leads to a more homogeneous porous structure, and a lower spinning rate accompanied by slower evaporation leads to higher droplet coalescence. Furthermore, the pores become smaller as the rotation speed increases. This is because the water droplets have a shorter time to grow during faster evaporation. In general, the use of spin-coating techniques results in irregular patterns in comparison to the air-flow or drop-casting techniques. Moreover, pores created in this way tend to be less spherical and more elongated [72]. During the dip-coating technique, the withdrawal time from the polymer solution plays an important role. The faster the extraction from the solution, the smaller the resulting pores (see Figure 7) [52]. The pore dimension may enhance the cytocompatibility of the substrate [73], which will also be discussed in the next section. The viscous resistance of the solution increases as the draw-out speed increases. Thus, a higher pull-out speed leads to more solution on the sample. Moreover, to achieve phase separation and stabilization of the droplets, sufficient thickness of the solution layer is required. Thus, a higher pull-out speed leads to a higher homogeneity of the HCP layer [52]. A pressure drop can quicken the solvent evaporation, in the same way as the acceleration of air flowing over the surface of the solvent [74]. In the work of Kuroda et al., a way of creating arrangements of porous fields on the surface layer of the polymer using a vacuum is presented. The degree of vacuum is shown to affect the size of the pores that are formed [75].

5.2. Polymer

Additionally, the properties and type of polymer used affect the resulting morphology of the layer. For example, larger polymers are produced, with an increase in the polymeric molecular weight [76]. Deepak et al., showed the effect of polymeric substrate functionalization on pore formation [77]. In the case of a polymer mixture, the hydrophobic/hydrophilic component can dramatically affect the structure of the pore [50]. An important parameter controlling the morphology of the layer is the concentration of dissolved polymer in the solution. If the concentration of the polymer in the solution is large enough to keep the water droplets stabilized and prevent them from coalescing, then low concentrations usually lead to larger pores. If the polymer concentration in the solution is too high, unevenly distributed and shaped pores are formed. This is due to the high viscosity, which affects the arrangement of the water droplets forming the pores [78].

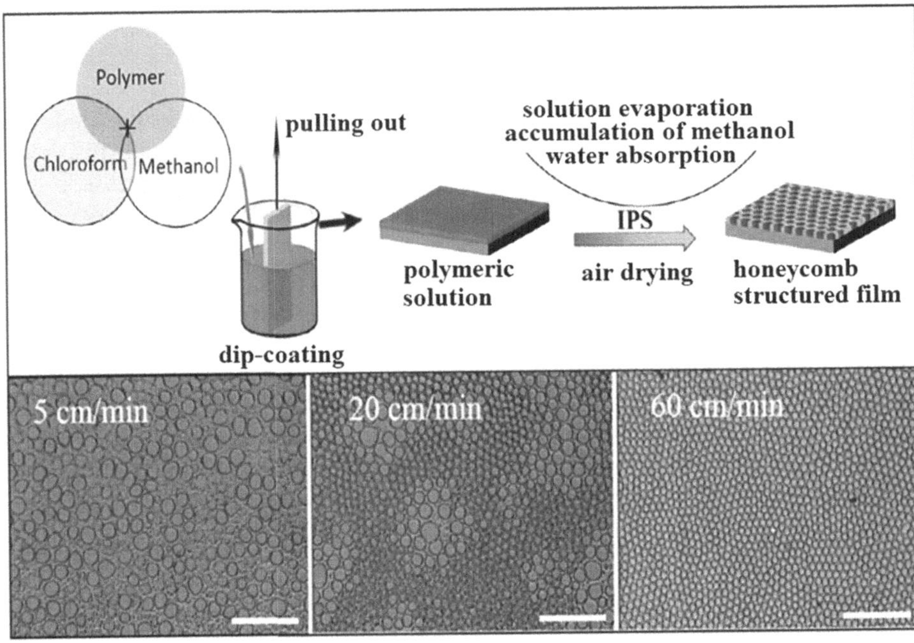

Figure 7.: The effect that the speed of the substrate's withdrawal from the solution using the dip-coating technique has on the morphology of the HCP layers [52].

5.3. Solvent

The choice of solvent is a very important factor affecting the morphology of HCP films. Inherent solvent properties, such as volatility (the associated evaporation time) and solute solubility, surface tension between water and solution, and the thermodynamic affinity between the solute/solvent system and the boiling point of the solvent, govern the regularity and pore size. High volatility, and insolubility in water mean that water droplets do not mix with the polymer solution; a low boiling point, polymer precipitation, and good compatibility with the polymer are the basic requirements when choosing a solvent in the BF process [3]. High-volatility solvents usually lead to the creation of smaller pores due to the rapid evaporation of the solvent and the short time needed for water droplet growth [26]. During the evaporation of the solvent, the water droplet is stabilized by high interfacial tension. [79]. However, if the solvent has too high a boiling point, the solvent evaporates more slowly and evaporative cooling is not sufficient to condense water droplets from the atmosphere; for condensation to occur, there must be a difference in temperature between the surface of the solution and the dew point of the atmosphere. When using solvents with a lower boiling point, the difference between the surface temperature of the solution and the dew point increases, and more water condenses on the surface of the solution [80]. Methanol and chloroform, used as a system solvent–non-solvent, show very effective sensitivity for controlling pore size, distribution and shape. The average pore size considerably increases with an increase in non-solvent content in the mixture. This is due to the accumulation of more methanol in the non-solvent droplets and their enlargement [39]. When a poor solvent is used in combination with a good solvent, it plays a dual role: it increases the number of nucleation sites and shortens the crystallization time of the polymer by increasing the extent of phase separation [81]. Bui et al., used water-miscible and water-immiscible mixtures of solvents and non-solvents to create and govern the morphology on the surface of the resulting porous PMMA films and compared their effects [64].

5.4. Substrate

To prove the universality of the BF method, HCP layers have been formed on several solid [52] and liquid substrates [82]. The porous film is influenced by the substrate through bulk and surface properties, such as thermal conductivity, surface chemistry, and morphology. Furthermore, the size of the formed pores was shown to be related to the wettability of the substrate. While the hydrophobic samples resulted in the formation of smaller pores, the hydrophilic samples produced larger pores [83]. A higher pore arrangement was achieved on a thin glass substrate compared to a thick glass substrate. For this reason it is assumed that substrate thermal conductivity is involved in the film formation mechanism. It is important to use substrates that are not eroded by solvents, as organic solvents that are immiscible in water are mostly used. In most studies, authors use smooth substrates; however, they can also be prepared on various curved and structured surfaces, as the polymer solution copies the shape of the surface [84,85]. The total volume of condensed water droplets also increases by using a substrate with a higher surface roughness, which corresponds well with the increasing number of nucleation droplets [86].

6. APPLICATIONS IN MEDICINE

The possible uses of polymer HCP materials are very diverse. In biomedical uses, research is usually focused on increasing the efficiency and biocompatibility of these materials, which is an important aspect of tissue substitutes, scaffolds in tissue engineering [87,88,89], biosensors, gene carriers, and drugs [90,91]. Films with a honeycomb structure have chemical stability, good compatibility, controllable porosity, and flexibility in functionalization, which enables them to serve as excellent biological carriers. Furthermore, porous films are applied in the field of optical devices [92], ion separation [93], in the production of surface-enhanced Raman scattering (SERS) elements [94], superhydrophobic surfaces [95], and as electronic components [96].

6.1. Tissue Engineering

The field of tissue engineering mainly requires the regeneration or replacement of damaged tissue or organ. So-called "scaffolds" play a critical role in the development of new tissue in the interaction with human cells. These are material structures that have been designed to encourage cells to adhere, spread, migrate, proliferate and differentiate to form the desired tissue structure. Therefore, the chemical, mechanical and topological surface properties of scaffolds are important for regulating cell behavior [97]. The porous nature and unique surface topography of polymeric HCP layers make them suitable samples for investigating the various effects of cell topography behavior and for the production of biomimetic materials for use in tissue engineering [98]. The term "biomimetic polymers" refers to polymer-made materials that are focused on mimicking a structure, biological environment, or function to elicit the desired cellular responses. This definition refers to various design principles and materials from ligands interacting with receptors on the surface of cells to material structures with micro- and nanometer dimensions that cells can perceive and respond to in a specific way [99]. The many advantages that porous films provide include the increased permeability necessary for in vivo and in vitro tests, topographical stimuli influencing cell behavior, and the increased, spatially controlled adsorption of proteins. Another function of the pores can be as a protective and storage site for growth factors and bioadhesive molecules, or its use in the creation of site-specific immobilization sites for drug-delivery systems (DDS) [100].

The edges and pores themselves strengthen the cell's ability to attach to the substrate during cultivation (increased cell adhesion) and can also mimic the essential microtopographic properties of the native extracellular matrix (ECM), further improving contact and interconnection in the horizontal direction and, thus, the possible formation of 3D structures. The pores' presence throughout the structure can facilitate the flow of oxygen, nutrients, and waste products. An adequate flow of substances is an extremely important prerequisite for the survival of a cell and its function in cell culture media and transplantation [80,101]. Different pore sizes can serve as an effective tool to control cell growth or select different cell types [102].

An excellent review of studies dealing with polymeric porous structures as scaffolds in the field of tissue engineering has been presented by Calejo [98]. Biodegradable and amphiphilic polymers, including copolymers containing PCL, PLGA, PLA, PS, and others, were used for the preparation of scaffolds. Many types of eukaryotic cells were investigated, including stem cells, cancer cells, and normal cells, whose attachment to HCP films is shown in Figure 8 for polystyrene and U2-OS cells, while the fluorescent images of the same cell type on PMMA is displayed in Figure 9. It has been shown that HCP structures can support the adhesion of most normal cells, such as cardiac myocytes [102], endothelial cells [103], chondrocytes [104], osteoblasts [105], hepatocytes [72], fibroblasts and keratinocytes [106,107], NIH-3T3 cells [71], and some cancer cells, such as a cervical epithelial carcinoma cell line [108] and lung cancer cells [109]. Regular surface topography induced the differentiation of neural and mesenchymal stem cells [110,111]. Stem cells, as undifferentiated cells, have great potential in the field of tissue engineering, especially in cell therapy and disease modeling. This is due to their ability to differentiate into specialized cells in response to adequate signals and their high regenerative capacity. Surface topography can be applied in the field of cell culture to guide differentiation in the desired direction or maintain self-renewal without the use of any harmful chemicals [112].

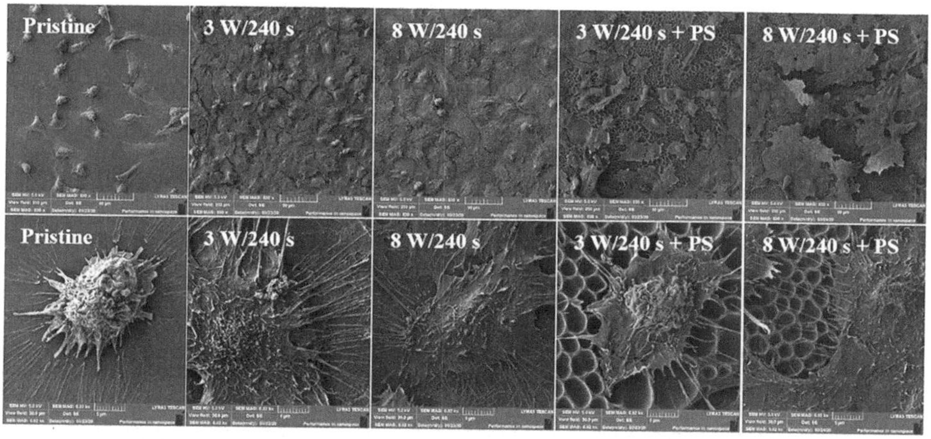

Figure 8.: *Scanning electron microscopy images of human osteoblasts (U-2 OS) six days post-seeding on FEP (pristine), FEP matrices treated by plasma with 3 W and 8 W (240 s) and subsequently coated with a honeycomb-like pattern formed from PS. The upper line represents a 250 × 250 μm2 scan, the bottom line a detailed scan with the area of 30 × 30 μm2. [27].*

6.2. Drug Delivery

Not only can structured films, thanks to their topography, control the "fate" of cells, they can also release therapeutics in a controlled manner into the surrounding medium (or into cells cultured on the material) or create reservoirs for various drug-delivery systems [113]. Due to its large specific surface area, the porous surface can accommodate a larger amount of therapeutics, even drugs with low solubility, which can be well dispersed in hydrophobic polymer matrices [114]. In addition, studies have shown that the encapsulation of drugs in porous layers contributes to a reduction in the so-called "burst release", the negative effects of which are pharmacologically dangerous and economically inefficient [115]. Due to the highly porous nature of HCP films, the release characteristics of encapsulated pharmaceutical compounds may be significantly modified compared to non-porous films of the same material. It is, therefore, important to characterize such releases, especially as the polymer film degrades over time. Salicylic acid (highly soluble in water) and ibuprofen (slightly soluble in water) were selected as model drug compounds to characterize the release rate. The release rate of both compounds from a porous PLGA thin film was higher compared to an equivalent non-porous film [116].

Porous coatings with antifibrotic drugs, used for glaucoma drainage implants in the treatment of intraocular pressure, have shown very promising results [117,118]. The success rate of these uncoated implants is relatively low because the resulting fibrosis during the wound-healing process blocks fluid drainage.

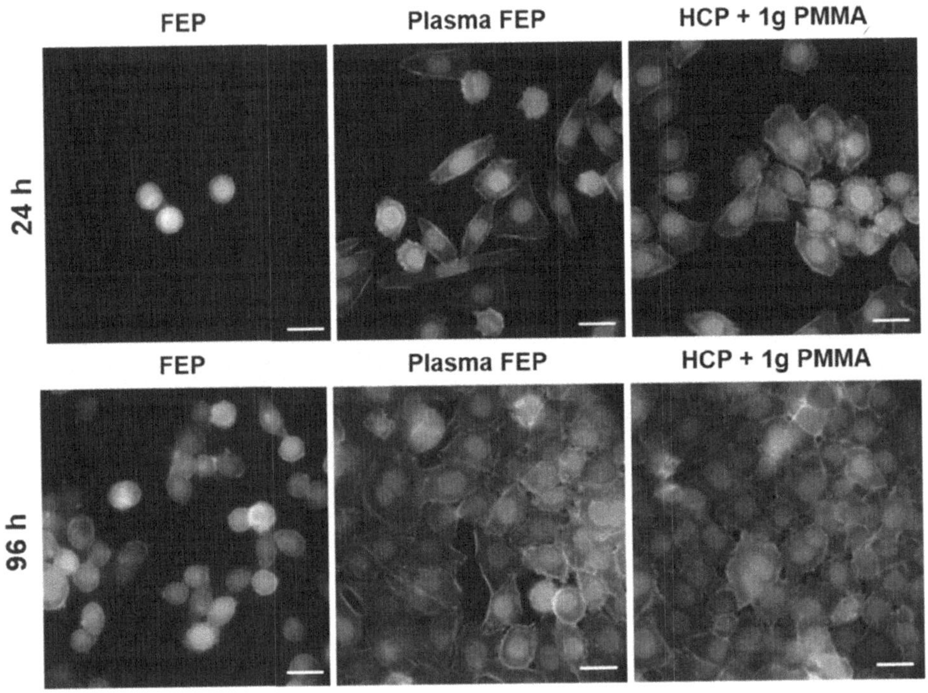

***Figure 9.**: Fluorescence microscopy images of VAMPIRO U-2 OS cells (human cells derived from osteosarcoma, stably transfected with tFP602) growing on pristine fluorinated ethylene propylene (FEP) and FEP treated with 8 W plasma for 240 s and subsequently coated with a honeycomb-like pattern formed from polymethyl methacrylate (PMMA) solution (1 g), 24 h and 96 h post-seeding, respectively. Scale bar represents 20 microns [24].*

The porous coating prevents the formation of fibrosis, and thus reduces the occurrence of post-operative complications. Another study has shown the potential of double-layer porous PLGA platelets loaded with fenofibrate to successfully fight the growth of glioblastoma, a brain tumor. These platelets enabled a constant and steady release of the drug that actively inhibited the growth of tumor cells more effectively than a drug administered in a single dose or at high concentrations [119]. The wide distribution of pores and cavities can also serve as an ideal space for drug-delivery systems (DDS) [120]. For example, polymer microspheres [121] and nanoparticles [122] can be dispersed in a very simple way using gravity inside the pores of HCP films. Physical capture or electrostatic interaction can also be used. The advantage of the presence of secondary DDS lies in the greater load capacity of the system and the modulation of drug release, or in the possibility of achieving controlled drug release according to external stimuli such as temperature, light, ionic strength and pH [121,122]. Other research has shown that it is possible for porous polymer films whose cavities act as protein reservoirs to be used as a glucose-responsive insulin delivery tool, which may have potential as a controlled-release drug-delivery system for the treatment of diabetes [100].

6.3. Biosensors

Porous films are promising materials for the fabrication of highly sensitive biosensors. One example is a nanoelectrode formed by a polymer thin HCP film for the electrochemical detection of uric acid [123]. This biosensor can isolate the anodic peak in uric acid from that of ascorbic acid in cyclic voltammograms, which, without HCP layers, shows overlapping responses. The measurement of uric acid in human physiological fluids is used in the diagnosis and treatment of many renal and metabolic disorders, including renal failure, gout, leukemia, starvation, or other wasting conditions, and in the treatment of patients receiving cytotoxic drugs [124]. Chen reported the design of a patterned polymer film containing phenylboronic acid, exhibiting a high sensitivity for glucose detection [125]. The phenylboron part is hydrophilic and aggregates on the surface of water droplets, forming pore walls after evaporation. The BF method indicates the specific location of polar groups on the pore surface. Thanks to this, HCP films can sense biomolecules, such as proteins, through the presence of sugars on the surface of the pores [126,127]. Flexible touch sensors with high sensitivity are promising candidates for the production of artificial "electronic skin" for prosthetic devices or robotics. Such applications require inexpensive and scalable methods for producing pixels (units of sensing capacity) that sense medium and low pressures. For this reason, the BF method was chosen to create micro-structured and pattern-compressible dielectric layers for capacitive pressure sensors. The authors prepared porous polystyrene layers that served as molds for structuring polydimethylsiloxane dielectrics [128].

6.4. Antibacterial Layers

Bacteria that adhere to the surface of materials, multiply and colonize to form biofilms often need to be controlled, especially on the surface of medical implants. The formation of a biofilm on an implant surface is one of the most common causes of failure [129]. The porous structure leads, in most cases, to an improvement in cell proliferation and may, therefore, raise concerns about increased bacterial colonization. It is essential to find effective strategies to prevent bacterial retention. Microporous surfaces have interesting properties that can enhance their antimicrobial effect [130]. This is very often achieved using antibacterial functional groups [131,132], organic and inorganic molecules [133], or different pore shapes and sizes. For example, films with pores with a diameter of 5–11 μm effectively prevent the attachment of Pseudomonas aeruginosa bacteria, a severe source of infection at present [134]. Vargas-Alfredo et al. [135] proposed the production of highly effective antibacterial substrates that are selective for bacteria, i.e., non-cytotoxic to mammalian cells. The proposed strategy is based on the different sizes of bacteria (1–4 μm) compared to mammalian cells (above 20 μm), which allow for the bacteria to contact the interior of the micrometer-sized pores containing antimicrobial functional groups. On the contrary, mammalian cells, which are larger, stay on the upper layer, thereby reducing unwanted cytotoxic effects and improving the biocompatibility of the substrates. Recently, Huang et al., used the same strategy and proposed a novel hybrid BF method for the preparation of functionalized HCP Petri dishes for cell culture [136]. The structure of mPEG, with its amphiphilic structure, activates OH groups that are located inside the pores and can covalently bind with other functional groups. Another effective strategy is doping polymer films with metal nanoparticles. One successful option is nanoparticles of silver (AgNPs), as they have good antifungal, antiviral and antibacterial properties (Figure 10) [137,138,139]. Ag, in its ionic form at low concentrations, acts as an antibacterial agent, although, for the Ag0 form, no significant antibacterial effect has been found. One example is an HCP film made of PLLA functionalized with Ag nanoparticles: the results show that the activity of Ag nanoparticles is more effective against E. coli bacteria in structured films compared to smooth films [140]. The addition of ferrocene to the PS solution showed an efficient way to control the Ag-functionalization and morphology of the surface of the fabricated HCP porous layers [141].

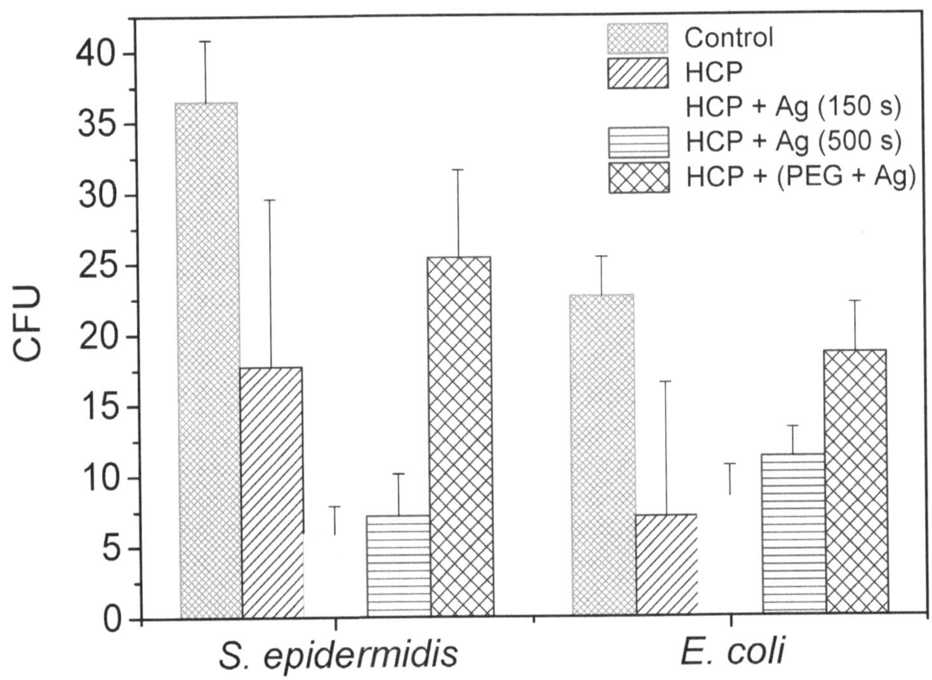

Figure 10.: *The number of colony-forming units (CFU) of S. epidermidis and E. coli were applied on the surface of the prepared samples of fluorinated ethylene propylene (FEP) for 2 h. The samples were tested after each phase of preparation: (1) A honeycomb-like (HCP) structure formation; (2) sputtering 150 s and 500 s of an Ag layer on the HCP surface; (3) sputtering of Ag into polyethylene glycol (PEG) and embedding the mixture into a solution of HCP structure. Bacterial suspension incubated only with phosphate-buffered saline without any sample addition served as a control. The samples were performed in triplicate [138].*

7. MODIFICATION OF HCP STRUCTURES

In the previous section, the possible use of polymer layers with a porous structure in medicine was discussed. Modification by physical and chemical means, or a combination thereof, is an effective way to prepare HCP films with new properties. For example, in the field of tissue engineering, the most common goal is to create a surface that is attractive to cells, and increase their adhesion and proliferation on the surface of the carrier [142]. This can be achieved by optimizing the wettability and surface roughness, chemical composition, charge, and mechanical strength [143,144]. Some medical applications require antibacterial properties on the polymer surface, which can be prepared, for example, by adding metal nanoparticles or a PEG polymer [145]. The morphology of the surface and its nanostructuring are particularly important for the differentiation of stem cells and biosensors [146,147].

7.1. Physical Modification

Physical modifications are implemented on already-created HCPs and can be split into cross-linking and surface treatments [3]. The purpose of surface treatments can be, for example, the improvement in adhesion properties, change in hydrophobicity, the introduction of special functional groups to the surface, or the modification of surface morphology [148] (see Figure 11).

One of the surface treatment methods that allows for the chemical and physical properties of polymer surfaces to be changed without affecting their bulk properties is treatment by plasma discharge [149]. Exposure of the polymer surface to plasma produces three main effects: surface ablation or etching, polymerization, and modification of the chemical structure of the exposed substrate. These phenomena lead to changes in wettability, surface energy, roughness, and chemistry, thereby contributing to the incorporation of functional groups or the attachment of proteins and cells to the polymeric material [150]. Yang et al., showed the incorporation of amino groups by plasma treatment, which promoted cell proliferation and attachment on a smooth PLA film [151]. In their next study, they combined the advantages of plasma treatment when incorporating amino functional groups and the excellent surface properties of HCP-patterned PLA to efficiently produce a biocompatible and cheap material [152]. Due to the presence of an external stimulus, such as a thermal or mechanical stress, osmotic pressure, or electric field, so-called "wrinkling instabilities" may appear, especially on the surface of polymers, which arise as a result of the system's reaction to the loss of equilibrium. As a result, different surface morphologies can be observed with specific shapes and geometries at the nano- and micro-scale [153]. If the instabilities are created in a controlled manner, then periodically structured surfaces in polymeric material with modified properties of technological importance, such as adhesion, friction or wetting, can be obtained. Various laser techniques have been studied to create surface instabilities leading to the formation of surface topography. An example of laser technology is "laser-induced periodic surface structure" (LIPSS) [154]. Wrinkling is caused by repeated irradiation of the polymer surface with pulsed lasers. The probability of generating LIPSS on the polymer surface mainly depends on the polymer effectively absorbing the laser light [155]. Excimer laser exposure can, in special cases, lead to the formation of a globular structure on the surface of the polymer material [155].

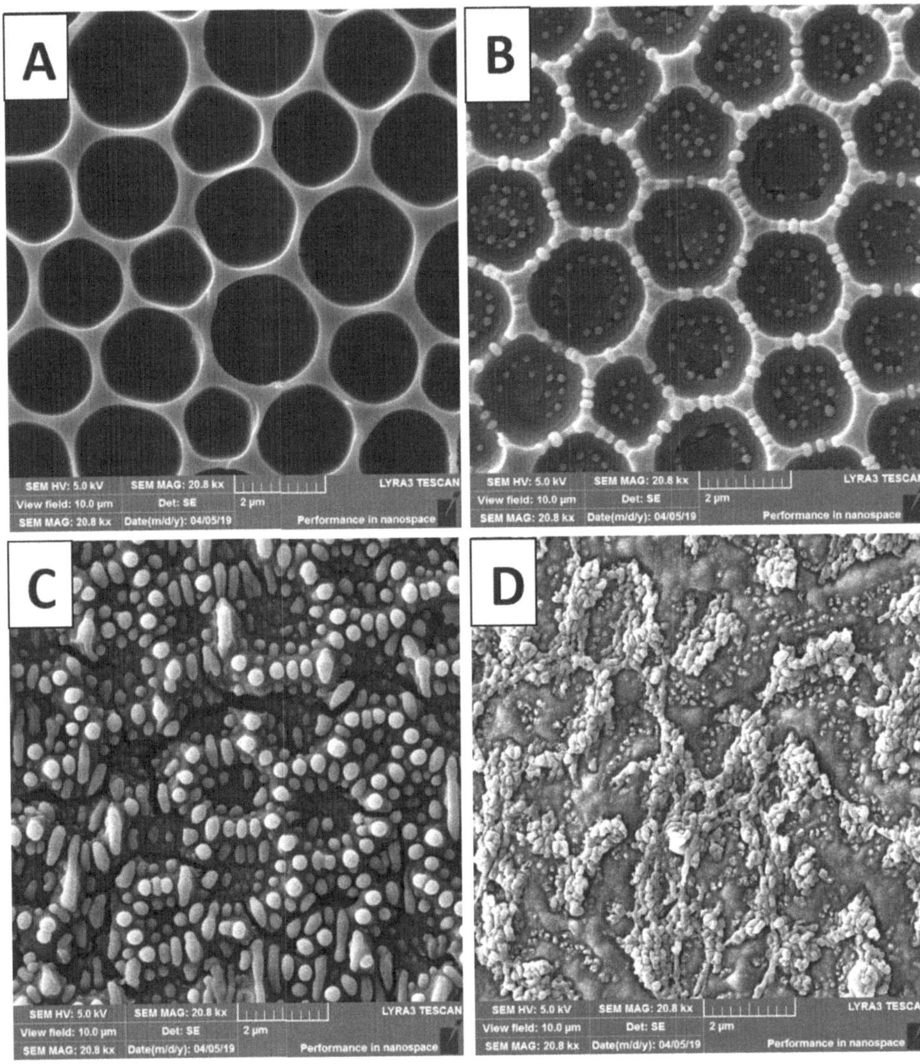

Figure 11.: *The different stages of modification of honeycomb-like PS patterns prepared on a FEP substrate plasma exposed to 8 W and 240 s (2 g of PS, ratio 90/10). The polystyrene HCP structures were exposed to an excimer laser with 6000 counts and laser fluence: (A)—0, (B)—6 (C)—10 and (D)—20 mJ.cm^{-2} [22].*

Layers with the HCP pattern show hydrophobic behavior in most cases, due to their surface structure and the hydrophobic polymers from which they are prepared. Surface treatment of PS honeycomb films with UV-ozone caused the oxidation of surface hydrocarbons and thus increased surface wettability [156]. Another physical treatment can be the deposition of a thin organic (e.g., carbon) [157] or inorganic (e.g., metal, ceramic) [158,159] layer on the HCP surface. For example, a metal thin layer can be deposited on the substrate surface by sputtering in a vacuum [160]. In this way, a surface is created that has different chemical and structural properties to the original film. Not all foils produced by the BF method have good mechanical strength, which is desirable in often harsh environments. There is also a significant limitation in their use in some materials and their bioanalytical applications due to their solubility in organic solvents. Cross-linking is a commonly used strategy for modifying the obtained HCP films, especially when influencing their mechanical properties, and increasing thermal and chemical resistance. Cross-linking processes can be divided into three types: photochemical cross-linking, chemical cross-linking, and thermal cross-linking, all of which try to not damage the regularity of the films [3]. After chemical cross-linking, the films showed excellent resistance to high temperatures and various organic solvents and good mechanical strength [161]. Photochemical processing is another way to activate the cross-linking of porous films, principally when the casting solution contains double bonds. For example, in one study, films became chemically resistant to a variety of solvents and stable at 350 °C after 1 h of UV exposure.

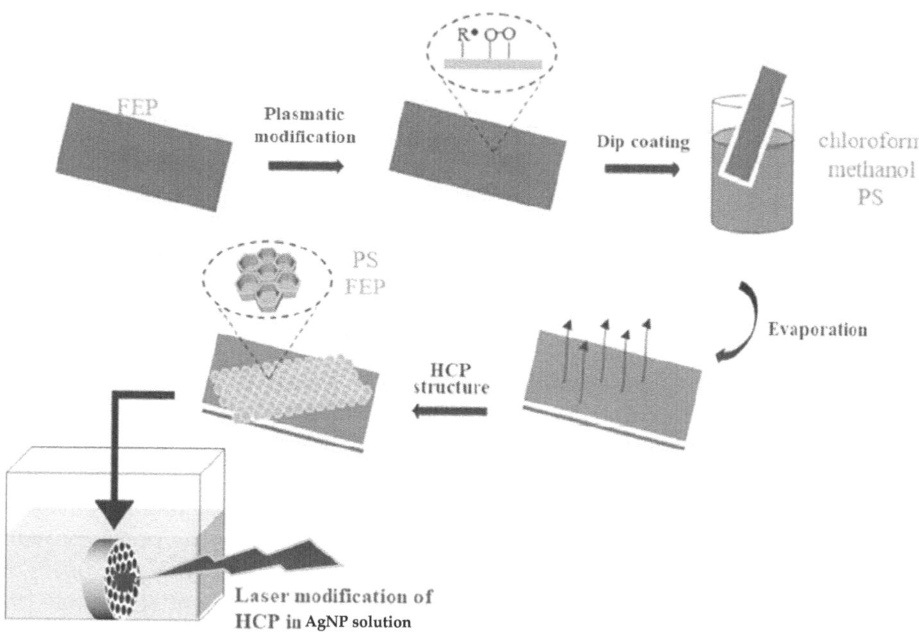

Figure 12.: Scheme of the preparation of a honeycomb pattern immobilized with Ag nanoparticles. [164].

Additionally, photochemical cross-linking led to the formation of polar groups that converted the surface from hydrophobic to hydrophilic [109]. Microporous-structured thin films were prepared on functionalized PDLLA with photoreactive groups. Subsequent photo-crosslinking rendered the film insoluble in chloroform [162]. Thermal cross-linking is a one-step, effective method for treating polymeric HCP films. Through high-temperature processing, the groups with temperature-sensitive properties are cross-linked and part of the polymer microstructures collapse [3]. To preserve the microstructure, Bolognesi came up with a two-step heat treatment [163]. Physical modification can also include the targeted mechanical deformation of prepared HCP structures. Porous layers formed from elastic and viscoelastic polymers can expand or contract. The deformed shapes of the porous structures are characterized by the stretch/shrinkage ratio and the stretch/shrinkage direction [80]. The physical immobilization of noble metal nanoparticles from the solution can be simply realized by the exposure to an excimer laser in liquid (Figure 12), when the 248 nm wavelength is applied, a honeycomb composite is formed [164].

7.2. Chemical Modification

The in situ functionalization of HCP surfaces is often limited by polymers with an amphiphilic structure that facilitate the formation of a porous structure using the BF method. This section describes various chemical reactions and grafting methods that are used for ex situ modification of the chemical functionality of already-prepared porous layers [2]. These are substrate reactions with the solution when the chemical composition inside the pores changes by introducing a wide range of functional groups, or by linking organic or inorganic substances (molecules, nanoparticles, etc.). In particular, the immobilization of bioactive ligands has attracted intense interest due to the potential use of protein microarrays, which are not only suitable for the study of cell adhesion but also as biosensors and a material for tissue engineering [165,166]. Another challenge in microarray preparation is the site-specific immobilization of proteins to ensure that protein activity is not restricted [167]. Min et al., anchored biotin to the pore surface, which selectively interacts with the protein streptavidin [168]. A similar approach, using avidin–biotin non-covalent interactions, was described by Ke [169] and Nyström [170]. Linear polystyrene was mixed with amine-bonded polystyrene to prepare porous layers in which the functionality of the cavities could be modified. After the formation of BF, the hydrophilic amino groups were oriented toward the pore surface. Proteins were attached to the pore surface by cross-linking to amino groups using glutaraldehyde. The turning of hydrophilic amino groups towards the pore surface occurred after the formation of BF. On the surface of the pores, proteins were cross-linked to amino groups using glutaraldehyde [171]. The grafting of polymers is primarily an interesting method of chemical treatment of the material. The techniques of controlled radical polymerization provide a broad spectrum of polymers with narrow polydispersity on surfaces and controlled chain length using either a "grafting from" or "grafting onto" approach [2,172].

8. CONCLUSIONS

For the appropriate selection of the material and the effective use of its properties, it is necessary to study and understand the material's surface and internal structure. We focused on the on formation of honeycomb microstructures and their applications, which include tissue engineering, antibacterial materials, replication processes or sensors. Thanks to this knowledge of the structure, it is possible to modify the existing properties of the material according to its intended application. Furthermore, surface nanostructuring provides a number of possibilities to modify or improve the properties of the studied material. We described experiments involving the breath figure process and the improved phase separation for honeycomb pattern formation, as well as subsequent surface modifications of the pattern.

Author Contributions

Conceptualization, P.S. and N.S.K.; validation, P.S. and V.Š.; formal analysis, B.F., N.S.K. and V.Š.; resources, P.S.; writing—original draft preparation, D.F., K.F. and B.F.; writing—review and editing, K.F., B.F. and P.S.; supervision, P.S.; funding acquisition, P.S. All authors have read and agreed to the published version of the manuscript.

REFERENCES

1. Tóth, L.F. What the bees know and what they do not know. Bull. Am. Math. Soc. 1964, 70, 468–481.
2. Munoz-Bonilla, A.; Fernandez-Garcia, M.; Rodriguez-Hernandez, J. Towards hierarchically ordered functional porous polymeric surfaces prepared by the breath figures approach. Prog. Polym. Sci. 2014, 39, 510–554.
3. Dong, C.; Hao, J. Honeycomb films with ordered patterns and structures. In Comprehensive Supramolecular Chemistry II (Volume 9 Nanotechnology); Atwood, J.L., Ed.; Elsevier: Amsterdam, The Netherlands, 2017; pp. 207–229. ISBN 978-0-12-803199-5.
4. Cilliers, L.; Retief, F.P. Bees, honey and health in antiquity. Akroterion 2008, 53, 7–19.
5. Bulmer-Thomas, I. Selections Illustrating the History of Greek Mathematics; Harvard University Press: Cambridge, MA, USA, 1939; Volume 2.
6. Kepler, J. The Six-Cornered Snowflake; Paul Dry Books: Philadelphia, PA, USA, 2010.
7. Hales, T.C. The honeycomb conjecture. Discret. Comput. Geom. 2001, 25, 1–22.
8. Gallo, V.; Chittka, L. Cognitive aspects of comb-building in the honeybee? Front. Psychol. 2018, 9, 900.
9. Wang, Z.; Zhang, Y.; Jiefu, L. Comparison between Five Typical Reinforced Honeycomb Structures; Atlantic Press: Dordrecht, The Netherlands, 2015; pp. 704–707.
10. Langstroth, L. Langstroth's The Hive and the Honey–Bee; Dover Publications Inc.: Mineola, NY, USA, 2004; originaly published in 1853.
11. Weaire, D.; Phelan, R. Optimal design of honeycombs. Nature 1994, 367, 123.
12. Zhang, Q.; Yang, X.; Li, P.; Huang, G.; Feng, S.; Shen, C.; Han, B.; Zhang, X.; Jin, F.; Xu, F. Bioinspired engineering of honeycomb structure–Using nature to inspire human innovation. Prog. Mater. Sci. 2015, 74, 332–400.
13. MacDonald, W.L. The Pantheon: Design, Meaning, and Progeny; Harvard University Press: Cambridge, MA, USA, 2002.
14. Hooke, R. Micrographia; BoD–Books on Demand: Mumbai, India, 2020.

15. Wittenauer, J.; Norris, B. Structural honeycomb materials for advanced aerospace designs. JOM 1990, 42, 36–41.
16. Bitzer, T. Honeycomb Technology: Materials, Design, Manufacturing, Applications and Testing; Springer Science & Business Media: Berlin/Heidelberg, Germany, 1997.
17. Höfler, R.; Renyi, S. Plattenförmiger Baukörper; DE355036; CompositesPRESS: Mennecy, France, 1914.
18. Lee, S.M. Handbook of Composite Reinforcements; John Wiley & Sons: Hoboken, NJ, USA, 1996.
19. Heng, L.; Wang, B.; Li, M.; Zhang, Y.; Jiang, L. Advances in fabrication materials of honeycomb structure films by the breath-figure method. Materials 2013, 6, 460–482.
20. Male, U.; Jo, E.J.; Park, J.Y. Surface functionalization of honeycomb-patterned porous poly (ε-caprolactone) films by interfacial polymerization of aniline. Polymer 2016, 99, 623–632.
21. Rodríguez-Hernández, J.; Bormashenko, E. Hierarchically Ordered Microporous Surfaces. In Breath Figures; Springer: Berlin/Heidelberg, Germany, 2020; pp. 169–187.
22. Slepička, P.; Neznalová, K.; Fajstavr, D.; Švorčík, V. Nanostructuring of honeycomb-like polystyrene with excimer laser. Prog. Org. Coat. 2020, 145, 105670.
23. Neznalová, K.; Sajdl, P.; Švorčík, V.; Slepička, P. Cellulose acetate honeycomb-like pattern created by improved phase separation. eXPRESS Polym. Lett. 2020, 14, 1078–1088.
24. Hurtuková, K.; Juřicová, V.; Fajstavrová, K.; Fajstavr, D.; Slepičková Kasálková, N.; Rimpelová, S.; Švorčík, V.; Slepička, P. Cytocompatibility of Polymethyl Methacrylate Honeycomb-like Pattern on Perfluorinated Polymer. Polymers 2021, 13, 3663.
25. Slepička, P.; Neznalová, K.; Fajstavr, D.; Slepičková Kasálková, N.; Švorčík, V. Honeycomb-like pattern formation on perfluoroethylenepropylene enhanced by plasma treatment. Plasma Processes Polym. 2019, 16, 1900063.
26. Peng, J.; Han, Y.; Yang, Y.; Li, B. The influencing factors on the macroporous formation in polymer films by water droplet templating. Polymer 2004, 45, 447–452.
27. Fajstavrová, K.; Rimpelová, S.; Fajstavr, D.; Švorčík, V.; Slepička, P. Cell behavior of primary fibroblasts and osteoblasts on plasma-treated fluorinated polymer coated with honeycomb polystyrene. Materials 2021, 14, 889.
28. Stenzel, M.H.; Barner Kowollik, C.; Davis, T.P. Formation of honeycomb-structured, porous films via breath figures with different polymer architectures. J. Polym. Sci. Part A Polym. Chem. 2006, 44, 2363–2375.
29. Yin, H.; Feng, Y.; Billon, L. Directed Self-Assembly in "Breath Figure" Templating of Melamine-Based Amphiphilic Copolymers: Effect of Hydrophilic End-Chain on Honeycomb Film Formation and Wetting. Chem. Eur. J. 2018, 24, 425–433.
30. Yabu, H.; Shimomura, M. Surface properties of self-organized honeycomb-patterned films. Mol. Cryst. Liq. Cryst. 2006, 445, 125–129.
31. Nurmawati, M.H.; Ajikumar, P.K.; Renu, R.; Valiyaveettil, S. Hierarchical Self-Organization of Nanomaterials into Two-Dimensional Arrays Using Functional Polymer Scaffold. Adv. Funct. Mater. 2008, 18, 3213–3218.

32. Falconnet, D.; Csucs, G.; Grandin, H.M.; Textor, M. Surface engineering approaches to micropattern surfaces for cell-based assays. Biomaterials 2006, 27, 3044–3063.
33. Norman, J.J.; Desai, T.A. Methods for fabrication of nanoscale topography for tissue engineering scaffolds. Ann. Biomed. Eng. 2006, 34, 89–101.
34. Xue, L.; Han, Y. Pattern formation by dewetting of polymer thin film. Prog. Polym. Sci. 2011, 36, 269–293.
35. Xue, L.; Zhang, J.; Han, Y. Phase separation induced ordered patterns in thin polymer blend films. Prog. Polym. Sci. 2012, 37, 564–594.
36. Xie, Y.; Kocaefe, D.; Chen, C.; Kocaefe, Y. Review of research on template methods in preparation of nanomaterials. J. Nanomater. 2016, 2016, 2302595.
37. Wang, D.M.; Lai, J.Y. Recent advances in preparation and morphology control of polymeric membranes formed by nonsolvent induced phase separation. Curr. Opin. Chem. Eng. 2013, 2, 229–237.
38. Guillen, G.R.; Pan, Y.; Li, M.; Hoek, E.M. Preparation and characterization of membranes formed by nonsolvent induced phase separation: A review. Ind. Eng. Chem. Res. 2011, 50, 3798–3817.
39. Bui, V.T.; Ko, S.H.; Choi, H.S. A surfactant-free bio-compatible film with a highly ordered honeycomb pattern fabricated via an improved phase separation method. Chem. Commun. 2014, 50, 3817–3819.
40. Liu, Q.; Tang, Z.; Ou, B.; Liu, L.; Zhou, Z.; Shen, S.; Duan, Y. Design, preparation, and application of ordered porous polymer materials. Mater. Chem. Phys. 2014, 144, 213–225.
41. Hsueh, H.Y.; Ho, R.M. Bicontinuous ceramics with high surface area from block copolymer templates. Langmuir 2012, 28, 8518–8529.
42. Rodríguez-Hernández, J. Nano-microporous structured surfaces prepared by the breath figures approach and their biorelated applications. In Surface Chemistry of Nanobiomaterials; Elsevier: Amsterdam, The Netherlands, 2016; pp. 107–133.
43. Aitken, J. Breath figures. Proc. R. Soc. Edinb. 1895, 20, 94–97.
44. Dou, Y.; Jin, M.; Zhou, G.; Shui, L. Breath figure method for construction of honeycomb films. Membranes 2015, 5, 399–424.
45. Li, M.; Xu, S.; Kumacheva, E. Convection in polymeric fluids subjected to vertical temperature gradients. Macromolecules 2000, 33, 4972–4978.
46. Wan, L.S.; Zhu, L.W.; Ou, Y.; Xu, Z.K. Multiple interfaces in self-assembled breath figures. Chem. Commun. 2014, 50, 4024–4039.
47. Wong, K.H.; Hernández-Guerrero, M.; Granville, A.M.; Davis, T.P.; Barner-Kowollik, C.; Stenzel, M.H. Water-assisted formation of honeycomb structured porous films. J. Porous Mater. 2006, 13, 213–223.
48. Eslamian, M.; Soltani-Kordshuli, F. Development of multiple-droplet drop-casting method for the fabrication of coatings and thin solid films. J. Coat. Technol. Res. 2018, 15, 271–280.

49. Yabu, H.; Tanaka, M.; Ijiro, K.; Shimomura, M. Preparation of honeycomb-patterned polyimide films by self-organization. Langmuir 2003, 19, 6297–6300.
50. Madej, W.; Budkowski, A.; Raczkowska, J.; Rysz, J. Breath figures in polymer and polymer blend films spin-coated in dry and humid ambience. Langmuir 2008, 24, 3517–3524.
51. Pilati, F.; Montecchi, M.; Fabbri, P.; Synytska, A.; Messori, M.; Toselli, M.; Grundke, K.; Pospiech, D. Design of surface properties of PET films: Effect of fluorinated block copolymers. J. Colloid Interface Sci. 2007, 315, 210–222.
52. Bui, V.T.; Tran, Q.C.; Nguyen, V.T.; Dao, V.D.; Choi, J.S.; Choi, H.S. Ordered honeycomb biocompatible polymer films via a one-step solution-immersion phase separation used as a scaffold for cell cultures. Chem. Eng. J. 2017, 320, 561–569.
53. Iqbal, M.; Dinh, D.K.; Abbas, Q.; Imran, M.; Sattar, H.; Ul Ahmad, A. Controlled surface wettability by plasma polymer surface modification. Surfaces 2019, 2, 349–371.
54. Morent, R.; De Geyter, N.; Desmet, T.; Dubruel, P.; Leys, C. Plasma surface modification of biodegradable polymers: A review. Plasma Processes Polym. 2011, 8, 171–190.
55. Cheng, Z.; Teoh, S.H. Surface modification of ultra thin poly (ε-caprolactone) films using acrylic acid and collagen. Biomaterials 2004, 25, 1991–2001.
56. Zelzer, M.; Scurr, D.; Abdullah, B.; Urquhart, A.J.; Gadegaard, N.; Bradley, J.W.; Alexander, M.R. Influence of the plasma sheath on plasma polymer deposition in advance of a mask and down pores. J. Phys. Chem. B 2009, 113, 8487–8494.
57. Shen, H.; Hu, X.; Yang, F.; Bei, J.; Wang, S. Combining oxygen plasma treatment with anchorage of cationized gelatin for enhancing cell affinity of poly (lactide-co-glycolide). Biomaterials 2007, 28, 4219–4230.
58. Nishikawa, T.; Ookura, R.; Nishida, J.; Arai, K.; Hayashi, J.; Kurono, N.; Sawadaishi, T.; Hara, M.; Shimomura, M. Fabrication of honeycomb film of an amphiphilic copolymer at the air– water interface. Langmuir 2002, 18, 5734–5740.
59. Wan, L.S.; Li, J.W.; Ke, B.B.; Xu, Z.K. Ordered microporous membranes templated by breath figures for size-selective separation. J. Am. Chem. Soc. 2012, 134, 95–98.
60. Farbod, F.; Pourabbas, B.; Sharif, M. Direct breath figure formation on PMMA and superhydrophobic surface using in situ perfluoro-modified silica nanoparticles. J. Polym. Sci. Part B: Polym. Phys. 2013, 51, 441–451.
61. Huang, C.; Kamra, T.; Chaudhary, S.; Shen, X. Breath figure patterns made easy. ACS Appl. Mater. Interfaces 2014, 6, 5971–5976.
62. Wang, Y.; Liu, Z.; Huang, Y.; Han, B.; Yang, G. Micropatterned polymer surfaces induced by nonsolvent. Langmuir 2006, 22, 1928–1931.
63. Hernández-Guerrero, M.; Stenzel, M.H. Honeycomb structured polymer films via breath figures. Polym. Chem. 2012, 3, 563–577.
64. Bui, V.T.; Choi, H.S. Surface morphology and wettability control of polymer Substrates: A comparison of water-miscible and water-immiscible mixture solvents. Eur. Polym. J. 2017, 93, 158–166.
65. Brien, F. Biomaterials & scaffolds Every day thousands of surgical procedures are performed to replace. Mater. Today 2011, 14, 88–95.

66. Chen, S.; Gao, S.; Jing, J.; Lu, Q. Designing 3D Biological Surfaces via the Breath-Figure Method. Adv. Healthc. Mater. 2018, 7, 1701043.
67. Nishikawa, T.; Nonomura, M.; Arai, K.; Hayashi, J.; Sawadaishi, T.; Nishiura, Y.; Hara, M.; Shimomura, M. Micropatterns based on deformation of a viscoelastic honeycomb mesh. Langmuir 2003, 19, 6193–6201.
68. Zhang, A.; Bai, H.; Li, L. Breath figure: A nature-inspired preparation method for ordered porous films. Chem. Rev. 2015, 115, 9801–9868.
69. Du, M.; Zhu, P.; Yan, X.; Su, Y.; Song, W.; Li, J. Honeycomb self-assembled peptide scaffolds by the breath figure method. Chem. Eur. J. 2011, 17, 4238–4245.
70. Li, J.; Peng, J.; Huang, W.; Wu, Y.; Fu, J.; Cong, Y.; Xue, L.; Han, Y. Ordered honeycomb-structured gold nanoparticle films with changeable pore morphology: From circle to ellipse. Langmuir 2005, 21, 2017–2021.
71. Wu, X.; Wang, S. Regulating MC3T3-E1 cells on deformable poly (ε-caprolactone) honeycomb films prepared using a surfactant-free breath figure method in a water-miscible solvent. ACS Appl. Mater. Interfaces 2012, 4, 4966–4975.
72. Kasuya, J.; Sudo, R.; Tamogami, R.; Masuda, G.; Mitaka, T.; Ikeda, M.; Tanishita, K. Reconstruction of 3D stacked hepatocyte tissues using degradable, microporous poly (d, l-lactide-co-glycolide) membranes. Biomaterials 2012, 33, 2693–2700.
73. Fajstavrová, K.; Rimpelová, S.; Fajstavr, D.; Švorčík, V.; Slepička, P. PLLA honeycomb-like pattern on fluorinated ethylene propylene as a substrate for fibroblast growth. Polymers 2020, 12, 2436.
74. Bunz, U.H.F. Breath figures as a dynamic templating method for polymers and nanomaterials. Adv. Mater. 2006, 18, 973–989.
75. Kuroda, K.; Caputo, G.A.; DeGrado, W.F. The role of hydrophobicity in the antimicrobial and hemolytic activities of polymethacrylate derivatives. Chem. Eur. J. 2009, 15, 1123–1133.
76. Lin, C.L.; Tung, P.H.; Chang, F.C. Synthesis of rod-coil diblock copolymers by ATRP and their honeycomb morphologies formed by the 'breath figures' method. Polymer 2005, 46, 9304–9313.
77. Deepak, V.; Asha, S. Self-organization-induced three-dimensional honeycomb pattern in structure-controlled bulky methacrylate polymers: Synthesis, morphology, and mechanism of pore formation. J. Phys. Chem. B 2006, 110, 21450–21459.
78. Wu, X.; Jones, M.D.; Davidson, M.G.; Chaudhuri, J.B.; Ellis, M.J. Surfactant-free poly (lactide-co-glycolide) honeycomb films for tissue engineering: Relating solvent, monomer ratio and humidity to scaffold structure. Biotechnol. Lett. 2011, 33, 423–430.
79. Fukuhira, Y.; Yabu, H.; Ijiro, K.; Shimomura, M. Interfacial tension governs the formation of self-organized honeycomb-patterned polymer films. Soft Matter 2009, 5, 2037–2041.
80. Yabu, H. Fabrication of honeycomb films by the breath figure technique and their applications. Sci. Technol. Adv. Mater. 2018, 19, 802–822.

81. Bui, V.T.; Ko, S.H.; Choi, H.S. Large-scale fabrication of commercially available, non-polar linear polymer film with a highly ordered honeycomb pattern. ACS Appl. Mater. Interfaces 2015, 7, 10541–10547.
82. Nishikawa, T.; Nishida, J.; Ookura, R.; Nishimura, S.I.; Scheumann, V.; Zizlsperger, M.; Lawall, R.; Knoll, W.; Shimomura, M. Web-structured films of an amphiphilic polymer from water in oil emulsion: Fabrication and characterization. Langmuir 2000, 16, 1337–1342.
83. Ferrari, E.; Fabbri, P.; Pilati, F. Solvent and substrate contributions to the formation of breath figure patterns in polystyrene films. Langmuir 2011, 27, 1874–1881.
84. Jiang, X.; Zhang, T.; Xu, L.; Wang, C.; Zhou, X.; Gu, N. Surfactant-induced formation of honeycomb pattern on micropipette with curvature gradient. Langmuir 2011, 27, 5410–5419.
85. Connal, L.A.; Vestberg, R.; Hawker, C.J.; Qiao, G.G. Fabrication of reversibly cross-linkable, 3-dimensionally conformal polymeric microstructures. Adv. Funct. Mater. 2008, 18, 3315–3322.
86. Ucar, I.O.; Erbil, H.Y. Dropwise condensation rate of water breath figures on polymer surfaces having similar surface free energies. Appl. Surf. Sci. 2012, 259, 515–523.
87. Lee, J.; Cuddihy, M.J.; Kotov, N.A. Three-dimensional cell culture matrices: State of the art. Tissue Eng. Part B Rev. 2008, 14, 61–86.
88. Gloria, A.; De Santis, R.; Ambrosio, L. Polymer-based composite scaffolds for tissue engineering. J. Appl. Biomater. Biomech. 2010, 8, 57–67.
89. Yuan, M.-S.; Xu, W.; He, Q.-G.; Cheng, J.G.; Fu, Y.-Y. Research progress of breath figure method in device application. Chin. J. Anal. Chem. 2022, 50, 44–52.
90. Shiohara, A.; Prieto-Simon, B.; Voelcker, N.H. Porous polymeric membranes: Fabrication techniques and biomedical applications. J. Mater. Chem. B 2021, 9, 2129–2154.
91. Davis, M.E. Non-viral gene delivery systems. Curr. Opin. Biotechnol. 2002, 13, 128–131.
92. Wu, L.W.; Wan, L.S.; Ou, Y.; Zhu, L.W.; Xu, Z.K. Fabrication of transferable perforated isoporous membranes on versatile solid substrates via the breath figure method. Adv. Mater. Interfaces 2015, 2, 1500285.
93. Yu, B.; Cong, H.; Li, Z.; Yuan, H.; Peng, Q.; Chi, M.; Yang, S.; Yang, R.; Ranil Wickramasinghe, S.; Tang, J. Fabrication of highly ordered porous membranes of cellulose triacetate on ice substrates using breath figure method. J. Polym. Sci. Part B Polym. Phys. 2015, 53, 552–558.
94. Ou, Y.; Wang, L.-Y.; Zhu, L.-W.; Wan, L.-S.; Xu, Z.-K. In-situ immobilization of silver nanoparticles on self-assembled honeycomb-patterned films enables surface-enhanced Raman scattering (SERS) substrates. J. Phys. Chem. C 2014, 118, 11478–11484.
95. Yabu, H.; Shimomura, M. Single-step fabrication of transparent superhydrophobic porous polymer films. Chem. Mater. 2005, 17, 5231–5234.
96. Vohra, V.; Bolognesi, A.; Calzaferri, G.; Botta, C. Multilevel organization in hybrid thin films for optoelectronic applications. Langmuir 2009, 25, 12019–12023.
97. Biswal, T. Biopolymers for tissue engineering applications: A review. Mater. Today Proc. 2021, 41, 397–402.

98. Calejo, M.T.; Ilmarinen, T.; Skottman, H.; Kellomäki, M. Breath figures in tissue engineering and drug delivery: State-of-the-art and future perspectives. Acta Biomater. 2018, 66, 44–66.

99. Bovey, F. Macromolecules: An Introduction to Polymer Science; Elsevier: Amsterdam, The Netherlands, 2012.

100. Liang, J.; Ma, Y.; Sims, S.; Wu, L. A patterned porous polymer film for localized capture of insulin and glucose-responsive release. J. Mater. Chem. B 2015, 3, 1281–1288.

101. Yamazaki, H.; Kohashi, S.; Ito, K.; Ijiro, K.; Shimomura, M. Production technology and applications of honeycomb films. Polym. J. 2022, 54, 107–120.

102. Arai, K.; Tanaka, M.; Yamamoto, S.; Shimomura, M. Effect of pore size of honeycomb films on the morphology, adhesion and cytoskeletal organization of cardiac myocytes. Colloids Surf. A Physicochem. Eng. Asp. 2008, 313, 530–535.

103. Tanaka, M.; Takayama, A.; Ito, E.; Sunami, H.; Yamamoto, S.; Shimomura, M. Effect of pore size of self-organized honeycomb-patterned polymer films on spreading, focal adhesion, proliferation, and function of endothelial cells. J. Nanosci. Nanotechnol. 2007, 7, 763–772.

104. Eniwumide, J.O.; Tanaka, M.; Nagai, N.; Morita, Y.; De Bruijn, J.; Yamamoto, S.; Onodera, S.; Kondo, E.; Yasuda, K.; Shimomura, M. The morphology and functions of articular chondrocytes on a honeycomb-patterned surface. BioMed Res. Int. 2014, 2014, 710354.

105. Birch, M.A.; Tanaka, M.; Kirmizidis, G.; Yamamoto, S.; Shimomura, M. Microporous "honeycomb" films support enhanced bone formation in vitro. Tissue Eng. Part A 2013, 19, 2087–2096.

106. Clement, A.L.; Moutinho, T.J., Jr.; Pins, G.D. Micropatterned dermal–epidermal regeneration matrices create functional niches that enhance epidermal morphogenesis. Acta Biomater. 2013, 9, 9474–9484.

107. McMillan, J.R.; Akiyama, M.; Tanaka, M.; Yamamoto, S.; Goto, M.; Abe, R.; Sawamura, D.; Shimomura, M.; Shimizu, H. Small-diameter porous Poly (€-Caprolactone) films enhance adhesion and growth of human cultured epidermal keratinocyte and dermal fibroblast cells. Tissue Eng. 2007, 13, 789–798.

108. Chen, S.; Lu, X.; Hu, Y.; Lu, Q. Biomimetic honeycomb-patterned surface as the tunable cell adhesion scaffold. Biomater. Sci. 2015, 3, 85–93.

109. Li, L.; Chen, C.; Li, J.; Zhang, A.; Liu, X.; Xu, B.; Gao, S.; Jin, G.; Ma, Z. Robust and hydrophilic polymeric films with honeycomb pattern and their cell scaffold applications. J. Mater. Chem. 2009, 19, 2789–2796.

110. Tsuruma, A.; Tanaka, M.; Yamamoto, S.; Shimomura, M. Control of neural stem cell differentiation on honeycomb films. Colloids Surf. A Physicochem. Eng. Asp. 2008, 313, 536–540.

111. Liu, Y.; Xu, J.; Zhou, Y.; Ye, Z.; Tan, W.-S. Layer-by-layer assembled polyelectrolytes on honeycomb-like porous poly (ε-caprolactone) films modulate the spatial distribution of mesenchymal stem cells. Mater. Sci. Eng. C 2017, 78, 579–588.

112. Farzaneh, Z.; Pournasr, B.; Ebrahimi, M.; Aghdami, N.; Baharvand, H. Enhanced functions of human embryonic stem cell-derived hepatocyte-like cells on three-dimensional nanofibrillar surfaces. Stem Cell Rev. Rep. 2010, 6, 601–610.

113. Jeon, G.; Yang, S.Y.; Kim, J.K. Functional nanoporous membranes for drug delivery. J. Mater. Chem. 2012, 22, 14814–14834.

114. Velayudhan, S.; Kumar, P.; Nair, P.D. A novel, single step, highly sensitive in-vitro cell-based metabolic assay using honeycomb microporous polymer membranes. J. Biomed. Nanotechnol. 2015, 11, 590–599.

115. Huang, X.; Brazel, C.S. On the importance and mechanisms of burst release in matrix-controlled drug delivery systems. J. Control. Release 2001, 73, 121–136.

116. Ponnusamy, T.; Lawson, L.B.; Freytag, L.C.; Blake, D.A.; Ayyala, R.S.; John, V.T. In vitro degradation and release characteristics of spin coated thin films of PLGA with a "breath figure" morphology. Biomatter 2012, 2, 77–86.

117. Ponnusamy, T.; Yu, H.; John, V.T.; Ayyala, R.S.; Blake, D.A. A novel antiproliferative drug coating for glaucoma drainage devices. J. Glaucoma 2014, 23, 526–534.

118. Dai, Z.; Yu, X.; Hong, J.; Liu, X.; Sun, J.; Sun, X. Development of a novel CsA-PLGA drug delivery system based on a glaucoma drainage device for the prevention of post-operative fibrosis. Mater. Sci. Eng. C 2016, 66, 206–214.

119. Grabacka, M.; Waligorski, P.; Zapata, A.; Blake, D.; Wyczechowska, D.; Wilk, A.; Rutkowska, M.; Vashistha, H.; Ayyala, R.; Ponnusamy, T. Fenofibrate subcellular distribution as a rationale for the intracranial delivery through biodegradable carrier. J. Physiol. Pharmacol. 2015, 66, 233.

120. Zhuang, C.; Shi, C.; Tao, F.; Cui, Y. Honeycomb structural composite polymer network of gelatin and functional cellulose ester for controlled release of omeprazole. Int. J. Biol. Macromol. 2017, 105, 1644–1653.

121. De León, A.S.; Molina, M.; Wedepohl, S.; Muñoz-Bonilla, A.; Rodríguez-Hernández, J.; Calderón, M. Immobilization of stimuli-responsive nanogels onto honeycomb porous surfaces and controlled release of proteins. Langmuir 2016, 32, 1854–1862.

122. Su, Y.; Dang, J.; Zhang, H.; Zhang, Y.; Tian, W. Supramolecular Host–Guest Interaction-Enhanced Adjustable Drug Release Based on β-Cyclodextrin-Functionalized Thermoresponsive Porous Polymer Films. Langmuir 2017, 33, 7393–7402.

123. Wang, C.; Liu, Q.; Shao, X.; Yang, G.; Xue, H.; Hu, X. One step fabrication of nanoelectrode ensembles formed via amphiphilic block copolymers self-assembly and selective voltammetric detection of uric acid in the presence of high ascorbic acid content. Talanta 2007, 71, 178–185.

124. Wang, Q.; Wen, X.; Kong, J. Recent progress on uric acid detection: A review. Crit. Rev. Anal. Chem. 2020, 50, 359–375.

125. Chen, P.C.; Wan, L.S.; Ke, B.B.; Xu, Z.K. Honeycomb-patterned film segregated with phenylboronic acid for glucose sensing. Langmuir 2011, 27, 12597–12605.

126. Ting, S.S.; Min, E.H.; Escale, P.; Save, M.; Billon, L.; Stenzel, M.H. Lectin recognizable biomaterials synthesized via nitroxide-mediated polymerization of a methacryloyl galactose monomer. Macromolecules 2009, 42, 9422–9434.

127. Munoz-Bonilla, A.; Ibarboure, E.; Bordegé, V.; Fernández-García, M.; Rodríguez-Hernández, J. Fabrication of honeycomb-structured porous surfaces decorated with glycopolymers. Langmuir 2010, 26, 8552–8558.

128. Miller, S.; Bao, Z. Fabrication of flexible pressure sensors with microstructured polydimethylsiloxane dielectrics using the breath figures method. J. Mater. Res. 2015, 30, 3584–3594.

129. Hall-Stoodley, L.; Costerton, J.W.; Stoodley, P. Bacterial biofilms: From the natural environment to infectious diseases. Nat. Rev. Microbiol. 2004, 2, 95–108.

130. Yang, M.; Ding, Y.; Ge, X.; Leng, Y. Control of bacterial adhesion and growth on honeycomb-like patterned surfaces. Colloids Surf. B Biointerfaces 2015, 135, 549–555.

131. Zhao, Y.; Shang, Q.; Yu, J.; Zhang, Y.; Liu, S. Nanostructured 2D diporphyrin honeycomb film: Photoelectrochemistry, photodegradation, and antibacterial activity. ACS Appl. Mater. Interfaces 2015, 7, 11783–11791.

132. Muñoz-Bonilla, A.; Cuervo-Rodríguez, R.; López-Fabal, F.; Gómez-Garcés, J.L.; Fernández-García, M. Antimicrobial porous surfaces prepared by breath figures approach. Materials 2018, 11, 1266.

133. Wang, Y.; Liu, Y.; Li, G.; Hao, J. Porphyrin-based honeycomb films and their antibacterial activity. Langmuir 2014, 30, 6419–6426.

134. Manabe, K.; Nishizawa, S.; Shiratori, S. Porous surface structure fabricated by breath figures that suppresses Pseudomonas aeruginosa biofilm formation. ACS Appl. Mater. Interfaces 2013, 5, 11900–11905.

135. Vargas-Alfredo, N.; Santos-Coquillat, A.; Martínez-Campos, E.; Dorronsoro, A.; Cortajarena, A.L.; Del Campo, A.; Rodríguez-Hernández, J. Highly efficient antibacterial surfaces based on bacterial / cell size selective microporous supports. ACS Appl. Mater. Interfaces 2017, 9, 44270–44280.

136. Huang, C.; Shen, X.; Liu, X.; Chen, Z.; Shu, B.; Wan, L.; Liu, H.; He, J. Hybrid breath figure method: A new insight in Petri dishes for cell culture. J. Colloid Interface Sci. 2019, 541, 114–122.

137. Slepička, P.; Fajstavr, D.; Krejčová, M.; Rimpelová, S.; Slepičková Kasálková, N.; Kolská, Z.; Švorčík, V. Biopolymer Composites with Ti / Au Nanostructures and Their Antibacterial Properties. Pharmaceutics 2021, 13, 826.

138. Hurtuková, K.; Fajstavrová, K.; Rimpelová, S.; Vokatá, B.; Fajstavr, D.; Slepičková Kasálková, N.; Siegel, J.; Švorčík, V.; Slepička, P. Antibacterial properties of a honeycomb-like pattern with cellulose acetate and silver nanoparticles. Materials 2021, 14, 4051.

139. Tormena, R.P.I.; Rosa, E.V.; de Fátima Oliveira Mota, B.; Chaker, J.A.; Fagg, C.W.; Freire, D.O.; Martins, P.M.; da Silva, I.C.R.; Sousa, M.H. Evaluation of the antimicrobial activity of silver nanoparticles obtained by microwave-assisted green synthesis using Handroanthus impetiginosus (Mart. ex DC.) Mattos underbark extract. RSC Adv. 2020, 10, 20676–20681.

140. Jiang, X.; Zhang, T.; He, S.; Ling, J.; Gu, N.; Zhang, Y.; Zhou, X.; Wang, X.; Cheng, L. Bacterial adhesion on honeycomb-structured poly (L-lactic acid) surface with Ag nanoparticles. J. Biomed. Nanotechnol. 2012, 8, 791–799.

141. Kim, Y.W.; Modigunta, J.K.R.; Male, U. Effect of ferrocene on the fabrication of honeycomb-patterned porous polystyrene films and silver functionalization of the film. Polymer 2019, 166, 55–62.

142. Mirotsou, M.; Abe, M.; Lanza, R. Corneal replacement tissue. In Principles of Tissue Engineering; Elsevier: Amsterdam, The Netherlands, 2020; pp. 1135–1143.

143. Slepicka, P.; Slepickova Kasalkova, N.; Siegel, J.; Kolska, Z.; Bacakova, L.; Svorcik, V. Nano-structured and functionalized surfaces for cytocompatibility improvement and bactericidal action. Biotechnol. Adv. 2015, 33, 1120–1129.

144. Slepička, P.; Siegel, J.; Lyutakov, O.; Slepičková Kasálková, N.; Kolská, Z.; Bačáková, L.; Švorčík, V. Polymer nanostructures for bioapplications induced by laser treatment. Biotechnol. Adv. 2018, 36, 839–855.

145. Toncheva, A.; Mincheva, R.; Kancheva, M.; Manolova, N.; Rashkov, I.; Dubois, P.; Markova, N. Antibacterial PLA / PEG electrospun fibers: Comparative study between grafting and blending PEG. Eur. Polym. J. 2016, 75, 223–233.

146. Yim, E.K.; Pang, S.W.; Leong, K.W. Synthetic nanostructures inducing differentiation of human mesenchymal stem cells into neuronal lineage. Exp. Cell Res. 2007, 313, 1820–1829.

147. Xia, L.; Wei, Z.; Wan, M. Conducting polymer nanostructures and their application in biosensors. J. Colloid Interface Sci. 2010, 341, 1–11.

148. Michaljaničová, I.; Slepička, P.; Rimpelová, S.; Slepičková Kasálková, N.; Švorčík, V. Regular pattern formation on surface of aromatic polymers and its cytocompatibility. Appl. Surf. Sci. 2016, 370, 131–141.

149. Slepička, P.; Michaljaničová, I.; Slepičková Kasálková, N.; Kolská, Z.; Rimpelová, S.; Ruml, T.; Švorčík, V. Poly-l-lactic acid modified by etching and grafting with gold nanoparticles. J. Mater. Sci. 2013, 48, 5871–5879.

150. Kasálková Slepičková, N.; Slepička, P.; Kolská, Z.; Hodačová, P.; Kučková, Š.; Švorčík, V. Grafting of bovine serum albumin proteins on plasma-modified polymers for potential application in tissue engineering. Nanoscale Res. Lett. 2014, 9, 161.

151. Yang, Y.W.; Wu, J.Y.; Liu, C.T.; Liao, G.C.; Huang, H.Y.; Hsu, R.Q.; Chiang, M.H.; Wu, J.S. Fast incorporation of primary amine group into polylactide surface for improving C2C12 cell proliferation using nitrogen-based atmospheric-pressure plasma jets. J. Biomed. Mater. Res. Part A 2014, 102, 160–169.

152. Cheng, K.Y.; Chang, C.H.; Yang, Y.W.; Liao, G.C.; Liu, C.T.; Wu, J.S. Enhancement of cell growth on honeycomb-structured polylactide surface using atmospheric-pressure plasma jet modification. Appl. Surf. Sci. 2017, 394, 534–542.

153. González-Henríquez, C.; Rodríguez-Hernández, J. Wrinkled Polymer Surfaces: Strategies, Methods and Applications; Springer: Berlin / Heidelberg, Germany, 2019.

154. Slepička, P.; Neděla, O.; Sajdl, P.; Kolská, Z.; Švorčík, V. Polyethylene naphthalate as an excellent candidate for ripple nanopatterning. Appl. Surf. Sci. 2013, 285, 885–892.

155. Fajstavr, D.; Slepička, P.; Švorčík, V. LIPSS with gold nanoclusters prepared by combination of heat treatment and KrF exposure. Appl. Surf. Sci. 2019, 465, 919–928.

156. Yabu, H.; Matsui, J.; Matsuo, Y. Site-Selective Wettability Control of Honeycomb Films by UV–O3-Assisted Sol–Gel Coating. Langmuir 2020, 36, 12023–12029.
157. Lišková, J.; Slepičková Kasálková, N.; Slepička, P.; Švorčík, V.; Bačáková, L. Heat-treated carbon coatings on poly (L-lactide) foils for tissue engineering. Mater. Sci. Eng. C 2019, 100, 117–128.
158. Yabu, H.; Hirai, Y.; Shimomura, M. Electroless plating of honeycomb and pincushion polymer films prepared by self-organization. Langmuir 2006, 22, 9760–9764.
159. Yabu, H.; Inoue, K.; Shimomura, M. Multiple-periodic structures of self-organized honeycomb-patterned films and polymer nanoparticles hybrids. Colloids Surf. A Physicochem. Eng. Asp. 2006, 284, 301–304.
160. Slepička, P.; Rimpelová, S.; Slepičková Kasálková, N.; Fajstavr, D.; Sajdl, P.; Kolská, Z.; Švorčík, V. Antibacterial properties of plasma-activated perfluorinated substrates with silver nanoclusters deposition. Nanomaterials 2021, 11, 182.
161. Kabuto, T.; Hashimoto, Y.; Karthaus, O. Thermally stable and solvent resistant mesoporous honeycomb films from a crosslinkable polymer. Adv. Funct. Mater. 2007, 17, 3569–3573.
162. Karikari, A.S.; Williams, S.R.; Heisey, C.L.; Rawlett, A.M.; Long, T.E. Porous thin films based on photo-cross-linked star-shaped poly (D, L-lactide) s. Langmuir 2006, 22, 9687–9693.
163. Bolognesi, A.; Galeotti, F.; Moreau, J.; Giovanella, U.; Porzio, W.; Scavia, G.; Bertini, F. Unsoluble ordered polymeric pattern by breath figure approach. J. Mater. Chem. 2010, 20, 1483–1488.
164. Slepička, P.; Siegel, J.; Šlouf, M.; Fajstavr, D.; Fajstavrová, K.; Kolská, Z.; Švorčík, V. The Functionalization of a Honeycomb Polystyrene Pattern by Excimer Treatment in Liquid. Polymers 2022, 14, 4944–4955.
165. Xu, W.Z.; Zhang, X.; Kadla, J.F. Design of functionalized cellulosic honeycomb films: Site-specific biomolecule modification via "click chemistry". Biomacromolecules 2012, 13, 350–357.
166. Nishida, J.; Nishikawa, K.; Nishimura, S.-I.; Wada, S.; Karino, T.; Nishikawa, T.; Ijiro, K.; Shimomura, M. Preparation of self-organized micro-patterned polymer films having cell adhesive ligands. Polym. J. 2002, 34, 166–174.
167. Cha, T. Surface Chemical Modification for the Immobilization of Biomolecules; University of Minnesota: Minneapolis, MN, USA, 2005.
168. Min, E.; Wong, K.H.; Stenzel, M.H. Microwells with patterned proteins by a self-assembly process using honeycomb-structured porous films. Adv. Mater. 2008, 20, 3550–3556.
169. Ke, B.B.; Wan, L.S.; Xu, Z.K. Controllable construction of carbohydrate microarrays by site-directed grafting on self-organized porous films. Langmuir 2010, 26, 8946–8952.
170. Nystrom, D.; Malmstrom, E.; Hult, A.; Blakey, I.; Boyer, C.; Davis, T.P.; Whittaker, M.R. Biomimetic surface modification of honeycomb films via a "grafting from" approach. Langmuir 2010, 26, 12748–12754.
171. Zhang, Y.; Wang, C. Micropatterning of Proteins on 3D Porous Polymer Film Fabricated by Using the Breath-Figure Method. Adv. Mater. 2007, 19, 913–916.

172. Hernández-Guerrero, M.; Min, E.; Barner-Kowollik, C.; Müller, A.H.; Stenzel, M.H. Grafting thermoresponsive polymers onto honeycomb structured porous films using the RAFT process. J. Mater. Chem. 2008, 18, 4718–4730.

10 IN SITU OBSERVATION OF ZNO NANOPARTICLE FORMATION BY A COMBINATION OF TIME-RESOLVED X-RAY ABSORPTION SPECTROSCOPY AND X-RAY DIFFRACTION

Franz Eckelt [1], Patrick Rothweiler [1], Frederic Braun [1], Lukas Voss [1], Ankica Šarić [2], Martina Vrankić [2] and Dirk Lützenkirchen-Hecht [1]

[1]*Faculty of Natural Sciences, University of Wuppertal, Gauss-Str. 20, 42119 Wuppertal, Germany*

[2]*Centre of Excellence for Advanced Materials and Sensing Devices, Division of Materials Physics, Ruđer Bošković Institute, HR-10002 Zagreb, Croatia*

ABSTRACT

The formation of ZnO nanomaterials from different Zn acetylacetonate precursor solutions was studied in situ by employing simultaneous, time-resolved X-ray diffraction (XRD) and X-ray absorption spectroscopy (EXAFS) at the Zn K-edge. The precursor solutions were heated from room temperature to the desired reaction temperatures in a hermetically sealed cell dedicated to X-ray experiments. In general, the first indications for the formation of hexagonal ZnO were found for elevated temperatures of about 80 °C both by XRD and EXAFS, and the contributions increase with temperature and time. However, no reaction intermediates could be proved in addition to the Zn precursors and the formed hexagonal ZnO materials. Furthermore, the results show that the efficiency of the reaction, i.e., the conversion of the precursor material to the ZnO product, strongly depends on the solvent used and the reaction temperature. ZnO formation is accelerated by an increased temperature of 165 °C and the use of 1-octanol, with a conversion to ZnO of more than 80% after only a ca. 35 min reaction time according to a detailed analysis of the EXAFS data. For comparison, an identical concentration of Zn acetylacetonate in water or dilute alkaline NaOH solutions and a reaction temperature of around 90 °C leads to a smaller conversion of approximately 50% only, even after several hours of reaction. The particle size determined from XRD for different orientations shows a preferred orientation along the c-direction of the hexagonal crystal system, as well in accordance with scanning electron microscopy. The LaMer model explained this highly non-uniform growth of needle-like ZnO crystallites.

Keywords: *Zn-oxide synthesis; in situ investigation; solution chemistry; X-ray absorption spectroscopy; X-ray diffraction*

1. INTRODUCTION

Zinc oxide (ZnO) has many interesting applications in various fields, such as catalysis, energy harvesting and storage, electronics, sensors, biomedicine and cosmetics, to mention just a few. With the advances in nanotechnology, new opportunities for preparing ZnO nanoparticles with tailored morphology and physicochemical properties appeared. In general, the preparation of ZnO nanoparticles from solution-based processes appears advantageous due to a high level of controllability and the opportunity to scale the reaction to a technological scale. In particular, hydrothermal synthesis routes for ZnO employing Zn-containing precursors in aqueous solutions became popular due to their simplicity and variability [1,2]. Important parameters for the reaction are the concentration of the Zn-precursor, possible additives and the reaction temperature leading to entirely different morphologies of the ZnO particles [2]. However, as the reaction temperature is limited by the water solvent's boiling point (T_B), different solvents with increased T_B appear attractive. In particular, organic solvents thus have gained interest for such a solvothermal synthesis [3]. Due to the use of different additives and the variation of the solution pH, temperature, pressure and time, these strategies are well-suited to synthesize a broad range of ZnO nanomaterials with different micro- and nanostructures and morphologies, ranging from micro- and nanospheres [3,4], nanoparticles, nanoflowers, and nanopyramids [3,5,6] to needle-like deposits with hollow structures [7]. The influence of an alcoholic solvent on the ZnO particle growth and morphology has been proved experimentally and theoretically [6,8,9,10,11,12], and microwave-assisted growth has also been demonstrated [3,13,14]. A comprehensive review of different preparation strategies has recently appeared [15].

Though studies focusing on the influence of synthesis conditions on structure and morphology have made progress, there is still a long way to go before a comprehensive understanding of the effects of synthesis parameters on structure, particle size, and morphology can be obtained. Consequently, the tailored preparation of ZnO nanomaterials with optimized structures and morphologies for specific purposes is still challenging. An in situ analysis and control of ZnO particle formation processes would thus be valuable and helpful for a target-oriented growth of nanomaterials. Specifically, the investigation of the in situ processes of ZnO formation in the liquid phase and at elevated temperatures continue to face obstacles that must be overcome.

Therefore, it is necessary to use analytical techniques compatible with the harsh reaction environment (such as solvents, temperatures, pressures, reaction speed, etc.) Electron-based techniques, such as electron microscopies (scanning electron microscopy, SEM; transmission electron microscopy, TEM) and electron spectroscopies (e.g., X-ray or ultraviolet photoelectron spectroscopies, XPS and UPS) that can substantially contribute to the elucidation of the structure, the morphology and the electronic properties of nanomaterials are not suited for in situ investigations in the liquid phase. In particular, functional groups of the species in the liquid phase can be identified using a visible, microwave, or infrared spectroscopy, such as UV–Vis, IR, or Raman spectroscopy (see, e.g., [16,17]). If the concentration of the species of interest—such as, e.g., small ZnO nuclei in the present case—is small compared to, e.g., the solvent or additives in the reaction solution, then the quantitative results are challenging to obtain. Furthermore, in conventional IR spectroscopy, the capability to characterize nanoparticle surfaces and their chemistry is limited. Nevertheless, although a quantitative identification of hydrolyzed or condensed species is impossible to obtain even from a thorough UV–Vis spectroscopy analysis, the kinetics of the nanoparticle formation could be derived, i.e., the time domain for hydrolysis, condensation and aggregation of primary particles could be obtained [16,18].

In this context, the use and combination of X-ray-based, photon-in, and photon-out techniques, such as X-ray scattering [19] and X-ray absorption spectroscopy [20] appear favorable. While X-ray scattering probes the sample structure at a long range, providing detailed information about crystalline structures, their size and orientation [19], X-ray absorption spectroscopy techniques (EXAFS and XANES) are sensitive to the short-range order around the X-ray absorbing atom, its chemical valence and the bond geometry [20]. Thus, EXAFS and XANES are well-suited for investigating amorphous materials, molecular complexes, and liquids [20,21]. For the study mentioned above, EXAFS and XANES measurements recorded simultaneously with the UV–Vis data allowed for the identification of the titanium (Ti^{4+}) tetraisopropoxide precursor as well as intermediate dodecatitanate condensate species within a $Ti_{11}O_{13}$ or $Ti_{12}O_{16}$ molecular framework, and amorphous TiO_2 particles in later stages of the growth [18]. Depending on the temperature and the hydrolysis rate, the kinetics of the reactions substantially changed, as already indicated by the UV–Vis data [16,18].

Koziej et al. recently reported on the in situ investigation of the formation of spherical CoO nanocrystals from metal-organic precursors $Co(acac)_3$ in an alcoholic solution when applying temperatures of 150 °C and above [22]. EXAFS/XANES revealed the rapid reduction from the pristine Co^{3+} ($Co(acac)_3$) species to Co^{2+} in the form of $Co(acac)_2$ within a few minutes, followed by the formation of small CoO oxide nuclei. A deconvolution/linear combination fit allows for the determination of $Co(acac)_3$, $Co(acac)_2$ and CoO concentrations as a function of time, with decreasing contributions of $Co(acac)_3$, an intermediate $Co(acac)_2$ species and conversion to about 90% of CoO in later stages of the reaction after ca. 60 min at about 160 °C [22]. These examples exemplarily indicate the useful combination of different X-ray techniques for the in situ, time-resolved analysis of nanoparticle formation processes.

In the case of ZnO, numerous ex situ X-ray studies successfully dealt with the characterization of differently prepared ZnO nanomaterials. The structure of ZnO at ambient conditions adopts the space group symmetry $P6_{3mc}$ (i.e., the Wurtzite structure, W). For moderate pressures, this structure can be transformed into a cubic setting within an F-43m space group (i.e., rock salt structure, RS) [23]. In contrast, the growth of a third structural modification of ZnO, namely the cubic zincblende type ZnO, is still challenging [24,25]. Compared to the bulk oxide,

Wurtzite ZnO nanorods prepared by oxidation of metallic Zn in NaCl solutions exhibited more disorder and slightly larger Zn-Zn interlayer distances as investigated by EXAFS measurements [26]. Wurtzite nanoparticles with a size of about 3–5 nm prepared by hydrolysis from $Zn(acac)_2$ under ultrasonic excitation with organic additives revealed a contraction of the Zn-O bond according to the strong interaction of Zn and oxygen [27].

At first glance, one may think these minor structural differences are negligible; however, the electronic structure and resulting physicochemical properties sensitively rely on these structural details. For example, slightly reduced ZnO_x nanomaterials reveal an improved catalytic reactivity and a good selectivity for the propane dehydrogenation reaction [28]. The antibacterial properties of ZnO nanostructures depend on the growth conditions and the size and orientation of the particles on the substrate [29]. Similarly, the photocatalytic reduction of CO_2 by ZnO microspheres, microflowers and nanorods sensitively depends on the percentage of exposed [0001] facets and the size of the catalysts [30]. The EXAFS experiments revealed detectable differences in the performance of hollow tubular ZnO nanostructures for the photocatalytic degradation of organic dye pollutants [11], while XANES showed the ability to distinguish coordination changes and surface disorder of the Zn-O adsorption complexes [31].

The performance of hollow tubular ZnO nanostructures for the photocatalytic degradation of organic dye pollutants can be sensitively monitored by EXAFS experiments that reveal detectable differences [11]. Moreover, XANES has proven to differentiate the coordination changes and surface disorder of the Zn-O adsorption complexes [31]. Since the use of X-ray absorption spectroscopy at the Zn K-edge allows the coordination changes and surface specificities of ZnO-based nanostructures to be distinguished, the application of this technique with a complementary XRD approach is an efficient and unique method for the immediate monitoring of the in situ synthesis of ZnO nanomaterials during the processing, providing relevant results on structural and microstructural features essential for the definition of the optimal ZnO nanostructures, which have applications across a variety of fields, as illustrated in many studies ([5,11,15,28,29,30] and references therein).

In this contribution, we, therefore, demonstrate the simultaneous application of X-ray diffraction and X-ray absorption spectroscopy to follow the formation of ZnO nanoparticles from a metal-organic ($Zn(acac)_2$) precursor in different solutions in situ with a time resolution of 2 s for each spectrum/diffractogram, so a very detailed analysis of the ZnO nanomaterials formation dynamics is feasible. A dedicated cell suited to a chemical synthesis employing inflammable liquids at temperatures up to ca. 165 °C and simultaneous X-ray measurements were conducted and realized for the experiments.

2. MATERIALS AND METHODS

2.1. ZnO Synthesis Procedures

We employed Zinc acetylacetonate monohydrate ($Zn(acac)_2$, $Zn(C_5H_7O_2)_2H_2O$, Alfa Aesar, Haverhill, MA, USA), sodium hydroxide (NaOH, Sigma-Aldrich, St.

Louis, MO, USA), 1-octanol ($CH_3(CH_2)_7OH$, Merck Millipore, Burlington, MA, USA) and triple distilled water as starting materials for the synthesis of ZnO nanomaterials. The reactants were mixed at room temperature in desired amounts and stirred for about 10 min. Subsequently, the solution was poured into the reaction cell and carefully sealed. Finally, the precursor solution was heated to the desired temperature under continuous stirring. The sample notation and the experimental conditions used for the preparation of the selected samples are given in Table 1. In a typical synthetic procedure, zinc acetylacetonate monohydrate ($Zn(acac)_2$) in the desired amount was mixed with water, a well-defined volume of aqueous NaOH solution (10^{-3} M) or 1-octanol, respectively, following the procedures described by Šarić et al. [4,7]. Depending on the actual solvent, the solutions were heated to about 90 °C (samples A and W) and 165 °C (sample O) to initiate the formation of ZnO nanomaterials for different reaction times, as compiled in Table 1. More than 25 individual preparations were performed and analyzed to ensure the reproducibility of the experiments.

Table 1.: *Compilation of the sample preparations and sample identification discussed in this work.*

Sample Id.	$Zn(acac)_2H_2O$	Solvent	Amount	Temperature	Time
A	0.75 g	10^{-3} M NaOH	50 ml	90 °C	150 min
O	0.75 g	1-octanol	50 ml	165 °C	60 min
W	0.75 g	water	50 ml	90 °C	450 min

2.2. In Situ Reaction Cell

The reactions were performed in a dedicated, hermetically sealed reaction cell, as shown in Figure 1. The main body of the cell with outer dimensions of 53 mm (width), 48 mm (length in the direction of the X-ray beam) and 102 mm (height) consists of a PTFE container in which the liquid precursor solution (maximum ca. 70 mL) is filled. Tests have shown that the electric heater attached to one side of the cell can heat the liquid sample solutions to a maximum temperature of ca. 300 °C, as measured with a platinum resistance thermometer encased in PTFE directly in the solution, and a second thermocouple fixed to the heater with a precision of ± 1 °C. The X-ray beam passes through the liquid sample volume about 20 mm above the bottom of the cell. It enters the cell via a small circular polyimide (Kapton) entrance window of 10 mm diameter and 25 μm thickness. The entrance and exit windows were mounted on specially designed PTFE frames employing a temperature and chemical-resistant adhesive (CMC 70110, CMC Klebetechnik, Frankenthal, Germany). It is important to note that the exit-window frame has a conical shape to allow scattering angles of up to 40°. The frames with the windows, as well as the top flange of the cell, are screwed to the main body and sealed by Viton

O-rings of a suited diameter (see Figure 1). Moreover, the effective thickness of the probed sample can be varied by varying the length of the window frames so that a reasonable X-ray absorption is achieved on the one hand while the diffracted volume is kept small enough to reduce the broadening of the Bragg peaks on the other. To guarantee a homogenous temperature distribution within the cell and to avoid precipitation of reaction products at the cell walls and the windows, a magnetic stirrer actuated by a brushless motor in the cell holder below the liquid were employed. The stirrer, as well as the control of the heater and the temperature readouts, were remote-controlled via a Raspberry Pi with an Ethernet connection.

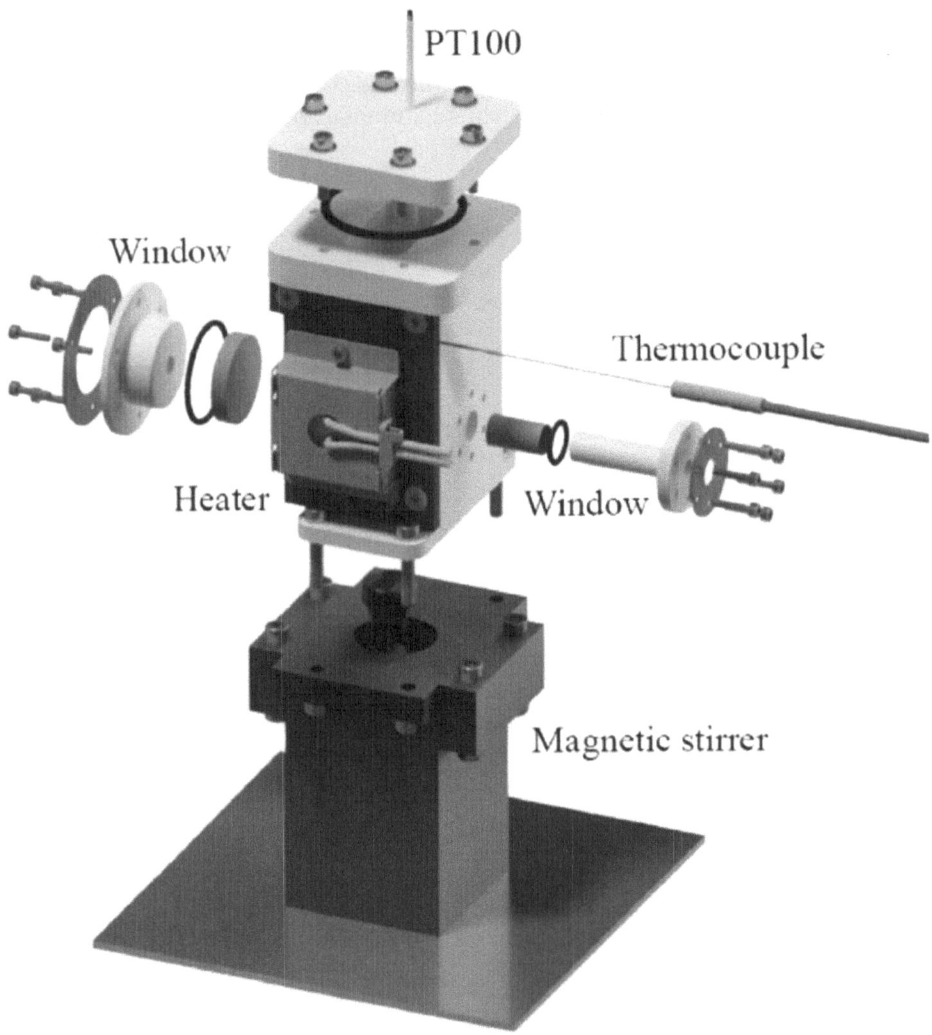

Figure 1.: *Schematic drawing of the reaction cell for hydrothermal/solvothermal preparation of ZnO nanomaterials. The X-ray beam passes from the right to the left.*

2.3. X-ray Experiments

The X-ray experiments were carried out at the advanced spectroscopy beamline P64 at the PETRA III storage ring (DESY, Hamburg, Germany). A cryo-cooled, Si(111) channel-cut monochromator suited for fast oscillations with up to 100 Hz (i.e., up to 200 spectra per second) was employed for the EXAFS measurements [32]. Gas-filled ionization chambers were used to monitor the incident (I_0) and transmitted (I_1) X-ray intensities. A third ionization chamber (I_2) was used to simultaneously investigate a Zn-metal reference foil with each sample spectrum for precise and reliable energy calibration, as depicted in Figure 2a. A Pilatus 100K pixel detector (DECTRIS, Dübendorf, Switzerland) was used to process the XRD images with forward-scattering geometry. While a sinusoidal excitation of the monochromator oscillation is usually employed for quick-scanning EXAFS experiments at P64 [32], it is important to perform X-ray diffraction studies with well-defined X-ray energy to obtain high-quality diffraction data. Thus, to avoid an interruption in the EXAFS data collection by stopping the oscillatory monochromator movements, we modified the sinusoidal movements by adding a signal of the third harmonic wave with suited amplitude so that the resulting oscillation obeys an energy plateau at the extremal points as can be seen in Figure 2b. With a trigger placed in this plateau by the monochromator control software, the illumination of the XRD detector can be initiated for a chosen period of time [33]. For the present experiments, a scan speed of 0.5 Hz was chosen, i.e., an EXAFS spectrum at the Zn K-edge (9.659 keV) covering the range from ca. 9.418 keV to 10.897 keV was taken every second, and an exposure of the XRD-detector was triggered every 2 s with an exposure time of 250 ms. It is important to note that the energy spread during the illumination of the XRD detector amounts to only 6.5 eV within this configuration, allowing for high-resolution XRD data to be obtained, as shown in Figure 2c. It should be noted that the used Pilatus detector has a rectangular active area of ca. 83.8×33.5 mm^2, so that in an upright position, a larger Bragg-angle range can be covered in comparison to a horizontally orientated detector. However, for this study, we chose the latter position because it allowed us to detect a large portion of the diffraction cones at the most prominent (100), (002), and (101) diffraction peaks of hexagonal ZnO, enabling the detection of subtle structural changes in the diffracting nanoparticles (see, e.g., refs. [4,11,12]).

Complementary, static in situ and ex situ X-ray experiments were performed at beamline 10 of the DELTA storage ring (Dortmund, Germany) [34]. Here, the reference spectra of Zn metal, polycrystalline Wurtzite ZnO powder, and zinc acetylacetonate monohydrate (Zn(acac)$_2$, Zn($C_5H_7O_2$)$_2$ H_2O) were recorded for comparison and their use in the linear combination fitting of the EXAFS data. EXAFS data reduction and analysis were performed using the Athena / Artemis software [35].

3. RESULTS AND DISCUSSION

3.1. X-ray Absorption Spectroscopy

Combined, simultaneous quick-EXAFS at the Zn K-edge and XRD measurements were measured during the synthesis of ZnO nanomaterials from different precursor solutions and for different temperatures and concentrations of the Zn(a-cac)$_2$ precursors.

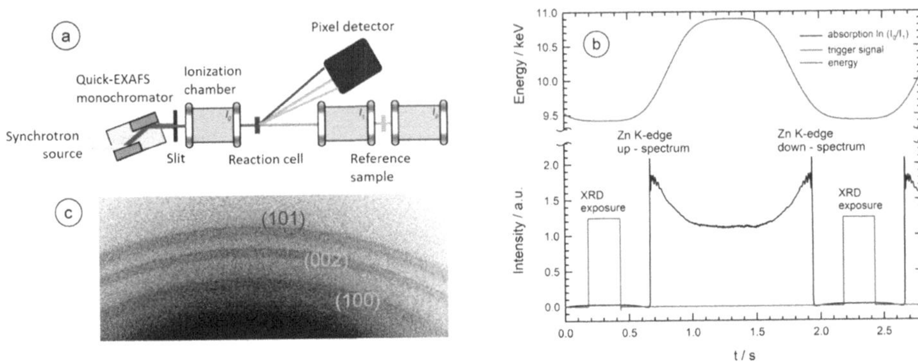

Figure 2.: *(a) Schematic representation of the experimental setup with the fast-scanning monochromator, the ionization chambers and the flat-panel detector arranged in forward-scattering geometry from the sample. (b) Continuous variation of the monochromator energy as a function of time, together with the measured absorption spectrum of the sample, and the trigger signal placed in the flat pre-edge region of the measurement. (c) Example of a diffraction pattern collected during the formation of ZnO nanoparticles in Zn(acac)$_2$ solution in 10^{-3}M NaOH (sample A) at a temperature of ca. 85 °C (t = 3200 s) within 250 ms of integration time. The (100)-reflection with a Bragg angle Θ_B of $\Theta_B \approx 27.0°$ is located at the bottom of the active area of the pixel detector, the (002) with $\Theta_B \approx 29.3°$ in the center, and the (101) in the upper region with $\Theta_B \approx 30.8°$, respectively, as also indicated by the colors in (a).*

As a typical example of quick-EXAFS data obtained during such a reaction, some selected spectra measured during the formation of ZnO in a solution containing 1-octanol as solvent (sample O, see Table 1) are presented in Figure 3. Although the spectra were collected with an acquisition time of only 1 s, the data were of sufficient quality to allow the extraction of k^3-weighted absorption fine structure oscillations $\chi(k)*k^3$ up to a photoelectron wave vector of 11 Å^{-1}, as shown in the inset of Figure 3. As can be seen, substantial changes in the spectra were observed when the stirred precursor solution was heated to about 100 °C and more. First, no changes in the absorption edge position defined as the inflection point of the spectra are observed. Thus, the chemical valence of the absorbing Zn-atoms does not change during the reaction, and the observed values of E_{Edge} = 9.663 keV ± 0.5 eV are typical for organic and inorganic Zn^{2+} species as expected [36,37]. While the white line intensity directly at the edge (9.67 keV) continuously decreases with increasing temperature, the absorption minimum at about 9.69 keV increases and slightly shifts to higher energies in the course of the reaction, and more prominent changes are observed for higher energies.

Most notable is the presence of isosbestic points in the absorption spectra at ca. 9.677 keV, 9.711 keV, 9.732 keV, 9.752 keV and 9.822 keV. In the $\chi(k)*k^3$ data (inset of Figure 3), these characteristic features are visible at 4.2 Å^{-1}, 5.6 Å^{-1}, 6.4 Å^{-1}, 7.1 Å^{-1} and 9.2 Å^{-1}, underlining the presence of only two different compounds in substantial concentration in the sample volume, i.e., on first glance, we cannot give any evidence for reaction intermediates as, e.g., proposed by calculations in reference [11].

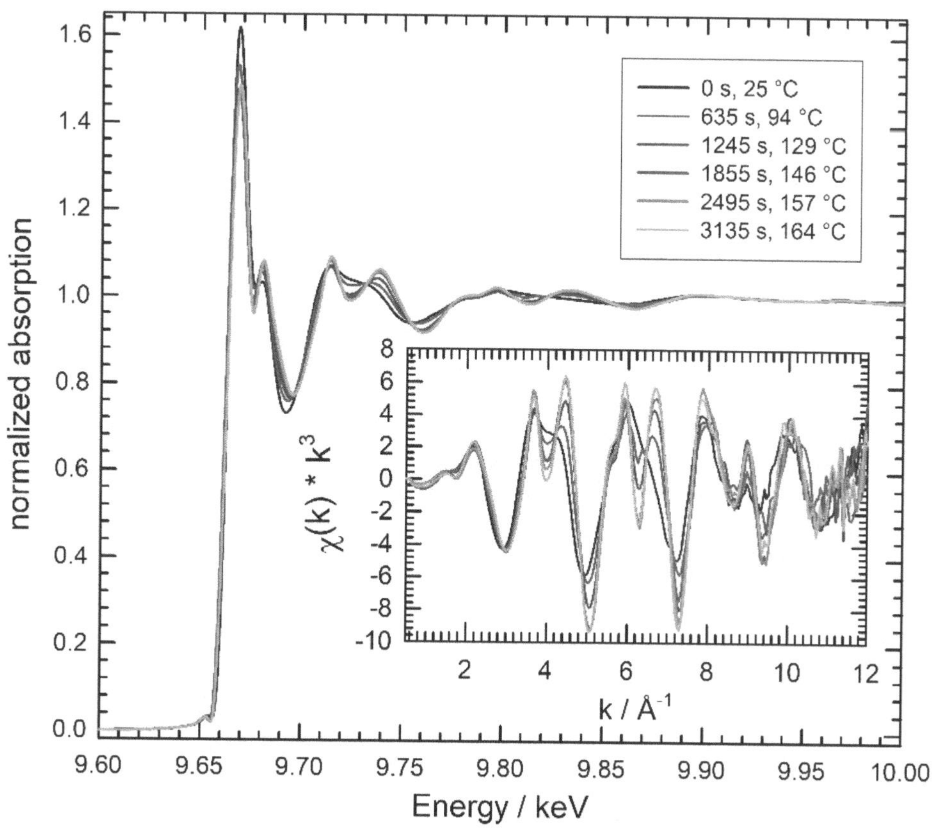

***Figure 3.**: Selection of characteristic Zn K-edge X-ray absorption data measured during the heating of a $Zn(acac)_2$ solution in 1-octanol from room temperature to ca. 165 °C as indicated (sample O). The k^3-weighted EXAFS $\chi(k)*k^3$ extracted from the data is depicted in the inset.*

The reaction course can be more sensitively followed in the Fourier-transformed X-ray absorption fine structure data, as displayed in Figure 4 for the three different reaction media. Peaks in these Fourier transforms correspond to the first few coordination shells around the X-ray absorbing Zn atom, i.e., the first peak at ca. 1.5 Å radial distance corresponds to the first Zn-O coordination of the reaction educts and products, while the peaks at a larger space are related to the Zn-Zn coordinations and multiple scattering events in the structures present at a particular time. It should be noted that the peaks in the Fourier transforms are generally shifted to smaller distances by about 0.4 Å compared to their crystallographic positions due to the phase shifts of the scattered photoelectrons causing these peaks [20]. For example, the first Zn-O and Zn-Zn bond distances in Wurtzite are 1.97 Å and 3.24 Å, respectively [11,38], while the related peaks in the Fourier transform appear at 1.5 Å and 2.9 Å.

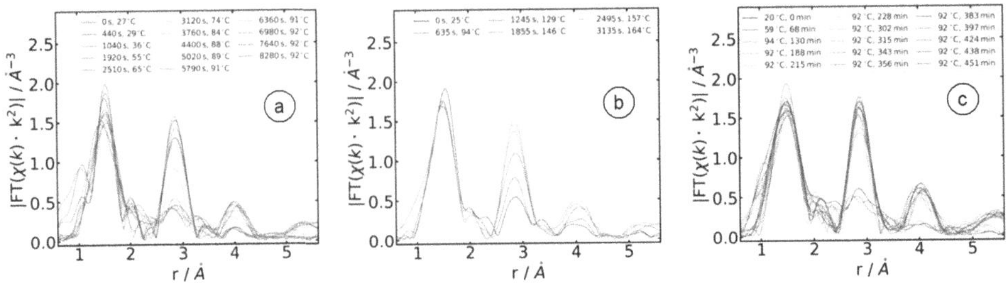

Figure 4.: *Magnitude of the k^2-weighted X-ray absorption fine structures $\chi(k)*k^2$ for the indicated times and temperatures (k-range for the FTs: 1.2 $Å^{-1} < k <$ 11.0 $Å^{-1}$, Kaiser–Bessel window functions). The peaks at ca. 1.5 Å belong to the Zn-O interactions, the peaks at ca. 2.9 Å are due to Zn-Zn bonds, respectively, and peaks at ca. 4 Å belong to multiple scattering events with larger path distances involving zinc and oxygen backscatters. (a) The reaction of Zn(acac)$_2$ solution in 10^{-3} M NaOH (sample A); (b) reaction of Zn(acac)$_2$ solution in 1-octanol (sample O); and (c) reaction of Zn(acac)$_2$ solution in distilled water (sample W).*

As can be seen in Figure 4, the position and the intensity of the peak at 1.5 Å remained almost constant as a function of the reaction time for all three preparations, while substantial differences were observed for the peaks at 2.9 Å and 4.0 Å, where a time- and temperature-dependent increase in intensity were observed. Comparing the Fourier-transform of Zn(acac)$_2$ and crystalline ZnO reference samples, it can be concluded that the peak at 1.5 Å arises from the Zn-O bonds in both compounds. Since Zn is 4-coordinated with oxygen in both sample systems and the Zn-O bond distances are very similar in both cases, there is not much change expected in the first peak during ZnO particle formation. The slight decrease in intensity, which can be observed in all samples, resulted from the increasing temperature and the resulting increase in thermal vibrations so that the amplitude of the EXAFS fine structures $\chi(k)*k^2$ and their Fourier-transform are reduced accordingly. The peak at 2.9 Å arises from the Zn-Zn shell in crystalline ZnO, which has a coordination number of 12 in hexagonal ZnO at a distance of 3.24 Å. Since this shell is not present in Zn(acac)$_2$, it forms only during the synthesis and is a consequence of the solvothermal formation of ZnO, as is the third peak at 4.0 Å, which is due to multiple scattering of the Zn-O-Zn pathways [11]. Therefore, the course of the ZnO reaction can be easily followed by examining these peaks. Obviously, the reaction kinetics of sample O (reaction in 1-octanol) are much faster compared to sample A (10^{-3} M NaOH), and especially to sample W (pure water).

For a more detailed analysis, the grown ZnO nanomaterials were extracted from the reaction cell after the completion of the reaction, filtered on paper, dried in air and investigated with EXAFS in transmission subsequently. The obtained experimental EXAFS data were fitted to a structural model based on the assumption of a hexagonal ZnO Wurtzite structure adopting a $P6_{3mc}$ space group symmetry, with unit cell metrics of a = b = 3.251 Å, c = 5.207Å, $\alpha = \beta = 90°$ and $\gamma = 120°$ [38]. All atoms up to a radius of 4.7 Å to the central Zn atom were considered for calculating the phase and amplitude functions using the FEFF 9 code [39], resulting in 14 coordination shells and a cluster size of 39 atoms.

Each coordination shell (i) considered is characterized by its distance (R_i), the number of atoms in the shell (N_i), the inner potential shift ($E_{0,i}$), the mean square displacement (σ_i^2) and the amplitude reduction factor ($S_{0,i}^2$). If each shell considered had been fitted with its own set of parameters, the results here would be a very large set of parameters with a statistically non-significant number of variables. The significance limit for the number of fit variables (N_{ipd}) can be estimated from the energy range measured during the experiment, the k-range used for the Fourier transform (here, the range from k_{min} = 1.2 Å^{-1} to k_{max} = 12.4 Å^{-1}), and the R-range in the FT used for the fit (here, R_{min} = 1.0 Å to R_{max} = 4.3 Å); for details see, e.g., ref. [40].

For the present experiments, N_{idp} had a value of $N_{idp} \approx 23$. To ensure this upper level of parameters was not exceeded, a simplified fit model was developed, which assumes identical values for S_0^2 and σ_i^2 on all the Zn-O shells. Similarly, the Zn-Zn scattering was described by a separate set of parameters. The distances R_i were varied individually for the first single scattering of Zn-O (R_1) and the Zn-Zn shells (R_2) to sensitively model small structural differences. All other distances were fitted with a single scaling factor (α) of the original ZnO Wurtzite lattice (R_{eff}), so the fitted distances for all paths with longer bond distances were given by $R_{eff}{}^{*}\alpha$. To minimize the influence of the correlation of the potential shift and all distances, the same value for E_0 was used for all shells. Thus, this model has only eight fit variables compared to the significance limit ($N_{idp} \approx 23$) and accordingly, the results are likely to have a high statistical significance. The fits were performed in R-space employing the k^3-weighted absorption fine structure data using Artemis software, and the fit results are compiled in Figure 5 and Table 2, where the R-factors of the fits are also shown.

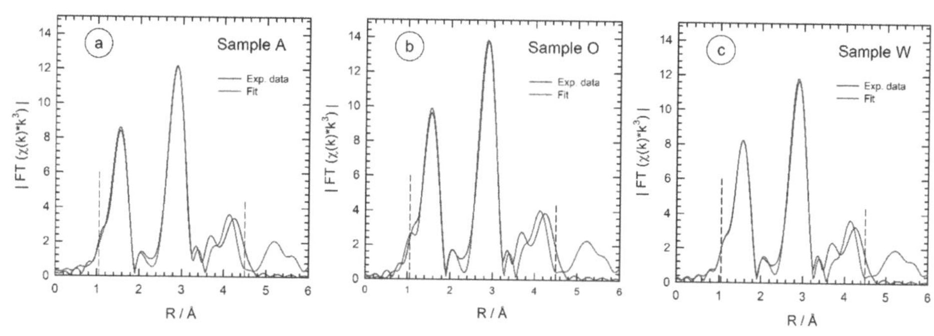

***Figure 5.:** Fits of the EXAFS data measured after extraction of the prepared ZnO nanomaterials from the reaction cell after finalization of the reaction. The data were fitted to a model structure consisting of hexagonal, wurtzite-type ZnO (k-range for the fits: 1.2 Å^{-1} < k < 12.4 Å^{-1}, Kaiser-Bessel-window functions, R-range 1.0 Å < R < 4.3 Å, dashed vertical blue lines). (a) The reaction of Zn(acac)$_2$ solution in 10^{-3} M NaOH (sample A); (b) reaction of Zn(acac)$_2$ solution in 1-octanol (sample O); and (c) reaction of Zn(acac)$_2$ solution in distilled water (sample W).*

The fit results show excellent agreement with the Wurtzite structure for all samples studied, regardless of the details of the preparations, i.e., all paths lead to Wurtzite-type nanomaterials, in good agreement with the literature.

Table 2.: *Compilation of quantitative fit results obtained from a detailed EXAFS data analysis for samples A, O and W after extraction of the synthesized ZnO nanomaterials from the reaction cell.*

Sample	Shell	N	S_0^2	R/Å	σ^2/Å^2	E_0/eV
Sample A (R-factor: 0.025)	Zn-O	4	0.793 ± 0.097	1.978 ± 0.007	0.0051 ± 0.0012	2.65 ± 0.69
	Zn-Zn	12	1.313 ± 0.210	3.247 ± 0.006	0.0122 ± 0.0012	2.65 ± 0.69
	Zn-O	1	0.793 ± 0.097	3.255 ± 0.007	0.0051 ± 0.0012	2.65 ± 0.69
	Zn-O-O	12	0.793 ± 0.097	3.600 ± 0.008	0.0051 ± 0.0012	2.65 ± 0.69
	Zn-O-Zn	24	1.053 ± 0.155	3.600 ± 0.008	0.0087 ± 0.0012	2.65 ± 0.69
	Zn-O	9	0.793 ± 0.097	3.808 ± 0.008	0.0051 ± 0.0012	2.65 ± 0.69
	Zn-O	4	0.793 ± 0.097	3.964 ± 0.009	0.0051 ± 0.0012	2.65 ± 0.69
	Zn-O-Zn-O	12	1.053 ± 0.155	3.964 ± 0.009	0.0087 ± 0.0012	2.65 ± 0.69
	Zn-O-O	6	0.793 ± 0.097	4.229 ± 0.009	0.0051 ± 0.0012	2.65 ± 0.69
	Zn-O-Zn	36	1.053 ± 0.155	4.516 ± 0.009	0.0087 ± 0.0012	2.65 ± 0.69
	Zn-O-O	36	0.793 ± 0.097	4.516 ± 0.009	0.0051 ± 0.0012	2.65 ± 0.69
	Zn-Zn-O	36	1.053 ± 0.155	4.516 ± 0.009	0.0087 ± 0.0012	2.65 ± 0.69
	Zn-Zn	6	1.313 ± 0.210	4.577 ± 0.010	0.0122 ± 0.0012	2.65 ± 0.69
	Zn-O	6	0.793 ± 0.097	4.599 ± 0.010	0.0051 ± 0.0012	2.65 ± 0.69
Sample O (R-factor: 0.030)	Zn-O	4	0.854 ± 0.114	1.971 ± 0.008	0.0047 ± 0.0012	3.12 ± 0.75
	Zn-Zn	12	1.573 ± 0.227	3.245 ± 0.006	0.0126 ± 0.0013	3.12 ± 0.75
	Zn-O	1	0.854 ± 0.114	3.250 ± 0.007	0.0047 ± 0.0012	3.12 ± 0.75
	Zn-O-O	12	0.854 ± 0.114	3.595 ± 0.008	0.0047 ± 0.0012	3.12 ± 0.75
	Zn-O-Zn	24	1.214 ± 0.096	3.595 ± 0.008	0.0087 ± 0.0013	3.12 ± 0.75
	Zn-O	9	0.854 ± 0.114	3.802 ± 0.009	0.0047 ± 0.0012	3.12 ± 0.75
	Zn-O	4	0.854 ± 0.114	3.958 ± 0.009	0.0047 ± 0.0012	3.12 ± 0.75
	Zn-O-Zn-O	12	1.214 ± 0.096	3.958 ± 0.009	0.0087 ± 0.0013	3.12 ± 0.75
	Zn-O-O	6	0.854 ± 0.114	4.223 ± 0.009	0.0047 ± 0.0012	3.12 ± 0.75
	Zn-O-Zn	36	1.214 ± 0.096	4.509 ± 0.010	0.0087 ± 0.0013	3.12 ± 0.75
	Zn-O-O	36	0.854 ± 0.114	4.509 ± 0.010	0.0047 ± 0.0012	3.12 ± 0.75
	Zn-Zn-O	36	1.214 ± 0.096	4.509 ± 0.010	0.0087 ± 0.0013	3.12 ± 0.75
	Zn-Zn	6	1.573 ± 0.277	4.570 ± 0.010	0.0126 ± 0.0013	3.12 ± 0.75
	Zn-O	6	0.854 ± 0.114	4.592 ± 0.010	0.0047 ± 0.0012	3.12 ± 0.75

Sample	Shell	N	S_0^2	R/Å	σ^2/Å^2	E_0/eV
Sample W (R-factor: 0.029)	Zn-O	4	0.735 ± 0.096	1.972 ± 0.008	0.0048 ± 0.0012	3.62 ± 0.74
	Zn-Zn	12	1.258 ± 0.222	3.244 ± 0.006	0.0123 ± 0.0013	3.62 ± 0.74
	Zn-O	1	0.735 ± 0.096	3.248 ± 0.007	0.0048 ± 0.0012	3.62 ± 0.74
	Zn-O-O	12	0.735 ± 0.096	3.593 ± 0.008	0.0048 ± 0.0012	3.62 ± 0.74
	Zn-O-Zn	24	0.997 ± 0.112	3.593 ± 0.008	0.0086 ± 0.0013	3.62 ± 0.74
	Zn-O	9	0.735 ± 0.096	3.800 ± 0.009	0.0048 ± 0.0012	3.62 ± 0.74
	Zn-O	4	0.735 ± 0.096	3.956 ± 0.009	0.0048 ± 0.0012	3.62 ± 0.74
	Zn-O-Zn-O	12	0.997 ±0.112	3.956 ± 0.009	0.0086 ± 0.0013	3.62 ± 0.74
	Zn-O-O	6	0.735 ± 0.096	4.220 ± 0.009	0.0048 ± 0.0012	3.62 ± 0.74
	Zn-O-Zn	36	0.997 ± 0.112	4.506 ± 0.010	0.0086 ± 0.0013	3.62 ± 0.74
	Zn-O-O	36	0.735 ± 0.096	4.506 ± 0.010	0.0048 ± 0.0012	3.62 ± 0.74
	Zn-Zn-O	36	0.997 ± 0.112	4.506 ± 0.010	0.0086 ± 0.0013	3.62 ± 0.74
	Zn-Zn	6	1.258 ± 0.222	4.567 ± 0.010	0.0123 ± 0.0013	3.62 ± 0.74
	Zn-O	6	0.735 ± 0.096	4.589 ± 0.010	0.0048 ± 0.0012	3.62 ± 0.74

There are only minor differences in the extracted short-range order structures; for example, in an overall trend, the distances are larger for sample A compared to samples O and W, while the intensity of the features (S_0^2) is maximal for sample O, i.e., the preparation in 1-octanol. For the disorder, no distinct differences are detectable, in agreement with recent EXAFS measurements [11].

Since the reaction products in all cases are nanocrystalline ZnO materials in a Wurtzite structure, and due to the fact that the isosbestic points observed in the raw data and in the extracted EXAFS functions ($\chi(k)*k^3$, see Figure 3), it is very likely that only two phases are present during the reaction, namely the Wurtzite ZnO and the $Zn(acac)_2$ precursor. Therefore, we performed linear combination fitting of the measured EXAFS data during the three preparation routes investigated here. The Zn K-edge spectra of the respective reference materials at room temperature were taken as constituting components, although the temperature during the reaction is substantially increased to about 90 °C for the reaction in 10^{-3} M NaOH (sample A) and distilled water (sample W) and to more than 160 °C for the reaction in 1-octanol (sample O). The fit results obtained for the different preparations are compiled in Figure 6 as a function of time, together with the measured temperature profiles of the reaction media.

There are clear differences depending on the solvent used. Namely, in the NaOH sample (sample A), the ZnO formation process starts after about 2000 s and a temperature of 55 °C. Up to about 5000 s and a temperature of about 90 °C, the ZnO phase fraction has increased to about 70% and remains at about this value until the end of the synthesis process considered. For sample A, the conversion process from $Zn(acac)_2$ to ZnO seems to be completed after about 5000 s and at a temperature of 90 °C.

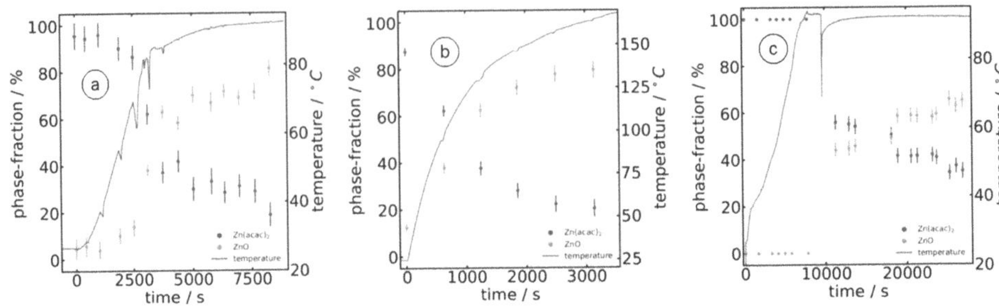

Figure 6.: *Results of linear combination fits of the EXAFS data measured in the course of the reaction in (a) $Zn(acac)_2$ solution in 10^{-3} M NaOH (sample A); (b) $Zn(acac)_2$ solution in 1-octanol (sample O); and (c) $Zn(acac)_2$ solution in distilled water (sample W), respectively. Reference spectra of crystalline ZnO and $Zn(acac)_2$ were used for the fits.*

A similar formation process can be observed with water as the solvent (sample W). It took about 120 min (ca. 7000 s) to detect the first evidence of the presence of ZnO, again at a temperature of ca. 90 °C. After about 200 min, i.e., another 3000 s at a temperature of 90 °C, the ZnO content increases to 45% and reaches about 65% at the end of the observation period after a total of 450 min. Here, too, the $Zn(acac)_2$ precursor has not yet been completely converted into ZnO after the period under consideration.

For the reaction in 1-octanol as solvent (sample O), ZnO is directly formed even at room temperature, i.e., ZnO is detectable at the initiation of the measurements with a fraction of ca. 10%. This proportion increases continuously with time and temperature and amounts to 35% after 600 s (90 °C), 60% after 1200 s (125 °C) and ca. 80% at the end of the observed process after 3200 s and a temperature of 165 °C. Thus, the opportunity to increase the reaction temperature to substantially more than 100 °C in the organic solvent increases the reaction rate and the conversion efficiency (i.e., the degree of transformation from $Zn(acac)_2$ to ZnO) of the process. It should be noted that similar results were found for the formation of ZnO nanocrystals by adding KOH solution to an ethanolic zinc oxy-acetate solution at a slightly elevated temperature of 40 °C by parallel X-ray absorption spectroscopy, small angle X-ray scattering, and UV–Vis spectroscopy [41]. Similarly, the conversion efficiency of such a hydrolysis-condensation process after a reaction time of several hours reaches about 80% [41]. The increase in reaction temperature obviously helps to overcome the activation energy of the reaction due to the rise of the thermal kinetic energies of the reactants, and it further increases the number of collisions between the reactants, leading to substantially larger reaction rates.

3.2. X-ray Diffraction

Some selected examples of two-dimensional XRD images recorded in parallel to the X-ray absorption measurements employing a photon energy of 9.423 keV

(X-ray wavelength $\lambda = 1.3158$ Å) are presented in Figure 7. The data collected on the different pixels of the X-ray detector were converted into conventional powder diffractograms and evaluated using the pyFAI software [42]. In comparison to a calculated diffractogram using the structure data of Wurtzite ZnO as already detailed above (PDF file no. 01-079-2205, [38]), it can be concluded that all samples only reveal the expected ZnO diffraction peaks, with the (100) peak at $\Theta_B = 27.0°$, the (002) reflection at $\Theta_B = 29.3°$ and the (101) with $\Theta_B = 30.8°$, which implies that the reaction product ZnO can also be detected via the XRD measurements. In agreement with the EXAFS measurements, the presence of ZnO took place with a slight time delay after the initiation of the reaction since a certain volume of crystalline ZnO must first be generated depending on the solution and the temperature. It was only in the reaction of 1-octanol (sample O) that very weak Scherrer rings were detected directly after initiating the reaction (see Figure 7a).

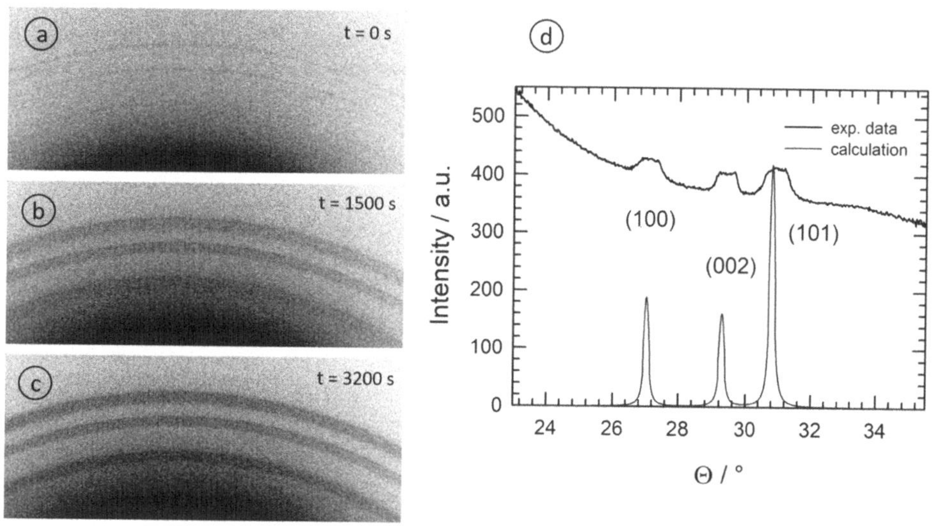

Figure 7.: *Selected diffraction patterns collected during the solvothermal reaction of $Zn(acac)_2$ in 1-octanol (sample O) for a photon energy of 9.423 keV; (a) t = 0 s; (b) t = 1500 s; (c) t = 3200 s. In (d), the diffractogram calculated by azimuthal integration is compared to a theoretical calculation for ZnO having Wurtzite structure without any preferential orientation (space group $P_{63}mc$, a = b = 3.251 Å, c = 5.207 Å; $\alpha = \beta = 90°$, $\gamma = 120°$, [38]).*

For further data analysis of the diffractograms, a polynomial background was subtracted from the diffraction data, followed by a peak fitting with three individual Gaussian peak functions. The evolution of the measured peak intensities of the (100), (002) and (101) reflections is displayed in Figure 8a for sample A, i.e., the reaction of $Zn(acac)_2$ in 10^{-3} M NaOH. The three peaks showed an identical trend, i.e., continuously increasing to a maximum intensity at about 6000 s ($T \approx 90$ °C). As can already be seen in Figure 7, the detected diffraction peaks exhibited considerable broadening due to the finite length of the X-ray path through the solution containing the nanoparticles, i.e., a particle located closer to the entrance window of the cell shows up at a position further away from the direct beam, while a particle located close to the exit window shows up at a correspondingly shorter distance. Assuming that the particles are homogeneously distributed in the liquid phase, a peak with a flat top can be expected. Such a trend is also experimentally observed by looking at the measured intensity distributions on the flat-panel detector and the extracted diffractogram (Figure 7d). In any case, this broadening must be considered carefully when the peak width is used for an estimation of the crystallite size L, e.g., by employing the Scherrer formula [43]

$$L = \frac{K\lambda}{\beta cos\Theta};$$

where $\lambda = 1.3158$ Å, β is the full width at half maximum of the considered Bragg peak, Θ is the position of the Bragg peak, and K is the Scherrer constant with a value of 0.9 [43]. The crystallite sizes for the above-mentioned orientations are accordingly determined and plotted in Figure 8b. Similar to the temporal evolution of the intensity, the crystallites grow to a size of about 25 nm ((100) and (101) orientation) and 30 nm ((002)-orientation). The temporal evolution of the diffraction peaks is well in accordance with the presence of small, hollow nanoparticles, which is usually observed in dilute alkaline solutions (see, e.g., refs. [5,10,11]).

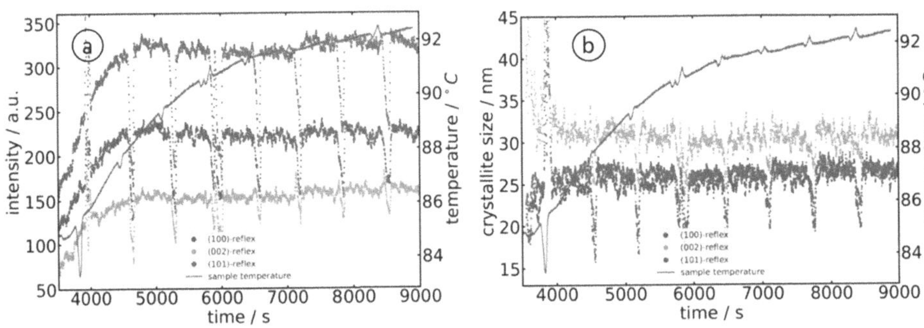

Figure 8.: *Evaluation of the diffractograms collected during the reaction of* $Zn(acac)_2$ *in* 10^{-3} *M NaOH (sample A) as a function of time and temperature: (a) temporal evolution of the diffraction peak intensities and (b) evolution of the crystallite size as calculated by the Scherrer equation. The periodic drops in intensity and crystallite size are due to the periodic interruption of the stirring process, so the particles in the solution tend to move out of the volume probed by the X-ray beam.*

In contrast, the use of organic solvents, such as triethanolamine [10], polyvinyl-pyrrolidone [27] or cetyl-trimethyl ammonium bromide [12], lead to needle- or flower-like nanocrystallites with stronger contributions of the otherwise weaker (002) Bragg reflection, and very intense (101) peaks compared to the ZnO reference without a preferred orientation (see the calculation in Figure 7d). This exact behavior was observed in the present study using 1-octanol as a solvent.

Comparing Figure 8a and Figure 9a, we notice the lower intensity of the (002) reflection in the case of the alkali compared to the (100) peak, as well as the dominant (101) reflection and the nearly equal intensities for (100) and (002) for the organic solvent (sample O). It is interesting to note that the intensity (peak area) of the (101) reflection exactly follows the course of the temperature during the reaction, thereby offering an opportunity to manipulate the growth easily. Such behavior is also well in-line with linear growth behavior [44,45]. The evaluation of the crystallite size further shows that the crystallites obey a larger and faster growth along the c-axis ((002)-reflection) in the first ca. 1500 s of the reaction, which is in agreement with the presence of needle-like crystallites. The growth of the (100)- and (101)-oriented crystallites was rather slower.

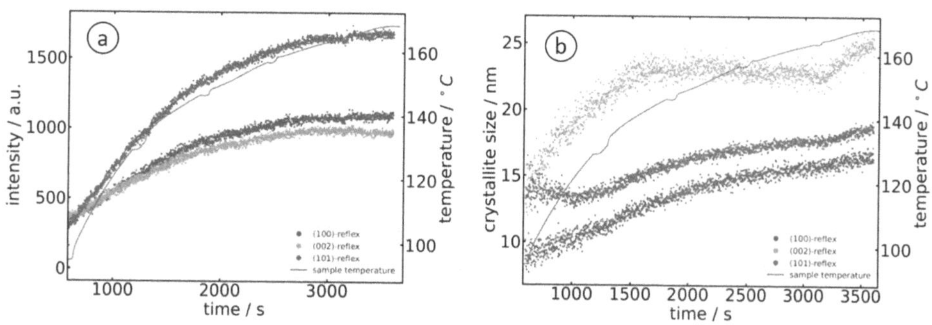

Figure 9.: Evaluation of the diffractograms collected during the reaction of Zn(acac)2 in the 1-octonal (sample O) as a function of time and temperature: (a) temporal evolution of the diffraction peak intensities and (b) evolution of the crystallite size as calculated by the Scherrer equation.

The comparison of the crystallite size found for the two experiments shows that crystallites are slightly larger for the reaction of $Zn(acac)_2$ in 10^{-3} M NaOH (sample A), with values of 31 nm ± 2 nm for the (002)-direction and slightly smaller values for (100) and (002) orientations, i.e., 25 nm ± 2 nm after about 4500 s (75 min). When prepared in 1-octanol (sample O), the crystallites were smaller, with a maximum of about 25 nm for (002) and much smaller values for (100) and (101), but with a shorter reaction time of less than one hour. Thus, the particles produced in 1-octanol were significantly smaller than in the NaOH sample, even though identical $Zn(acac)_2$ concentrations of the precursor solutions were initially used. The smaller crystallite size in (100) and (101) correlates well with the preparation of hollow nanostructures [11]. Since the diffracted intensity of the diffraction pattern is proportional to the volume leading to the diffraction of the radiation, it can be concluded from the measured Bragg intensities and particle sizes that more, but smaller, nanoparticles are formed in 1-octanol (sample O) than in the alkaline solution (sample A).

Electron microscopy measurements on particles using similar synthesis processes involving, however, a longer synthesis time of several hours or even days showed particle sizes of several hundred nanometers [4,7,11,12]. This indicates that the growth of the ZnO nanoparticles is comprised of the formation of small hexagonal nuclei in the initial stages of growth and a subsequent, temperature-controlled, and eventually non-isotropic growth by diffusion, depending on the solvent. The agglomeration of particles is presumably responsible for the decreased growth rates observed in the alkaline solvent in the later stages of the growth, as all the precursor material in close vicinity of the grown crystallites was already consumed.

3.3. Comprehensive Discussion

As described in the sections above, time-resolved quick-scanning EXAFS spectroscopy and X-ray diffraction are able to monitor the formation of hexagonal ZnO nanoparticles during a solvothermal reaction from $Zn(acac)_2$. Although EXAFS and XANES are generally very sensitive to small concentrations of additional species of the X-ray absorbing atom, no contributions from the Zn-O intermediate phases could be determined, such as Zn-organometallic species, $ZnO\text{-}H_2O$, or thermodynamically quite stable tetrameric (ZnO-OH) species, both of which were predicted by density functional calculations (DFT) [10,11]. The situation is similar to the hydrolysis–condensation route of zinc oxy-acetate precursors in ethanolic solution, where only two species were identified by EXAFS and XANES [41]. This could be because the concentrations of these species are beyond the detection limit of the X-ray techniques used, which are on the order of fractions of a percent, or the fact that the short-range order structures of these intermediate complexes are very similar to the hexagonal ZnO structure, with four oxygen atoms in the first coordination sphere (see Table 2). The fact that the amplitude reduction factor S_0^2 needed to properly model the first Zn-Zn coordination amounts to values of typically 1.2 and more (see Table 2) may be an indication for an additional Zn-containing phase, as S_0^2 is usually approximately 1.0 at its maximum, as it is the case for all the Zn-O coordination spheres. Since the Zn-Zn-coordination number is smaller in the above-mentioned Zn-O complexes, a larger value of S_0^2 is needed to compensate for this zinc deficiency in the model fit. Thus, a more detailed data analysis should be performed, including the synthesis, isolation and preparation of suited reference compounds. It should also be mentioned that the hydrolysis of an ethanolic Zn precursor solution with LiOH leads to zinc hydroxide double salts in later stages of the reaction, i.e., after several hours of the reaction [46], so that the presence of additional Zn-O-species would in general not be unexpected. Furthermore, the amorphous Zn-O phases are also reported in the literature (e.g., ref. [47,48]).

Ex situ electron microscopy measurements on the particles derived from similar synthesis processes showed particle sizes of up to several hundred nanometers (e.g., [4,7,9,10]). This leads to the general conclusion that the growth of the ZnO nanoparticles is a sequence of nucleation, diffusion, and subsequent agglomeration and ripening since the measured growth rates decrease rapidly and sharply (see Figure 8 and Figure 9). The growth of the particles could therefore be well described by employing the LaMer model [49].

With increasing time and temperature, there is a continuous increase in the ZnO monomer concentration until a critical concentration is reached, and sudden nucleation may occur. This nucleation can be traced by the sharply increasing intensities of the diffraction peaks and the ZnO concentration detected by EXAFS linear combination fits (Figure 6). The subsequent growth of the particles is diffusion-limited; thus, particle growth is substantially faster for an increased reaction temperature. However, due to the better solubility of $Zn(acac)_2$ in 1-octanol, the critical concentration is reached more quickly and many Zn nuclei are immediately available for particle formation and growth (sample O). In NaOH and pure water, the critical concentration needed for nucleation is reached with a delay in time, and fewer Zn cations are available for the growth process (samples A and W). Since the nuclei formed only become stable above a certain size, nuclei may dissolve again because there are not enough Zn cations available in the close vicinity of the nucleus to reach the required critical size. This may explain the Bragg peaks observed, almost constant intensities with time. If those small aggregates agglomerate in later stages of growth, the size of the individual crystallites remains more or less unchanged, and only a slight tendency for an Ostwald ripening is observed.

4. CONCLUSIONS

In this work, the formation of ZnO nanoparticles by hydrothermal synthesis of $Zn(acac)_2$ in distilled water and dilute NaOH and solvothermal synthesis in 1-octanol was followed in situ by time-resolved, simultaneous X-ray absorption (EXAFS/XANES) and X-ray diffraction (XRD) measurements, respectively. The simultaneous combination of the two techniques has proven to be very powerful in exploring details of the growth mechanisms of ZnO nanomaterials, as both techniques can investigate different aspects of the reaction. While EXAFS/XANES is suitable for studying both reaction products and reactants as well as possible reaction intermediates with high sensitivity, XRD can provide parallel details about the microstructure of the crystalline reaction products. Depending on the precursor solution and the applied temperature, the formation of hexagonal ZnO could be detected in all solvents by EXAFS and XRD and most effectively in 1-octanol with a conversion of more than 80% after about 35 min. The prepared nanoparticles showed preferential growth along the c-axis of the hexagonal lattice structure, which is most pronounced in 1-octanol with needle-like deposits.

The growth of the nanoparticles, in particular the observed anisotropy, can be explained using the LaMer model. After a critical ZnO cation concentration has been reached, there is a sudden, instantaneous nucleation with subsequent diffusion-controlled growth. Consequently, the solution is depleted in the $Zn(acac)_2$-precursor in the vicinity of the formed ZnO nuclei. Thus, stable nuclei may build up smaller aggregates, so the final particle size is determined by agglomeration and Ostwald ripening. The different degrees of solubility of $Zn(acac)_2$ in the various solvents used here obviously affect the ZnO growth processes. $Zn(acac)_2$ dissolves most readily in 1-octanol, which is why the critical concentration is already reached at room temperature and a large number of ZnO cations are available for particle growth, i.e., many small nuclei are formed compared to the dilute alkali (sample A) or distilled water (sample W) as solvents. The results of the present study allow us to easily determine the preparation conditions for tailored nanoparticles in an in situ experiment. By varying the solvent, the reaction temperature, the pressure in the cell, and the reaction time, particles with the desired shape, size, and crystallographic structure can be obtained.

For future investigations, it might be useful to additionally couple UV–Vis spectroscopy [18,41,50,51], Raman spectroscopy [50,52], small angle X-ray scattering [41,53] or FT-IR [54] to the combined X-ray measurements to obtain additional physicochemical properties of the formed particles and deeper insights into the reaction mechanisms and dynamics.

Furthermore, the present setup only allows monitoring a rather small angular range of the X-ray diffraction pattern. A larger detector with smaller pixels would facilitate measurements of more diffraction peaks and thus provide a more detailed insight into the microstructure of the grown particles. Regarding the detection of intermediate reaction products, the present experiments suggest that the reaction from the $Zn(acac)_2$ precursor to the ZnO occurs directly, i.e., without reaction intermediates. Quick-EXAFS with a faster repetition rate should be performed to exclude the presence of short-living intermediates. At present, the technical limitations of the method are around 5-10 ms for an individual spectrum [32,55], so the transient species with a subsecond lifetime should be feasible to detect.

Author Contributions

Conceptualization, F.E. and D.L.-H.; methodology, F.E., A.Š., M.V. and D.L.-H.; software, F.E., F.B., P.R. and L.V.; validation, F.E., F.B. and D.L.-H.; formal analysis, F.E.; investigation, F.E., P.R., F.B., L.V. and D.L.-H.; resources, A.Š., M.V. and D.L.-H.; data curation, A.Š., M.V. and D.L.-H.; writing—original draft preparation, F.E. and D.L.-H.; writing—review and editing, D.L.-H., F.E., and M.V.; visualization, F.E. and D.L.-H.; supervision, D.L.-H.; project administration, A.Š. and D.L.-H.; funding acquisition, A.Š. and D.L.-H. All authors have read and agreed to the published version of the manuscript.

Acknowledgments

We gratefully acknowledge the provision of beamtime at beamline P64 at PETRA III, DESY, Hamburg, Germany, and beamline 10, DELTA synchrotron facility, Dortmund, Germany. In particular, we would like to thank W. Caliebe, M. Görlitz and A. Tayal for their manifold help and technical support during the beamtimes.

REFERENCES

1. Darr, J.A.; Zhang, J.; Makwana, N.M.; Weng, X. Continuous Hydrothermal Synthesis of Inorganic Nanoparticles: Applications and Future Directions. Chem. Rev. 2017, 117, 11125–11238.
2. Baruah, S.; Dutta, J. Hydrothermal growth of ZnO nanostructures. Sci. Technol. Adv. Mater. 2009, 10, 013001.
3. Demazeau, G. Solvothermal Processes: Definition, Key Factors Governing the Involved Chemical Reactions and New Trends. Z. Naturforsch. 2010, 65, 999–1006.

4. Šarić, A.; Štefanić, G.; Dražić, G.; Gotić, M. Solvothermal synthesis of zinc oxide microspheres. J. Alloys Compd. 2015, 652, 91–99.
5. Al-Gaashani, R.; Radiman, S.; Daud, A.; Tabet, N.; Al-Douri, Y. XPS and optical studies of different morphologies of ZnO nanostructures prepared by microwave methods. Ceram. Int. 2013, 39, 2283–2292.
6. Zhang, W.; Chen, X.; Xu, N.; Xiang, R.; Zhu, Y.; Tang, Z. Morphology and Optical Property of ZnO Nanostructures Grown by Solvothermal Method: Effect of the Solution Pretreatment. J. Nanomater. 2013, 2013, 986157.
7. Music, S.; Šarić, A. Formation of hollow ZnO particles by simple hydrolysis of zinc acetylacetonate. Ceram. Int. 2012, 38, 6047–6052.
8. Cho, S.; Jang, J.-W.; Jung, S.-H.; Lee, B.; Oh, E.; Lee, K.-H. Precursor Effects of Citric Acid and Citrates on ZnO Crystal Formation. Langmuir 2009, 25, 3825–3831.
9. Razali, R.; Zak, A.K.; Majid, W.A.; Darroudi, M. Solvothermal synthesis of microsphere ZnO nanostructures in DEA media. Ceram. Int. 2011, 37, 3657–3663.
10. Šarić, A.; Despotović, I.; Štefanić, G.; Dražić, G. The Influence of Ethanolamines on the Solvothermal Synthesis of Zinc Oxide: A Combined Experimental and Theoretical Study. ChemistrySelect 2017, 2, 10038–10049.
11. Šarić, A.; Vrankić, M.; Lützenkirchen-Hecht, D.; Despotović, I.; Petrović, Ž.; Dražić, G.; Eckelt, F. Insight into the Growth Mechanism and Photocatalytic Behavior of Tubular Hierarchical ZnO Structures: An Integrated Experimental and Theoretical Approach. Inorg. Chem. 2022, 61, 2962–2979.
12. Zou, X.; Ke, J.; Hao, J.; Yan, X.; Tian, Y. A new method for synthesis of ZnO flower-like nanostructures and their photocatalytic performance. Phys. B 2022, 624, 413395.
13. Wojnarowicz, J.; Chudoba, T.; Lojkowski, W. A Review of Microwave Synthesis of Zinc Oxide Nanomaterials: Reactants, Process Parameters and Morphologies. Nanomaterials 2020, 10, 1086.
14. Mohammadi, E.; Aliofkhazraei, M.; Hasanpoor, M.; Chipara, M. Hierarchical and Complex ZnO Nanostructures by Microwave-Assisted Synthesis: Morphologies, Growth Mechanism and Classification. Crit. Rev. Solid State Mater. Sci. 2018, 43, 475–541.
15. Sauvik Raha, S.; Ahmaruzzaman, M. ZnO nanostructured materials and their potential applications: Progress, challenges and perspectives. Nanoscale Adv. 2022, 4, 1868–1925.
16. Soloviev, A.; Jensen, H.; Søgaard, E.G.; Kanaev, A.V. Aggregation kinetics of sol-gel process based on titanium tetraisopropoxide. J. Mater. Sci. 2003, 38, 3315–3318.
17. Kiefer, J.; Grabow, J.; Kurland, H.-D.; Müller, F.A. Characterization of Nanoparticles by Solvent Infrared Spectroscopy. Anal. Chem. 2015, 87, 12313–12317.
18. Stötzel, J.; Lützenkirchen-Hecht, D.; Frahm, R.; Santilli, C.V.; Pulcinelli, S.H.; Kaminski, R.; Fonda, E.; Villain, F.; Briois, V. QEXAFS and UV/Vis Simultaneous Monitoring of the TiO2-Nanoparticles Formation by Hydrolytic Sol–Gel Route. J. Phys. Chem. C 2010, 114, 6228–6236.
19. Zolotoyabko, E. Basic Concepts of X-ray Diffraction, 1st ed.; Wiley VCH: Weinheim, Germany, 2014.

20. Koningsberger, D.; Prins, R. X-ray Absorption. Principles, Applications and Techniques of EXAFS, SEXAFS, and XANES; John Wiley & Sons: New York, NY, USA, 1988.
21. Lützenkirchen-Hecht, D.; Frahm, R. The corrosion of Mo in KOH: Time resolved XAFS investigations. J. Phys. Chem. B 2001, 105, 9988–9993.
22. Grote, L.; Zito, C.A.; Frank, K.; Dippel, A.C.; Reisbeck, P.; Pitala, K.; Kvashnina, K.O.; Bauters, S.; Detlefs, B.; Ivashko, O.; et al. X-ray studies bridge the molecular and macro length scales during the emergence of CoO assemblies. Nature Commun. 2021, 12, 4429–4441.
23. Karzel, H.; Potzel, W.; Köfferlein, M.; Schiessl, W.; Steiner, M.; Hiller, U.; Kalvius, G.M.; Mitchell, D.W.; Das, T.P.; Blaha, P.; et al. Lattice dynamics and hyperfine interactions in ZnO and ZnSe at high external pressures. Phys. Rev. B 1996, 53, 11425–11438.
24. Ashrafi, A.; Jagadish, C. Review of zincblende ZnO: Stability of metastable ZnO phases. J. Appl. Phys. 2007, 102, 071101.
25. Ashrafi, A.; Ueta, A.; Kumano, H.; Suemune, I. Role of ZnS buffer layers in growth of zincblende ZnO on GaAs substrates by metalorganic molecular-beam epitaxy. J. Cryst. Growth 2000, 221, 435–439.
26. Park, S.-H.; Kwak, C.-H.; Seo, S.-Y.; Kim, S.-H.; Kim, B.-H.; Park, C.-I.; Park, Y.-W.; Han, S.-W. A Large Quantity of ZnO Nanorods Grown at Room Temperature. J. Korean Phys. Soc. 2009, 55, 94–97.
27. Guo, L.; Yang, S.; Yang, C.; Yu, P.; Ge, W.; Wong, G.K.L. Synthesis and Characterization of Poly(vinylpyrrolidone)-Modified Zinc Oxide Nanoparticles. Chem. Mater. 2000, 12, 2268–2274.
28. Zhao, D.; Tian, X.; Doronkin, D.E.; Han, S.; Kondratenko, V.A.; Grunwaldt, J.-D.; Perechodjuk, A.; Vuong, T.H.; Rabeah, J.; Eckelt, R.; et al. In situ formation of ZnOx species for efficient propane dehydrogenation. Nature 2021, 599, 234–241.
29. Shaban, M.; Mohamed, F.; Abdallah, S. Production and Characterization of Superhydrophobic and Antibacterial Coated Fabrics Utilizing ZnO Nanocatalyst. Sci. Rep. 2018, 8, 3925.
30. Liu, X.; Ye, L.; Liu, S.; Li, Y.; Ji, X. Photocatalytic Reduction of CO2 by ZnO Micro/nanomaterials with Different Morphologies and Ratios of {0001} Facets. Sci. Rep. 2016, 6, 38474.
31. Nelson, J. XANES reflects coordination change and underlying surface disorder of zinc adsorbed to silica. J. Synchrotron Radiat. 2021, 28, 1119–1126.
32. Bornmann, B.; Kläs, J.; Müller, O.; Lützenkirchen-Hecht, D.; Frahm, R. The quick EXAFS setup at beamline P64 at PETRA III for up to 200 spectra per second. AIP Conf. Proc. 2019, 2054, 040008.
33. Lützenkirchen-Hecht, D.; Bornmann, B.; Frahm, R.; Rothweiler, P.; von Polheim, S.; Eckelt, F.; Müller, O. Simultaneous quick-scanning X-ray absorption spectroscopy and X-ray diffraction. In Proceedings of the 14th International Conference on Synchrotron Radiation Instrumentation, Hamburg, Germany, 28 March–1 April 2022.
34. Lützenkirchen-Hecht, D.; Wagner, R.; Szillat, S.; Hüsecken, A.K.; Istomin, K.; Pietsch, U.; Frahm, R. The multi-purpose hard X-ray beamline BL10 at the DELTA storage ring. J. Synchrotron Radiat. 2014, 21, 819–826.

35. Ravel, B.; Newville, M. ATHENA and ARTEMIS: Interactive Graphical Data Analysis Using IFEFFIT. Phys. Scr. 2005, T115, 1007–1010.

36. Castorina, E.; Ingall, E.D.; Morton, P.L.; Tavakoli, D.A.; Lai, B. Zinc K-edge XANES spectroscopy of mineral and organic standards. J. Synchrotron Radiat. 2019, 26, 1302–1309.

37. Thomas, S.A.; Mishra, B.; Myneni, S.C.B. High Energy Resolution-X-ray Absorption Near Edge Structure Spectroscopy Reveals Zn Ligation in Whole Cell Bacteria. J. Phys. Chem. Lett. 2019, 10, 2585–2592.

38. Kisi, E.H.; Elcombe, M.M. u parameters for the wurtzite structure of ZnS and ZnO using powder neutron diffraction. Acta Crystallogr. Sect. C Cryst. Struct. Commun. 1989, 45, 1867–1870.

39. Rehr, J.J.; Kas, J.J.; Vila, F.D.; Prange, M.P.; Jorissen, K. Parameter-free calculations of X-ray spectra with FEFF9. Phys. Chem. Chem. Phys. 2010, 12, 5503–5513.

40. Ravel, B. Quantitative EXAFS Analysis. In X-ray Absorption and X-Ray Emission Spectroscopy: Theory and Applications; Van Bokhoven, J.A., Lamberti, C., Eds.; John Wiley & Sons: Chichester, UK, 2016.

41. Caetano, B.L.; Santilli, C.V.; Meneau, F.; Briois, V.; Pulcinelli, S.H. In Situ and Simultaneous UV−vis/SAXS and UV−vis/XAFS Time-Resolved Monitoring of ZnO Quantum Dots Formation and Growth. J. Phys. Chem. C 2011, 115, 4404–4412.

42. Kieffer, J.; Karkoulis, D. PyFAI, a versatile library for azimuthal regrouping. J. Phys. Conf. Ser. 2013, 425, 202012.

43. Warren, B.E. X-ray Diffraction; Addison-Wesley Publishing Co.: Boston, MA, USA, 1969.

44. Wong, E.M.; Bonevich, J.E.; Searson, P.C. Growth Kinetics of Nanocrystalline ZnO Particles from Colloidal Suspensions. J. Phys. Chem. B 1998, 102, 7770–7775.

45. Ludi, B.; Niederberger, M. Zinc oxide nanoparticles: Chemical mechanisms and classical and non-classical crystallization. Dalton Trans. 2013, 42, 12554–12568.

46. Briois, V.; Giorgetti, C.; Baudelet, F.; Blanchandin, S.; Tokumoto, M.S.; Pulcinelli, S.H.; Santilli, C.V. Dynamical Study of ZnO Nanocrystal and Zn-HDS Layered Basic Zinc Acetate Formation from Sol−Gel Route. J. Phys. Chem. C 2007, 111, 3253–3258.

47. Huang, Y.; Shen, Z.; Wu, Y.; Wang, X.; Zhang, S.; Shi, X.; Zeng, H. Amorphous ZnO based resistive random access memory. RSC Adv. 2016, 6, 17867–17872.

48. Bruncko, J.; Vincze, A.; Netrvalova, M.; Šutta, P.; Hasko, D.; Michalka, M. Annealing and recrystallization of amorphous ZnO thin films deposited under cryogenic conditions by pulsed laser deposition. Thin Solid Films 2011, 520, 866–870.

49. LaMer, V.K.; Dinegar, R.H. Theory, Production and Mechanism of Formation of Monodispersed Hydrosols. J. Am. Chem. Soc. 1950, 72, 4847–4854.

50. Briois, V.; Lützenkirchen-Hecht, D.; Villain, F.; Fonda, E.; Belin, S.; Griesebock, B.; Frahm, R. Time-Resolved Study of the Oxidation of Ethanol by Cerium(IV) Using Combined Quick-XANES, UV−Vis, and Raman Spectroscopies. J. Phys. Chem. A 2005, 109, 320–329.

51. Bauer, M.; Heusel, G.; Mangold, S.; Bertagnolli, H. Spectroscopic set-up for simultaneous UV-Vis/(Q)EXAFSin situ and in operando studies of homogeneous reactions under laboratory conditions. J. Synchrotron Radiat. 2010, 17, 273–279.
52. Tsakoumis, N.; Dehghan, R.; Johnsen, R.E.; Voronov, A.; van Beek, W.; Walmsley, J.C.; Borg, Ø.; Rytter, E.; Chen, D.; Rønning, M.; et al. A combined in situ XAS-XRPD-Raman study of Fischer–Tropsch synthesis over a carbon supported Co catalyst. Catal. Today 2013, 205, 86–93.
53. Just, J.; Coughlan, C.; Singh, S.; Ren, H.; Müller, O.; Becker, P.; Unold, T.; Ryan, K.M. Insights into Nucleation and Growth of Colloidal Quaternary Nanocrystals by Multimodal X-ray Analysis. ACS Nano 2021, 15, 6439–6447.
54. Rabeah, J.; Briois, V.; Adomeit, S.; La Fontaine, C.; Bentrup, U.; Brückner, A. Multivariate Analysis of Coupled Operando EPR/XANES/EXAFS/UV–Vis/ATR-IR Spectroscopy: A New Dimension for Mechanistic Studies of Catalytic Gas-Liquid Phase Reactions. Chem.–A Eur. J. 2020, 26, 7395–7404.
55. Müller, O.; Nachtegaal, M.; Just, J.; Lützenkirchen-Hecht, D.; Frahm, R. Quick-EXAFS setup at the SuperXAS beamline for in situ X-ray absorption spectroscopy with 10 ms time resolution. J. Synchrotron Radiat. 2016, 23, 260–266.

11 DEVELOPMENT OF 2.45 GHZ SEMICONDUCTOR MICROWAVE SYSTEM FOR COMBUSTION IGNITION ENHANCEMENT AND FAILURE ANALYSIS

Yuji Ikeda

i-Lab, Inc., #213 KIBC Bldg., 5-5-2 Minatojima-Minami, Chuo, Kobe 650-0047, Japan

ABSTRACT

We developed a semiconductor microwave system to improve the ignition process in a combustion system. Under atmospheric pressure conditions, large plasma was successfully ignited by a 2.45 GHz microwave, and it is characterized in comparison with standard spark plug ignition and laser ignition. The size of the microwave power source was also effectively reduced with the minimal size (100 × 60 mm²) that could fit in the palm of a hand. We then prototyped a microwave plug with a diameter of 4 mm, which is smaller than the standard spark plugs for passenger cars. The design and electric field strength are discussed in detail. Combustion experiments were conducted using a motorcycle engine and an actual light vehicle, and significant fuel efficiency improvement was experimentally obtained. We investigated the wear of the plug caused by continuous operation, and efficiently improved the endurance by swinging the resonance frequency between 2.4 and 2.5 GHz. In a passenger car engine experiment using a flat panel igniter, significant fuel efficiency improvement was confirmed. Further failure analysis revealed that the ceramic was severely damaged by a large current surge.

Keywords: *microwave chemistry; microwave ignition process; microwave control*

1. INTRODUCTION

Ignition is the most important part of the combustion process. Optimal emission control during ignition is indispensable for reducing CO2 and exhaust-gas emissions [1]. We are heading towards a decarbonized society, and a wide variety of fuels are being considered [2,3,4,5]. This is a shift towards a hydrogen-based society where hydrogen transport is actively conducted [6]. Other alternatives such as ammonia combustion ideally reduce emission pollutants, but viability issues have not been resolved [7].

Aside from emission pollutants, heat generation during ignition decreases overall efficiency. Most of the ignition in a combustion system, such as spark plug and laser ignition, locally generates a high-temperature heat source. Those represented by the spark plug generate heat transfer losses through cylinder walls [8,9]. Laser ignition solves this problem by moving the flame kernel away from the combustion chamber [10,11]. However, the feasibility of laser ignition decreases in terms of the system cost.

In general, the following are important parameters for a highly efficient combustion process: stable ignition, suppression of misfire, rapid combustion, suppression of unburned fuel, miniaturization of the combustor, high output, exhaust gas heat recirculation, use of catalyst, etc. [1]. Stable ignition requires high voltage, high-energy heat source breakdown, plasma initialization, flame kernel development, and flame propagation physicochemical processes. These phenomena are summarized in Figure 1. To see the breakdown phenomenon in detail [12,13,14], it is desirable to control the energy and time similarly to laser ignition. Breakdown plasma is formed, a shock wave is generated, and a flame kernel is created. Time changes from nanoseconds to microseconds, and when it comes to the flame surface, it is a phenomenon that happens after a microsecond period. We see multi-point breakdown, ionization, and cascade.

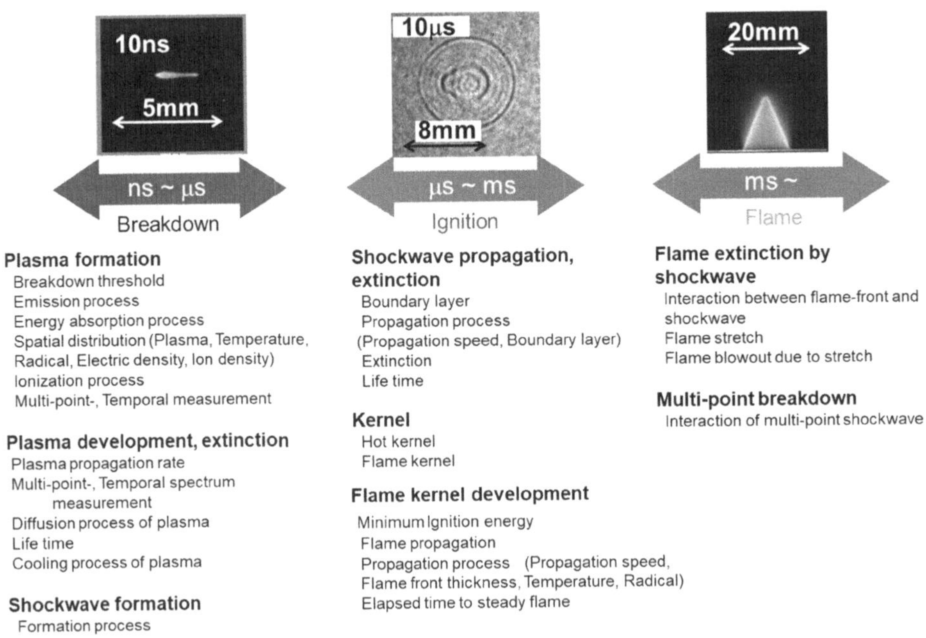

Figure 1.: *Physicochemical process of combustion.*

Further illustrations of laser-induced breakdown plasma, flame formation, and flame propagation are summarized in Figure 2. Here, significant enhancement and stabilization of the ignition process using microwaves are also shown. Plasma size is expanded by thousand-fold, and the plasma lifetime is prolonged by irradiating the laser-induced plasma with a microwave antenna.

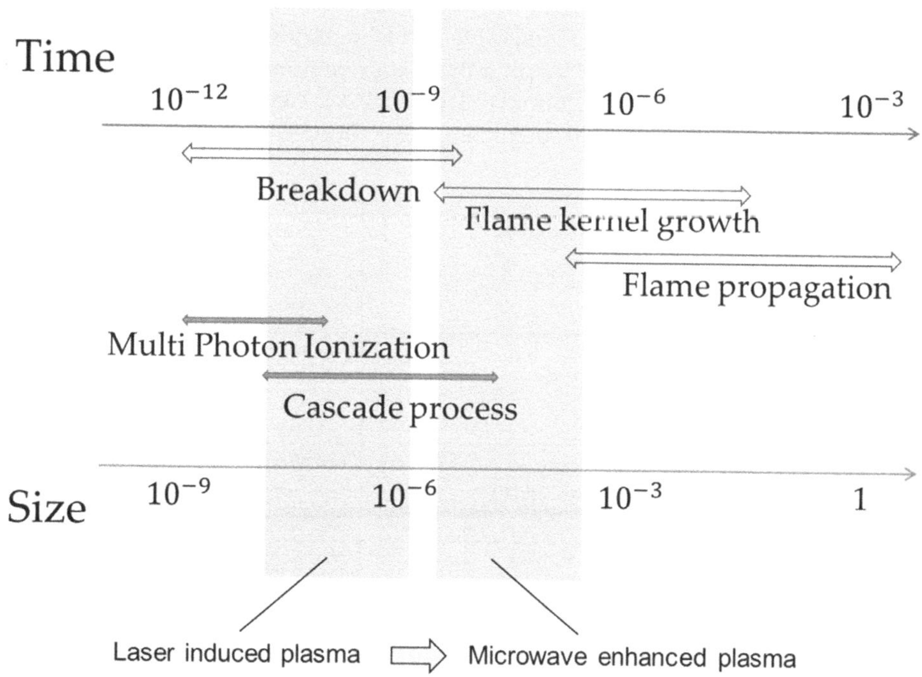

Figure 2.: *Temporal evolution of flame kernel in combustion process.*

Figure 3 summarizes the characteristics of laser ignition and microwave plasma ignition using a standard spark ignition with superimposed microwave input. The advantages of each type of ignition are highlighted (pink). The volume of the kernel produced by the microwave plasma ignition is ideally larger than the laser ignition due to the addition of radicals. This type of ignition can be considered to be volume ignition instead of local-point ignition generated by laser ignition. Non-thermal plasma is produced with microwave plasma ignition, while high-density thermal plasma is generated by laser ignition. A huge advantage with using laser ignition is the nonoccurrence of the heat shrinkage problem, which is often seen in spark plugs; thus, heat loss is small. However, a window for laser light is needed, and the size and cost of the power supply are problems that are still under consideration [15]. On the other hand, a spark source coupled with a microwave system is cheaper, but the size of the entire system would be large, and integrating them into an actual engine is not yet resolved. Therefore, we are developing a compact ignition source with a novel structure that initiates the breakdown by microwaves instead of a high-voltage source, sustains the expansion of the kernel, and stabilizes the ignition by further input of microwaves [16,17,18,19,20].

This paper reports on the development of miniaturization of microwave systems, the enhancement of ignition and flame propagation, and the improvement of engine combustion efficiency. We report on the breakdown phenomenon caused by microwaves, the control of microwaves, and the temporal sustainment of breakdown plasma using semiconductor devices [21,22,23]. We developed a circuit that transmits microwaves with a coaxial cable and a breakdown plug (microwave discharge ignitor (MDI)) [24,25,26] with a resonant circuit and an antenna structure on a ceramic plate (flat panel ignitor (FPI)) [27,28,29]. We also report on the limits of combustion experiments with actual engines, countermeasures against erosion problems, and failure results.

	Laser Ignition 	Microwave Plasma Ignition 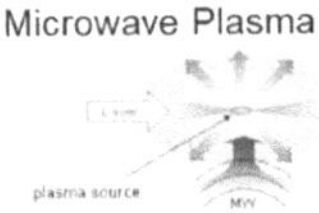
Principle	Thermal effect Local ignition + Flame propagation	Thermal effect + Radical addition Volume ignition + Flame propagation
Ignition	Local Point	Volume Ignition
	High Energy Density (LTE)	Non-Thermal Plasma
	Heat Source	Heat Source + Kinetic (Radicals) Effect
	Short Pulse (not so bad)	Wide Controllable Plasma Duration
Heat Loss	No Heat Loss (No electrode)	Less Heat Loss
Engine Modification	Optical window (Durability, Cleaning)	Commercial-Available Standard Plug
Power source	Laser, LED	Magnetron, Semiconductor
Stale ignition / Misfire	Limited by Laser Repetition Rate	Stale ignition / Less Misfire
Cost	Expensive (Laser, Fiber, Window...)	Total system < $300
Enhancement by Microwave	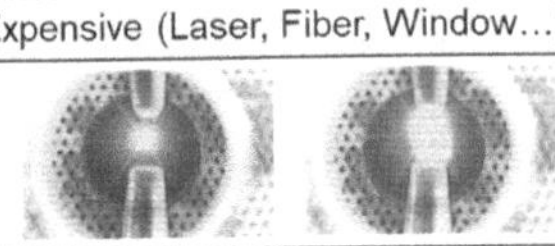	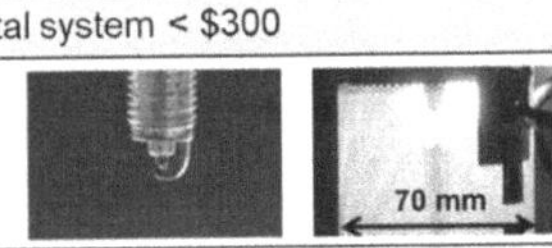

Figure 3.: *Laser ignition compared with microwave plasma ignition.*

2. EXPERIMENTAL SYSTEM

2.1. Laser Ignition and Spark Ignition System

Figure 4a shows the schematic diagram of the microwave-enhanced laser ignition system, and Figure 4b shows the actual image of the combustion chamber of the microwave-enhanced spark ignition. For microwave-enhanced laser ignition, an Nd–YAG laser is used as a light source (Quanta-Ray ND1–004; Spectra-Physics) that is focused inside the combustion chamber to initiate breakdown. For the microwave-enhanced spark ignition, a standard spark plug was used. A coaxial antenna was used to superimpose the microwaves for both types of ignitions. The antenna was powered by a semiconductor-controlled MW generator [21,22,23]. The pulse delay generator controls the pulse input of each piece of equipment. Actual images of the plasma generated by the microwave-enhanced laser and spark ignitions are also shown in comparison with the standard ignitions without the microwave. Plasma enlargements are shown with the superimposition of microwaves for both types of ignitions. We attempted both magnetrons and semiconductor devices as microwave sources [22]. The formed plasma was significantly larger as visibly seen by the human eye (see Figure 3). Plume size varied depending on the emission spectrum. This photo was taken without a wavelength filter.

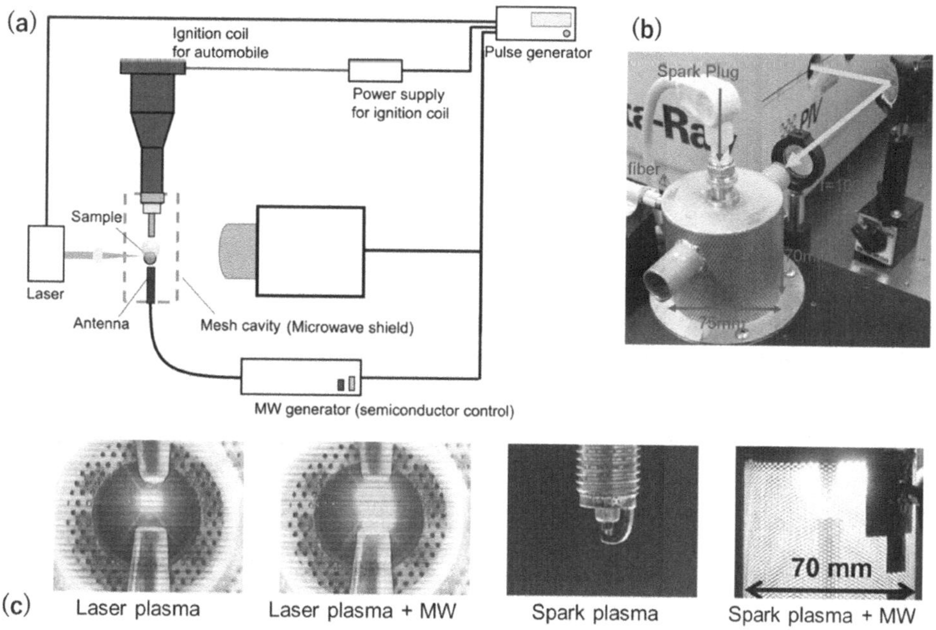

Figure 4.: *(a) Schematic diagram of microwave-enhanced laser ignition system; (b) actual image of combustion chamber of spark plug ignition plus microwave. (c) Enlargements of plasma plume with superimposition of microwave to laser ignition and spark ignition [22].*

2.2. Semiconductor Microwave Generator

Oscillation in semiconductor devices rather than magnetrons is essential for energy efficiency and reducing heat generation. Figure 5 shows the development in the reduction in semiconductor size. We developed this microwave oscillation system in 10 months with the latest model (100 × 195 mm2), achieving less than half the size of the previous model (530 × 230 mm2). The original model includes a power control unit, a driver amplifier oscillator, and the main amplifier generating a maximal output of 500 W. The latest model has a standard output of 1000 W, and a simple combination of 4 oscillators can generate a total of 4 kW. A simple arrangement of the circuitry downsized the product to a size of 60 × 100 mm2. This version has the size of the palm of a typical hand and fits inside the hood of a passenger car when mounted on an actual device. In the future, further improvements can reduce the production cost by using plastic packages instead of ceramic packages.

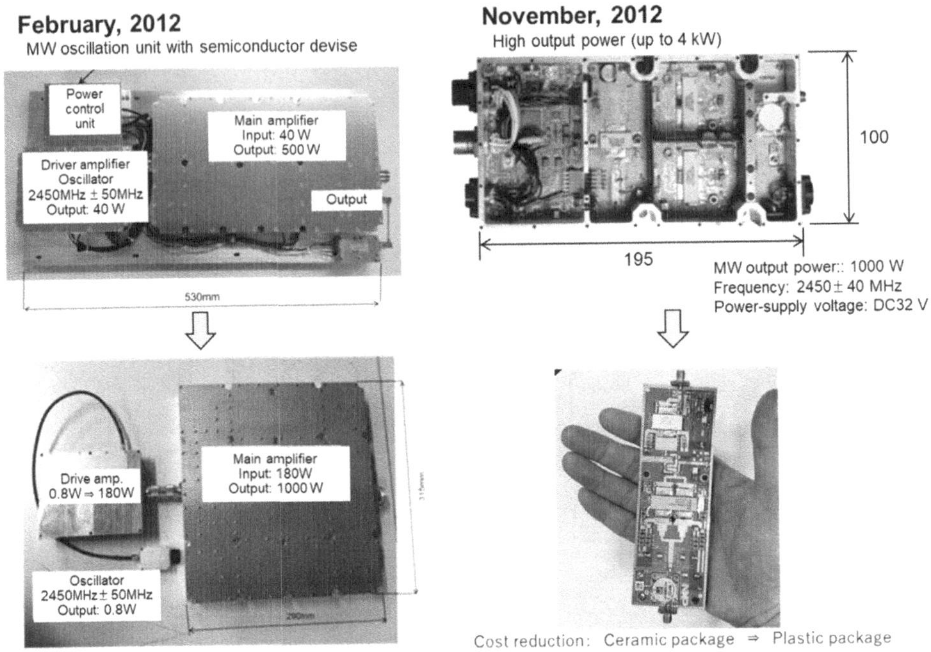

Figure 5.: Size reduction of microwave oscillator.

2.3. Antenna Characteristics

Figure 6 shows the comparison of the antenna that we developed for the superimposition of MW compared with a standard spark plug. The standard spark plug had small plasma generation, but it was the simplest and cheapest system. A modified spark plug sustained by an integrated microwave antenna (MW plasma ignitions) produce a significantly larger plasma compared with the standard spark plug, but it also requires a more expensive system due to the need for two power sources: (a) high voltage source for spark plug breakdown and (b) microwave source for sustaining the enlarged plasma. This led to the development of two major types of antennas that only require one microwave source for the breakdown and sustainment of the plasma: the microwave discharge igniter (MDI) and the flat panel igniter (FPI). The MDI transmits high frequencies from a microwave oscillator into a ceramic oscillator that concentrates the MW transmission in the central electrode where the plasma breakdown is initiated [28]. The FPI, on the other hand, uses a ceramic chip antenna where the antenna structure is printed in a ceramic plate. Plasma breakdown is initiated between the smallest distanced antenna prints. The single-point MDI and FPI compete with the standard igniter in terms of system simplicity and costs. Both types of ignitions have the advantage of creating larger plasma. Two variations of the MDI using triple igniters (3 ×× MDI and M12) only use one oscillation pattern for breakdown and sustaining the plasma. The 3 ×× MDI variation uses three separate MDI antennas that are placed linearly. The M12 antenna integrates the 3 MDI into one structure. Significant improvements in the plasma size are observed for both types of antennas. The multipoint ignition of gas fuel is also superior in terms of ignition characteristics and stability.

	Plasma Image	Device image
MW plasma ignition (SI + MW)	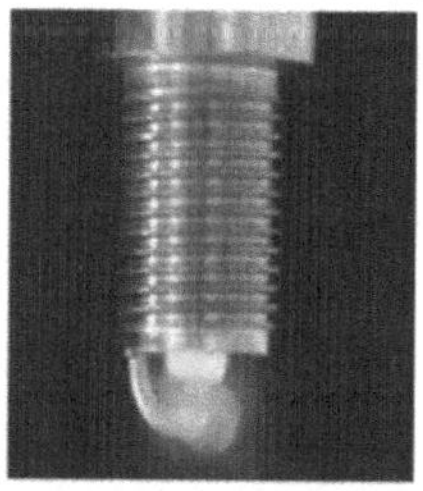	
MDI enhanced by MW (MW + MW)	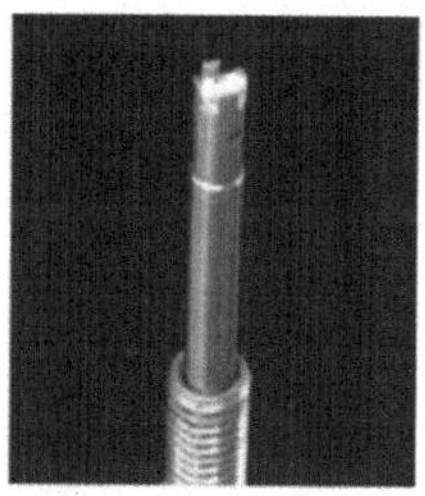	
Multi-MDI Ignition (MW only)		
Flat Panel Igniter (Ceramic chip)		

Figure 6.: *Comparison of spark plug, microwave discharge igniter (MDI), and flat panel igniter (FPI). There are multiple variations of the ignition process spark plug and MDI depending on the energy input.*

In this paper, we conducted an experiment using an M12 plug fitting the whole system in the hood of a passenger car. The cost, ignition control, and maintenance were comparable to those of standard spark plugs with a standard lifetime of about 100,000 km. The balance of effectiveness, function, material, and cost made the development challenging. We considered using the system not only for gasoline-powered vehicles, but also for diesel engines and gas engines.

2.4. MDI Detailed Design and Erosion

Regarding the development of MDI, we developed one that is smaller than the smallest spark plug diameter used in gasoline-powered vehicles. This is because the degree of freedom in designing the engine cylinder can be improved. Another advantage of using a smaller diameter spark plug is the increase in distance of the plasma from the engine cylinder that can lessen the heat loss. Figure 7a–c show the schematic diagram, equivalent circuit, and actual images of the MDI. Two concentric metal rods and a ceramic resonator formed the MDI. Breakdown occurs in the concentric space at the tip of the antenna, and pulsed microwaves are continuously sent to sustain the plasma temporally and spatially. The equivalent circuit shows how the capacitive values of the MW connection point and the resonant cavity affect the central electrode. With the optimum MW coupling and resonant capacitance, the ignition is observed as shown in Figure 7e. This produced plasma was larger, denser, and more intense compared with the spark produced in the standard spark plug in Figure 7d.

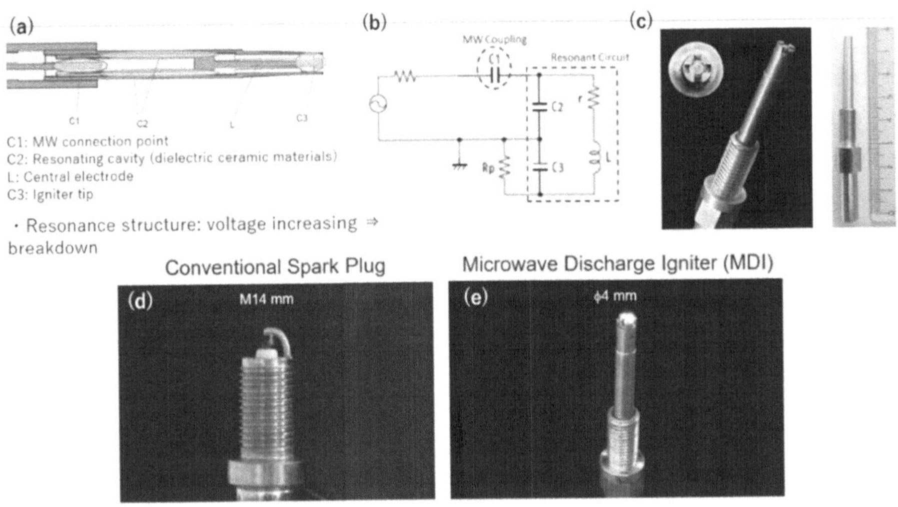

Figure 7.: (a) Schematic diagram, (b) equivalent circuit, (c) and actual image of microwave discharge igniter (MDI). Actual scale size is compared with that of a (d) conventional spark plug. (e) Actual ignition of MDI.

To further understand the ignition process in the MDI, electric-field analysis using HFSS is shown in Figure 8. The microwave was transmitted from the MW oscillator into the C1 structure. The MW creates a localized electric field in the whole MDI structure through proper coupling in structure C1. The C1 structure was optimized for generally low distribution of the electric field in the resonant structure C2 and L, and maximum electric field strength concentration in the discharge tip, C3. The high electric field strength at the tip causes the breakdown of plasma. The combination of C2 and C1 is the most difficult parameter tuned. The performance of MDI increases with a larger C2 diameter (data not shown). However, dimensions C2 and C3 were determined by the overall MDI diameter of 4 mm. The issue of misfire was not observed with this type of igniter. Impedance matching was not performed in MDI.

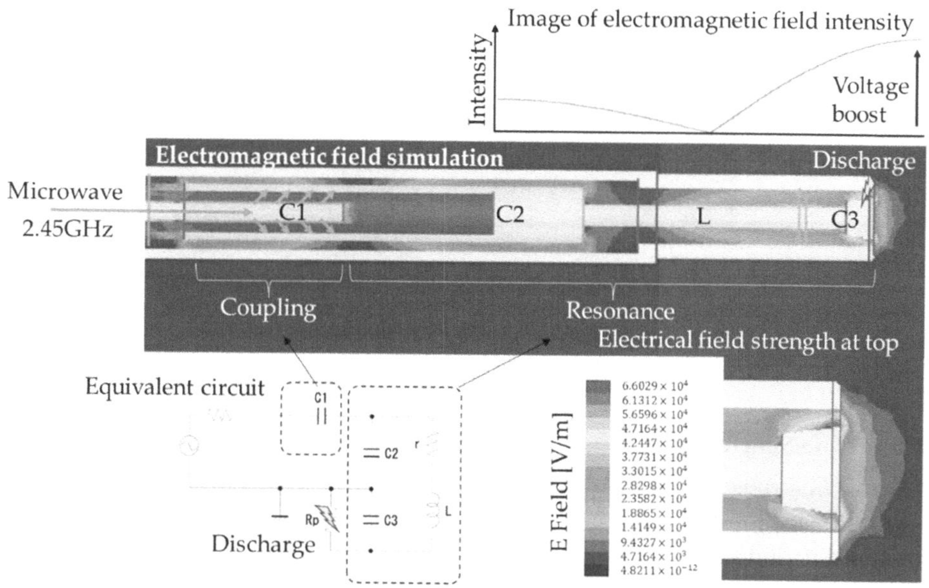

Figure 8.: *Electric-field strength simulation inside microwave discharge igniter (MDI). The capacitance of the microwave coupling point C1 was varied for maximum E-field in ignition tip C3. The resonant cavity C3 and inductance L comprised the resonant circuit.*

When injecting microwaves in a pulsed manner, a few microseconds of energy are applied for breakdown, which creates a pilot fire, shown on the lower left corner of Figure 9. When a pulsed microwave is injected there, the plasma expands, as shown in the photo on the right, moves between the electrodes, and is sustained.

It was necessary to produce multiple microwave pulses instead of a signal microwave pulse to optimize the duty ratio. Each pulse was set to 0.1 mJ or less, and pulse width was controlled to be 50–100 ns. The required energy for the breakdown and that to sustain plasma is very different. It takes a certain amount of time to form a flame after the breakdown. For that purpose, it was found that sustaining the plasma for a longer period is more efficient than creating a high-energy heat source. If the pulse interval is too short, the plasma cannot be sustained. If it is too long, the temperature of the plasma rises, resulting in unnecessary energy consumption and the material melting of the spark plug itself. If the ignition energy used for two different engines is compared to that of the existing spark plug, the total fuel consumption is improved. Therefore, as little energy as possible is required. It is necessary to sustain the plasma with short pulses of microwaves.

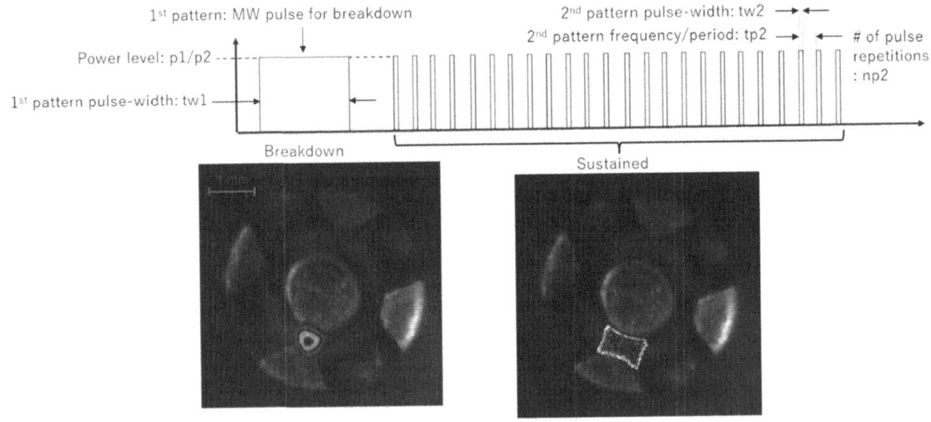

Figure 9.: Oscillation pattern of microwave input of microwave discharge igniter (MDI) initiates plasma breakdown and sustains ignition for a long period [30].

3. RESULTS

3.1. MW on Spark Plug

Two types of actual engines were used: 125 cc of gasoline for motorcycles, and 660 cc for light vehicles [29,31]. The used spark plug with microwave, shown in Figure 10, indicated the integration of the MW antenna into a high voltage spark plug for the breakdown of plasma. This system is a mixer unit type, in contrast to a system where the external microwave is added into a separate MW antenna. Spark energy was 20–30 mJ, and the microwave gave an energy of 500 W peak for 5.0 ms. The plasma emission in the atmosphere is expanded by microwaves. Using this system, we first conducted an experiment to expand the lean limit of a motorcycle engine. The coefficient of variation (COV) is reduced by about 5% to improve fuel efficiency and suppress the expansion of engine combustion fluctuations.

The engine speed of the gasoline engine was 4000 rpm, and the indicated mean effective pressure (IMEP) had a low load of 120 kPa. Results of constant microwave superposition of 2.5 ms for varied air/fuel (A/F) ratio are shown in Figure 11. In normal operation, the A/F is about 14.5, and there is stable operation that is stoichiometric. By inserting microwaves, this was great, and the dilution limit was expanded. Indicated specific fuel consumption (ISFC) improved by 11% in terms of fuel economy. These are the ones in which the flame propagation is accelerated, and the fuel consumption is improved by stable ignition and flame expansion.

Most motorcycle engines are air-cooled, and the temperature of the spark plug attached to the engine is higher than that of a water-cooled passenger car engine. An 11% improvement in indicated specific fuel consumption (ISFC) was obtained under this experimental condition, but it was not the same under other engine rotation conditions. The higher the engine speed was, the shorter the ignition period and the shorter the flame propagation time were. Furthermore, intake efficiency decreased, and the amount of air used for combustion decreased, so fuel efficiency tended to decrease.

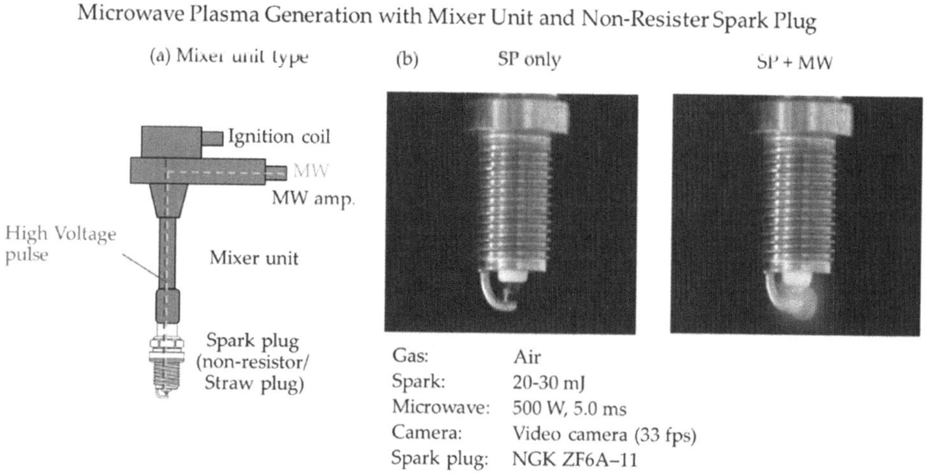

Figure 10.: *(a) Schematic diagram of microwave-assisted spark plug. (b) Actual images of plasma generated by spark plug with and without microwave input.*

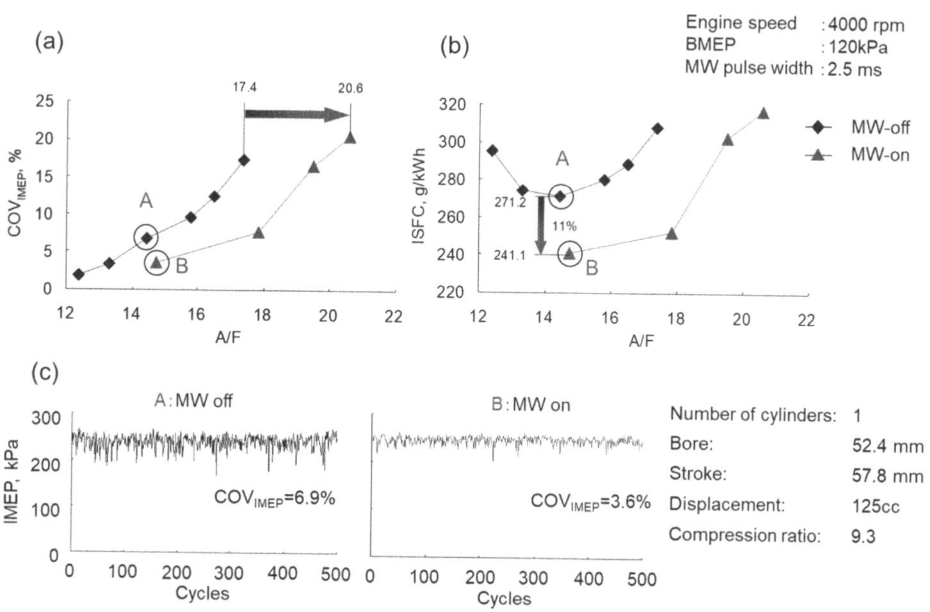

Figure 11.: *Effects of microwave to (a) coefficient of variation (COV) and (b) indicated specific fuel consumption (ISFC) in the spark plug. (c) Indicated mean effective pressure (IMEP) is the same with and without the MW input.*

3.2. Single vs. Multipoint MDI

Next, using a spark plug that integrates two sets of this MDI system with a gasoline engine for a light vehicle, we conducted an experiment to expand the combustion dilution limit. This engine had three cylinders, and a pressure sensor was inserted into each cylinder to measure combustion efficiency. Figure 12 shows the volatility COV of the engine output: IMEP obtained from the pressure of the first cylinder of the three cylinders. Combustion efficiency was examined with a COV value of 5% as a threshold. For a standard spark plug, the limit was an A/F ratio value of about 23. However, when one set of MDI was activated in the spark plug, this value was expanded to 28. Furthermore, when two sets of MDI were activated, it was greatly expanded to 31. This is an immense improvement in fuel efficiency. Combustion did not become unstable. The exhaust gas temperature of the engine also decreased. Furthermore, the NO value in the exhaust gas was greatly improved. This engine uses a three-way catalyst, but the fact that the NO value dropped to almost a single value is a remarkable achievement.

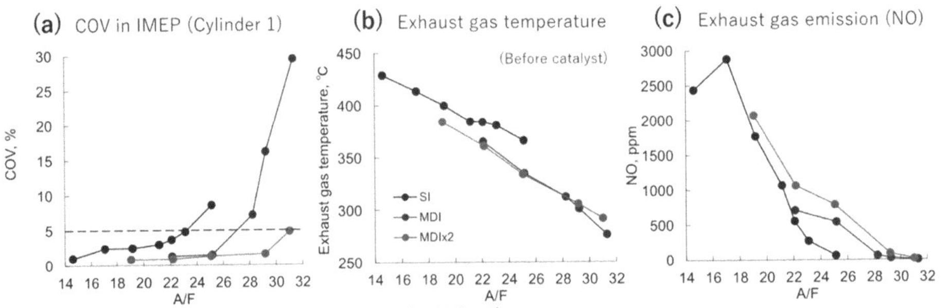

Engine specifications

Number of cylinder	3
Displacement	658 cc
Compression ratio	12.2

Test conditions

Engine speed	1460 rpm
Torque	20 Nm
EGR rate	0

Figure 12.: Comparison of (a) coefficient of variation (COV), (b) exhaust gas temperature, and (c) NO production in standard spark plugs, single-point microwave discharge igniter (MDI), and double point MDI. (d) Actual images of used multi-MDI.

3.3. Endurance Test of MDI

Figure 13 shows the endurance test result of this MDI. The photo shows a new product under wide throttle opening (WOT) conditions at 3000 rpm for 40 and 124 h, and 124 h means 22 million ignitions. The volatility COV of indicated mean effective pressure (IMEP) when the output was operated at 550 kPa for 100 h was 1% or less, which was excellent. Even under WOT conditions, fluctuation was about 2%, and there was no problem in practical operations. However, material erosion occurred. The distance between electrodes became narrow. Commercially available spark plugs can be operated for 100,000 km, and only 0.1 mm or less is allowed because, when the gap between electrodes becomes wide, discharge energy and its probability change, and stable ignition cannot be achieved. Wear due to electric discharge also occurs in MDI. When the gap is changed, the resonance wavelength changes. The resonance wavelength is shown in the figure. When the number of ignitions was taken on the horizontal axis and the resonance wavelength was measured, it was 2.41–2.45 GHz. Since the used microwave oscillator could freely control the transmission frequency up to 2.40–2.50 GHz, this problem of gap expansion due to wear can be solved, and the life of MDI cleared that of commercial passenger cars.

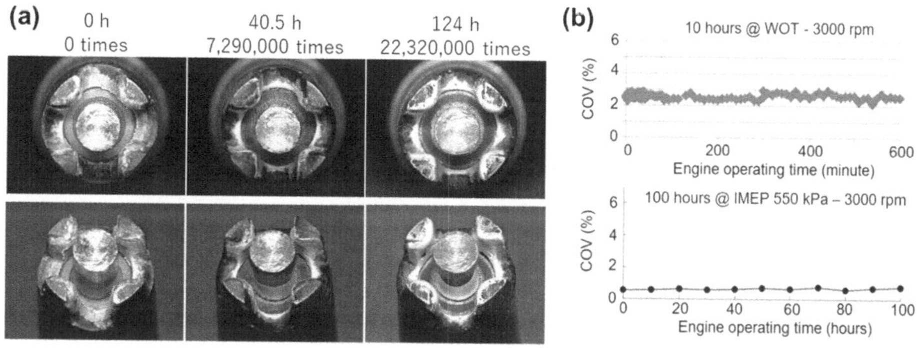

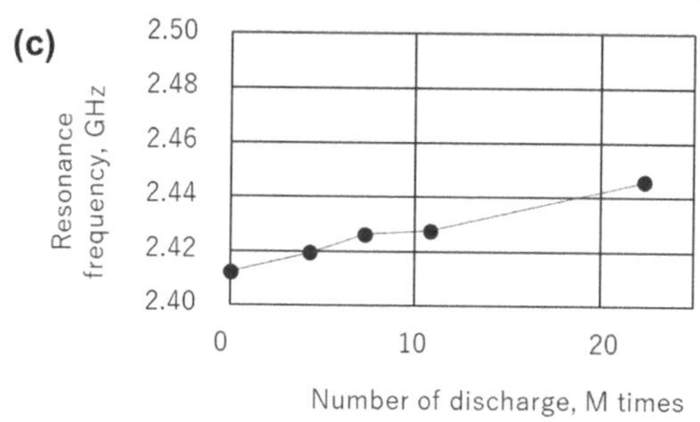

(d) Experimental conditions	
Number of discharge	Over 20 million times
Frequency of discharge	50Hz (equivalent to ignition interval of 6000rpm)
Gas	N_2 (non-reactive, no oxidation)
Pressure	0.5 MPa
Flow rate	< 1000 mL/min (for gas exchange)

Figure 13.: *(a) Actual images of microwave discharge igniter (MDI) under wide throttle opening (WOT) conditions at 3000 rpm for 0 h (0 ignitions), 40.5 h (7.29 ignitions), and 124 h (22 million ignitions). (b) Effects of engine operation time on coefficient of variation (COV) were insignificant. (c) Effects of number of discharges to resonance frequency for (d) these experimental conditions.*

3.4. FPI Combustion Promotion Effect and Component Destruction

Figure 14 shows the actual image of the FPI and the schematic diagram of the implementation of the antenna. Microwave power was transmitted using a coaxial cable from the microwave power supply to the antenna. An input of MW pulsed oscillation of 1.6 kW peak to peak was performed which generated a breakdown and plasma generation at the antenna and sustainment of nonequilibrium plasma. When the forward wave and the reflected wave were measured, about 2/3 input power contributed to the plasma generation.

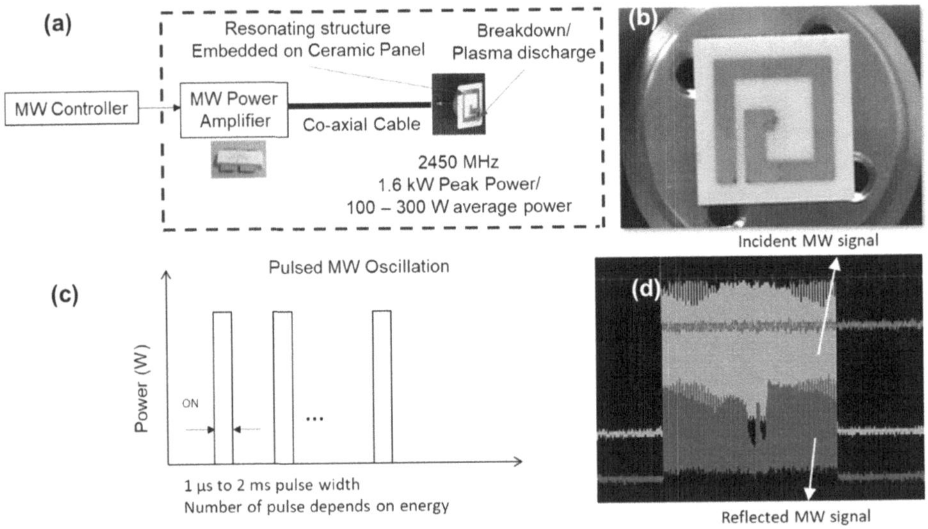

Figure 14.: *Flat panel igniter (FPI) (a) schematic operation, (b) actual image, (c) schematic oscillation pattern, (d) actual oscillation pattern.*

Figure 15 shows the experimental results of using the FPI in a constant combustion chamber [28] using propane. The flame propagation in the chamber was measured under atmospheric pressure and standard conditions. A comparative experiment was also conducted at 68 mJ using a standard spark plug. FPI energy results using the same input energy of 68 mJ and minimal ignition energy of 4.0 mJ are also shown. Compared to spark plugs, FPI 68 mJ forms flame nuclei faster and pressure rises earlier.

The magnitude change of the flame spread from 68 to 4.0 mJ is indicated by the radius of the flame kernel. In parameter BD10 (pressure increase of 10%) used for the flame spread of combustion, the larger the input energy was, the faster the flame spread. The initial flame propagation velocity at the minimal energy of 4 mJ was as small as 2 m/s. Compared to conventional spark plugs, this FPI has a relatively large degree of freedom in size, and shape can greatly contribute to the degree of freedom in designing cylinders, pistons, and plug positions.

3.5. Failure Analysis

An FPI destruction experiment was conducted. At the same level of input energy as a standard spark plug, ignition probability was 100%, there was no problem, and it can be used for life. Therefore, a large current was applied to try to destroy it. As a result of changing the current from 0.3 to 4.0 A in normal operation, it was destroyed, as shown in Figure 16. It glowed red and the ceramic was destroyed. Furthermore, cracks were generated in the remaining ceramic part. The energy was concentrated by the large current and was input to the plasma, and the temperature became high. This failure analysis is indispensable for applications to actual device development because it is vital to take measures against malfunction of the oscillation circuit due to individual differences, noise, overcurrent in an accident, lightning, etc. The thickness of the ceramic plate was 0.8 mm, but it should be increased to several millimeters, the antenna material should be tungsten, the reflected wave of microwaves should be monitored, and the safety circuit should be installed.

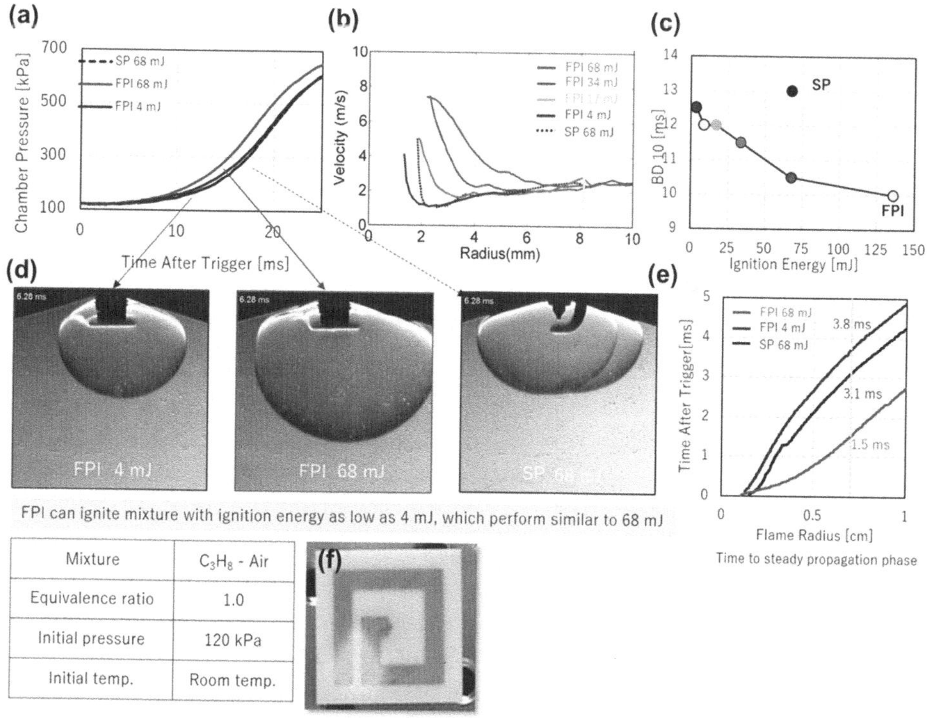

Mixture	C_3H_8 - Air
Equivalence ratio	1.0
Initial pressure	120 kPa
Initial temp.	Room temp.

Figure 15.: *Effects of (a) time after trigger on chamber pressure, (b) flame kernel radius on plume velocity, (c) BD10 (pressure increase of 10%) in flat panel igniter (FPI). (d) Actual images of flame kernel after 6.28 ms for the low energy (4 mJ) FPI, and high energy (68 mJ) FPI and spark plug (SP). (e) Correlation of flame radius vs. time after trigger shows a shorter time (1.5 ms) to enlarge the flame kernel in FPI. (f) Top view image of flame kernel [31].*

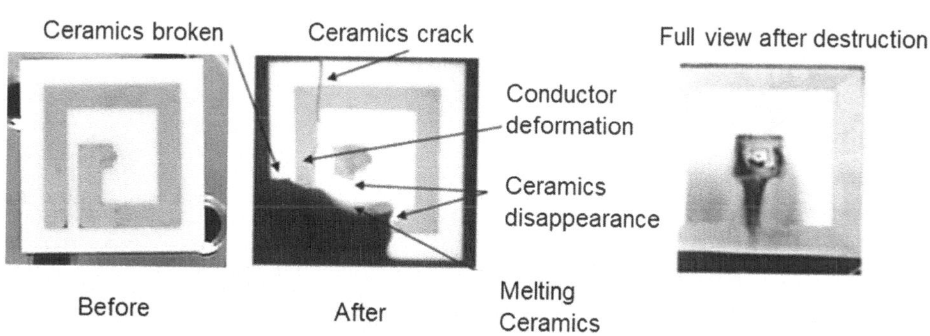

Figure 16.: *Failure analysis shows ceramic melting after 4 A current surge. Normal operations only require 0.3 A.*

4. CONCLUSIONS

A microwave semiconductor system was successfully developed to improve the ignition process in a combustion system. Large plasma was confirmed compared to a spark plug under atmospheric pressure conditions. A smaller microwave ignitor of 4 mm in diameter was developed, and a microwave generator was also developed in hand-carrying size. Practical experiments were demonstrated both in practical passenger car engines of a motorcycle engine 125 cc and small passenger gasoline engine 660 cc. Remarkable lean limit extensions were observed in both engines.

Erosion problems were tested in the engine and could be controlled by changing the resonant microwave frequency of 2.4 to 2.5 GHz. A ceramic plate antenna was designed and it demonstrated its excellent combustion ignition performance. Further failure analysis revealed that the ceramic itself was severely damaged by a large current when it broke.

REFERENCES

1. Bradley, D.; Sheppard, C.G.W.; Suardjaja, I.M.; Woolley, R. Fundamentals of high-energy spark ignition with lasers. Combust. Flame 2004, 138, 55–77.
2. Sahoo, B.B.; Sahoo, N.; Saha, U.K. Effect of engine parameters and type of gaseous fuel on the performance of dual-fuel gas diesel engines—A critical review. Renew. Sustain. Energy Rev. 2009, 13, 1151–1184.
3. Malenshek, M.; Olsen, D.B. Methane number testing of alternative gaseous fuels. Fuel 2009, 88, 650–656.
4. Nichols, R.J. 1993 Soichiro Honda Lecture: The Challenges of Change in the Auto Industry: Why Alternative Fuels? J. Eng. Gas Turbines Power 1994, 116, 727–732.
5. Seko, T.; Kuroda, E.; Hamano, Y. Methanol lean burn in an auto-ignition DI engine. SAE Trans. 1998, 107, 721–730.
6. Karim, G.A. Hydrogen as a spark ignition engine fuel. Int. J. Hydrogen Energy 2003, 28, 569–577.
7. Kobayashi, H.; Hayakawa, A.; Somarathne, K.K.A.; Okafor, E.C. Science and technology of ammonia combustion. Proc. Combust. Inst. 2019, 37, 109–133.
8. Nishiyama, A.; Le, M.K.; Furui, T.; Ikeda, Y. The relationship between in-cylinder flowfield near spark plug areas, the spark behavior, and the combustion performance inside an optical SI engine. Appl. Sci. 2019, 9, 1545.
9. Mariani, A.; Foucher, F. Radio frequency spark plug: An ignition system for modern internal combustion engines. Appl. Energy 2014, 122, 151–161.
10. Morsy, M.H. Review and recent developments of laser ignition for internal combustion engines applications. Renew. Sustain. Energy Rev. 2012, 16, 4849–4875.
11. Beduneau, J.L.; Ikeda, Y. Application of laser ignition on laminar flame front investigation. Exp. Fluids 2004, 36, 108–113.

12. Beduneau, J.L.; Kim, B.; Zimmer, L.; Ikeda, Y. Measurements of minimum ignition energy in premixed laminar methane/air flow by using laser induced spark. Combust. Flame 2003, 132, 653–665.
13. Kawahara, N.; Beduneau, J.L.; Nakayama, T.; Tomita, E.; Ikeda, Y. Spatially, temporally, and spectrally resolved measurement of laser-induced plasma in air. Appl. Phys. B 2007, 86, 605–614.
14. Beduneau, J.L.; Kawahara, N.; Nakayama, T.; Tomita, E.; Ikeda, Y. Laser-induced radical generation and evolution to a self-sustaining flame. Combust. Flame 2009, 156, 642–656.
15. Ikeda, Y.; Nishiyama, A.; Kaneko, M. Microwave enhanced ignition process for fuel mixture at elevated pressure of 1MPa. In Proceedings of the 47th AIAA Aerospace Sciences Meeting Including the New Horizons Forum and Aerospace Exposition, Orlando, FL, USA, 5–8 January 2009.
16. Cheng, X.; Zhang, X.; Wang, Z.; Wu, H.; Wang, Z.; Chen, J.Y. Effect of Microwave Pulses on the Morphology and Development of Spark-Ignited Flame Kernel. Energies 2021, 14, 6205.
17. Hwang, J.; Kim, W.; Bae, C. Influence of Plasma-Assisted Ignition on Flame Propagation and Performance in a Spark-Ignition Engine. Appl. Energy Combust. Sci. 2021, 6, 100029.
18. Kumar, P.; Yamaki, Y.; Lee, J.; Nakaya, S.; Tsue, M. Effects of Microwave Radiation on Laser Induced Plasma Ignition of N-Butane/Air Mixture under Atmospheric Conditions. Proc. Combust. Inst. 2021, 38, 6593–6603.
19. Liu, S.; Zhao, N.; Zhang, J.; Yang, J.; Li, Z.; Zheng, H. Experimental and Numerical Investigations of Plasma Ignition Charac-teristics in Gas Turbine Combustors. Energies 2019, 12, 1511.
20. Zhao, H.; Zhao, N.; Zhang, T.; Wu, S.; Ma, G.; Yan, C.; Ju, Y. Studies of Multi-Channel Spark Ignition of Lean n-Pentane/Air Mixtures in a Spherical Chamber. Combust. Flame 2020, 212, 337–344.
21. Ikeda, Y.; Nishiyama, A.; Wachi, Y.; Kaneko, M. Research and development of microwave plasma combustion engine (Part I: Concept of plasma combustion and plasma generation technique). In Proceedings of the SAE World Congress & Exhibition, Detroit Michigan, MI, USA, 16–19 March 2007.
22. Tsuruoka, R.; Ikeda, Y. Development of plasma source sustained by semiconductor microwaves. In Proceedings of the 51st AIAA Aerospace Sciences Meeting Including the New Horizons Forum and Aerospace Exposition, Grapevine, TX, USA, 7–10 January 2013.
23. Ikeda, Y. Plasma-Assisted Combustion in Automobile Engines Using Semiconductor-Oscillated Microwave Discharge Igniters. In RF Power Semiconductor Generator Application in Heating and Energy Utilization; Springer: Singapore, 2020; pp. 195–216.
24. Ikeda, Y.; Padala, S.; Makita, M.; Nishiyama, A. Development of innovative microwave plasma ignition system with compact microwave discharge igniter. In Proceedings of the 12th International Conference on Engines & Vehicles, Napoli, Italy, 13–17 September 2015.

25. Le, M.K.; Padala, S.; Nishiyama, A.; Ikeda, Y. Control of microwave plasma for ignition enhancement using microwave discharge igniter. In Proceedings of the 13th International Conference on Engines & Vehicles, Capri Napoli, Italy, 10–14 September 2017.

26. Padala, S.; Nagaraja, S.; Ikeda, Y.; Le, M.K. Extension of Dilution Limit in Propane-Air Mixtures Using Microwave Discharge Igniter. In Proceedings of the SAE 13th International Conference on Engines and Vehicles, Capri, Italy, 10–14 September 2017.

27. Padala, S.; Le, M.K.; Nishiyama, A.; Ikeda, Y. Ignition of propane-air mixtures by miniaturized resonating microwave flat-panel plasma igniter. In Proceedings of the 13th International Conference on Engines & Vehicles, Capri Napoli, Italy, 10–14 September 2017.

28. Le, M.K.; Nishiyama, A.; Ikeda, Y. Evaluation of a novel miniaturised microwave resonating igniter: The Flat Panel Igniter. Proc. Combust. Inst. 2019, 37, 5613–5620.

29. Ikeda, Y.; Nishihara, H.; Nakajima, T. Detailed Spectrum Analysis of Chemiluminescent Radicals at Flame Front in an SI Engine. In Proceedings of the International Symposium on Diagnostics and Modeling of Combustion in Internal Combustion Engines, Nagoya, Japan, 1–4 July 2001.

30. Shcherbanev, S.A.; Cherif, M.A.; Starikovskaia, S.M.; Ikeda, Y. Experimental study of pulsed microwave discharges at pressures ranging over five orders of magnitude. Plasma Sources Sci. Technol. 2019, 28, 045009.

31. Nishiyama, A.; Yuji, I. Improvement of lean limit and fuel consumption using microwave plasma ignition technology. In Proceedings of the SAE 2012 World Congress & Exhibition, Detroit, MI, USA, 24–26 April 2012.

INDEX

G

H

I

J

K

L

M

N

O

P

Q

R